If ȷ

Sustainable Waste Management

Proceedings of the International Symposium
held at the University of Dundee, Scotland, UK
on 9-11 September 2003

Edited by

Ravindra K. Dhir
Director, Concrete Technology Unit
University of Dundee

Moray D. Newlands
CPD / Consultancy Manager, Concrete Technology Unit
University of Dundee

and

Tom D. Dyer
Lecturer, Concrete Technology Unit
University of Dundee

 ThomasTelford

Published by Thomas Telford Publishing, Thomas Telford Ltd, 1 Heron Quay, London E14 4JD.
www.thomastelford.com

Distributors for Thomas Telford books are
USA: ASCE Press, 1801 Alexander Bell Drive, Reston, VA 20191-4400, USA
Japan: Maruzen Co. Ltd, Book Department, 3–10 Nihonbashi 2-chome, Chuo-ku, Tokyo 103
Australia: DA Books and Journals, 648 Whitehorse Road, Mitcham 3132, Victoria

First published 2003

The full list of titles from the 2003 International Symposium 'Advances in waste management and recycling'
and available from Thomas Telford is as follows

- Sustainable waste management. ISBN: 0 7277 3251 X
- Recycling and reuse of waste materials. ISBN: 0 7277 3252 8

A catalogue record for this book is available from the British Library

ISBN: 0 7277 3251 X

PREFACE

The issue of waste management is currently high on the list of priorities of many governments, organisations and agencies worldwide, and a variety of waste-related directives have been, or are being, implemented to deal with this. A common aim of these is to control resource use through regulation of waste disposal, with an emphasis of placing ultimate responsibility on producers. Concurrently, governments are attempting to reduce the burden on primary resources, often through more direct means e.g. taxation. Together, these will form the essential elements of a future sustainable society.

The Concrete Technology Unit at the University of Dundee has organised this International Symposium to review the latest advances in waste management and to provide a forum for discussion on how to embrace international commitments whilst maintaining a balance between ecological requirements, economic growth and social progress.

The event was organised in collaboration with six internationally recognised institutions: the Chartered Institution of Wastes Management, the Waste and Resources Action Programme (WRAP), the Scottish Environment Protection Agency, the Scottish Office, Remade Scotland and London Remade. The proceedings contain thirty-eight papers presented in five themes: (i) Global and International Commitments, (ii) European Waste Directives and Priorities, (iii) National Government Policy, (iv) Local Government Policy and (v) Assessing Environmental Impact, summarising the state-of-the-art in sustainable waste management.

The Opening Paper of the event was given by Mr Pierre Portas, Secretariat of the Basel Convention, United Nations Environment Programme, Switzerland and four Keynote Papers were presented by Dr Ana Filipa-Ferraz, PROCESL, Portugal, Sir Ken Collins, Scottish Environment Protection Agency, UK, Dr Vincent Cassar, Ministry for Resources and Infrastructure, Malta and Dr Prasad Modak, Environmental Management Centre, India.

The support of the Collaborating Institutions, Sponsoring Organisations and International Professional Associations was a major contribution to the success of the Symposium. The work of the event was an immense undertaking and all of those involved are gratefully acknowledged, in particular, the members of the Organising Committee for managing the event from start to finish; members of the Scientific and Technical Committee for advising on the selection and reviewing of papers; the Authors and the Chairmen of Technical Sessions for their invaluable contributions to the proceedings and the exhibiting organisations who took part in the Trade Fair.

All of the proceedings have been prepared directly from the camera-ready manuscripts submitted by the authors and editing has been restricted to minor changes where it was considered absolutely necessary.

Dundee
September 2003

Ravindra K Dhir
Moray D Newlands
Tom D Dyer

iii

ORGANISING COMMITTEE

Concrete Technology Unit

Professor R K Dhir OBE (Chairman)

Dr M D Newlands (Secretary)

Ms C Mitchell
Scottish Environment Protection Agency

Dr M R Jones

Dr M J McCarthy

Dr T D Dyer

Dr K A Paine

Dr L J Csetenyi

Dr J E Halliday

Dr L Zheng

Dr S Caliskan

Ms K Giannakou

Mr M C Tang

Mrs E Csetenyi

Ms P I Hynes (Symposia Assistant)

S R Scott (Unit Assistant)

SCIENTIFIC AND TECHNICAL COMMITTEE

Mr Stephen Aston,
Chartered Institution of Wastes Management, *UK*

Professor Hasan Belevi
University of Innsbruck, *Austria*

Professor Werner Bidlingmaier
Bauhaus University of Weimar, *Germany*

Mr Jeff Cooper
Environment Agency, *UK*

Mr Paul Dumpleton
SITA, *UK*

Dr Robert Eden
Organics, *UK*

Professor Salah El-Haggar
The American University in Cairo, *Egypt*

Professor Andy Fourie
University of Witwatersrand, *South Africa*

Ms Ana Filipa Ferraz
PROCESL - Engenharia Hidraulica e, *Portugal*

Dr Jane Gilbert
The Composting Association, *UK*

Mr Iain Gulland,
Alloa Community Enterprises Ltd, *UK*

Professor William Hogland,
Kalmar University, *Sweden*

Mr Stephen Jenkinson,
Greater Manchester Waste Group, *UK*

Dr Katia Lasaridi,
Harokopio University, *GREECE*

SCIENTIFIC AND TECHNICAL COMMITTEE

COLLABORATING INSTITUTIONS

Chartered Institution of Wastes Management

London Remade

Remade Scotland

Scottish Environment Protection Agency

Scottish Executive

Waste and Resources Action Programme

SPONSORING ORGANISATIONS WITH EXHIBITION

Angus and Dundee Convention Bureau

Building Research Establishment

Dundee City Council

Scottish Environment Protection Agency

Valpak Ltd

Waste and Resources Action Programme

SUPPORTING INSTITUTIONS

Association of Estonian Cities

ASSURE

Finnish Environment Institute

GFR Engineering Solutions

The Packaging Federation

CONTENTS

THEME 5 ASSESSING ENVIRONMENTAL IMPACT

Keynote Paper

SYMPOSIUM OPENING PAPER

FROM MAKERS TO BREAKERS:
A NEW DIMENSION IN WORLDWIDE WASTE MANAGEMENT

P Portas

Basel Convention UNEP

Switzerland

ABSTRACT. With the adoption of the Basel Convention on the Control of the Transboundary Movements of Hazardous Wastes and Their Disposal in 1989, the world community has imposed on itself to achieve the following far-reaching goals:

- To control transboundary movements of hazardous and other wastes and to reduce such movements to a minimum consistent with their environmentally sound management.

- To treat and dispose of such wastes as close as possible to their place of generation and in an environmentally sound manner.

- To minimize their quantity and hazardousness.

In the context of the Basel Convention, environmentally sound management means to take all practical steps to protect human health and the environment from the adverse effects of hazardous and other wastes. The environmentally sound management approach calls for a continuous improvement in waste management and reduction. Policy integration is a key element to reconcile and bring together approaches that normally exclude each other. The protection of the environment and human health from wastes requires comprehensive policies. Because the awareness of the dangers posed by wastes has come late in government agenda, there often is a lack of institutional integration. Waste management links to macro-economic and development policies, to poverty alleviation or community well-being being is often weak and loose. Environmentally sound management, as the underlying principle of the Basel Convention, is an approach that has the potential to bridge the makers to the breakers and thus contributing worldwide to the protection of human health and the environment. Environmentally sound management is both a global and universal concept; its ambition is to be accessible to everyone, at any time and any place. It calls for public and private partnership, for far-sighted policies and sustainable technical assistance and funding.

Keywords: Basel Convention; Environmentally sound management; Partnership, Recovery

Mr Pierre Portas is Senior Programme Officer at the Secretariat of the Basel Convention, United Nations Environment Programme.

INTRODUCTION

The Basel Convention was adopted in March 1989 and entered into force on 5 May 1992. As of today, the Convention has 156 Parties. The main purpose of the Convention is to protect human health and the environment from the dangers posed by hazardous and other wastes. To achieve this, the Convention aims at achieving three interrelated objectives, namely:

- To reduce transboundary movements of hazardous and other wastes to a minimum consistent with their environmentally sound and efficient management.

- To treat and dispose of such wastes as close as possible to their source of generation.

- To minimize both the quantity and hazardousness of these wastes.

The underlying principle is environmentally sound management. It means taking all practicable steps to ensure that hazardous and other wastes are managed in a manner which will protect human health and the environment from these wastes. The transboundary movements of hazardous and other wastes are subject to a strict control system under the Convention. The Basel Convention is a dynamic international instrument and can, through its Annexes in particular, adapt to scientific and technological changes. The implementation of the Convention is supported by the decisions adopted by the Conference of the Parties, its supreme legislative body. Building on the achievements of the first decade of the Convention, the Conference of the Parties at its fifth meeting in December 1999 adopted the Basel Declaration on Environmentally Sound Management. In this Declaration, the Parties and other States asserted the vision that the environmentally sound management of hazardous and other wastes should be accessible to them, emphasizing the minimization of such wastes and the strengthening of capacity building.

The Basel Declaration on Environmentally Sound Management has become both the foundation and framework for the development of the Strategic Plan for the Implementation of the Basel Convention (to 2010) adopted by the Conference of the Parties at its sixth meeting in December 2002. The Basel Declaration and the Strategic Plan recognize that the worldwide environmentally sound management of hazardous and other wastes requires action at all levels of society within the approach of the life-cycle management of materials, from makers to breakers. The sound management of chemicals and hazardous wastes was addressed by the World Summit on Sustainable Development that was held in Johannesburg from 26 August to 4 September 2002; its Plan of Implementation promotes efforts of UNEP and other intergovernmental organisations to develop a strategic approach to international chemicals management; this would include support for strengthening ongoing cooperation between the Stockholm Convention on Persistent Organic Pollutants, the Rotterdam Convention on the Prior Informed Consent Procedure for Certain Hazardous Chemicals and Pesticides in International Trade and the Basel Convention.

THE BURDEN OF UNCERTAINTY

There is growing pressure on policy-makers from different stakeholders, often having conflicting agendas. In the world of wastes, uncertainties prevail and affect the policy-making process. The use of best available technique, integrated pollution prevention control framework, as well as waste prevention and minimization are arguably considered as the best way forward in improving worldwide environmental quality and health safety. Science, economic analysis and regulations are tuned to address such issues; risks are assessed and prioritised and cost-benefit analysis are carried out with the purpose of more efficient use of resources with greater benefit to the environment. Technology is the system by which a society provides its members with those things needed or desired. Today economic resources and social systems are under the influence of technology; even, are guided by it. Cleaner technology or processes, product design, information technology are perceived as being critical tool to improve environmental protection.

Agenda 21 of UNCED reflected a global consensus towards changing production and consumption patterns and to use environmentally sound technology to reach this goal. More efficient energy production and use and wiser use of resources in general is therefore an essential component of this change. It also implies that society needs to move towards more environmentally sound lifestyles. Despite all these assumptions, stress on nature, natural resources or ecosystems is increasing. The nocivity of human activities seems greater than the benefits technology could bring to improve quality of life and sustainable lifestyles that are environmentally sound.

Mankind could be seen as a global ecological force that has not yet mastered knowledge of the globality of its acts on nature, natural resources, ecosystems and biogeophysical processes. Unrestrained technological optimism could therefore have unfortunate consequences. Although, there is much that technology can contribute to people's problems. The problem is not shortage of information or data but rather our inability to foresee many of the consequences of action or inaction in a global perspective. This is the burden of uncertainty.

LOCAL AND GLOBAL SIGNIFICANCE

Euphoria over globalization is being challenged by its sequels. Assured supplies of materials and energy are required to meet the increased needs of development. It is interesting to note that the largest known stock of metal is metal-in-use. Appropriate technology and product design can help in obtaining further economic benefits while protecting human health and the environment.

Waste issues are both local and global; they concern individuals and different sectors of society. Let's take as an example the dismantling of ships. There is an agreement by the international community that a global programme to address the issue is essential. The concerned intergovernmental organizations namely, the International Maritime Organization, the International Labour Organization and the Basel Convention are working together in the development of guidelines that cover the design of ships to be more friendly when they will be disposed of, the preparation (eg: decontamination) of ships destined for dismantling, the environmentally sound dismantling itself and the protection of workers in particular during the scrapping or dismantling and when steel is recycled. Every year, some 600-700 seavessels are taken out of service.

Old ships or other ships sold for breaking contain a wide range of toxic or hazardous materials (eg: used oils, asbestos). Most of these ships are broken down on beaches in Asia, threatening the health of workers and surrounding communities and polluting the environment severely. Shipping companies received an average price of US$ 2 million for each ship sold for scrap. The shipping industry and the countries where dismantling occur have taken step forwards improving the safety and environmental conditions of ship scrapping. But we are not yet where we should be. Million of dollars are required to up-grade dismantling facilities or create new facilities to ensure sound working practices and the respect of environmental standards in the yards to avoid having workers with lung problems, gas explosions or beaches polluted with chemicals and toxic substances or wastes; this is of paramount importance. But still, this will not be enough. It is critical to clean up vessels, to the extend possible, destined for dismantling before they are beached. But more is necessary. Indeed, there is a logic in designing ships in such a way that they are free of as many hazardous materials as possible and easier to scrap. This will greatly alleviate the burden at the disposal site.

Economics has a role to play because, often, places that are taxed less have more incentives to import ships for dismantling and recycling, but these places are not necessarily where ships are scrapped in an environmentally sound way. Sound environmental, social and economic policies are required both at the national and global levels to influence and guide all those stakeholders having a role from the makers to the breakers.

RECOVERY OF WASTES

Recovery[1] is a complex issue because of its environmental, trade and economic implications. It includes all the steps involved in recovering useable components of wastes so that they may be re-used, and the management of the unusable residues. This issue of transboundary movement of hazardous wastes destined for recovery operations could be viewed in the context of the comprehensive management of such wastes and the environmental costs of using the analogous virgin materials.

Waste management policies recognize the value of residues and wastes for their potential as economically useful materials or sources of energy and their potential for avoiding the environmental and economic costs from the extraction and processing of virgin materials. In future, cleaner production methods, low waste technology and hazardous waste avoidance may help to reduce waste generation at source. Recycling and recovery operations would generally be preferred to landfilling and incineration. Recovery operations have also disadvantages. There is always a danger of sham recycling-wastes moved for final disposal in the guise of a recovery operation. Genuine recovery operations can be polluting.

The availability of a cheap and easy recovery option can reduce the incentive to find and use other cleaner production options. There is a lack of sufficient quantified information or access to it about what hazardous wastes are exported to developing countries and what environmental problems have occurred as a result; in addition, the knowledge about activities of the informal sector, and their definitive impacts on health and the environment, in recovery (eg: of lead) is scarce. In contrast to this, a large body of information is available from authoritative sources in industrialized countries on the many aspects of recovery. This includes information about trends in world supply and demand, statistics on transboundary movements within industrialized countries, role of the secondary industrial sectors, priority waste streams more likely to be subject to recovery, the present extent of recovery, current proven and developing technologies and environmental and health impacts of recovery, processes or operations.

---

[1] Recovery includes resource recovery, recycling, reclamation, direct re-use or alternative uses

A FRAMEWORK FOR ACTION

The development of standards, the consideration of voluntary and regulatory approaches, the environment goals and economic and trade interest are useful elements to assess environmentally sound management. They would, however, not suffice for designing the framework. Other critical elements should be part of the equation; namely: the progressing harmonization of procedures and control systems for the transboundary movements of wastes, including hazardous wastes, compilation of data on generation of hazardous and other wastes and monitoring their import and export.

The level of enforcement, prevention and monitoring of illegal traffic, adequate waste management infrastructure and trained personnel, compliance with international obligations, liability and compensation regimes, reduction of pollutants, waste minimization programmes (voluntary or mandatory), cleaner production and prevention programmes and economic incentives are critical factors. Local, national, regional and international comparable knowledge, know-how and infrastructure are required to achieve environmentally sound management.

Environmentally sound management cannot be restricted to the measurement of the level of operation of one waste facility or industry dealing with wastes. Its implication is wider and more comprehensive. One can use a metaphor: it is not enough to drive well a solid and reliable car if the road is in a very poor condition and the other drivers do not know how to drive or have unreliable vehicles. Environmentally sound management is a concept that means different things for different people; factors such as geographical locations, level of economic development or professional activities would influence such perception. It will take time to build a common understanding of what constitutes environmentally sound management among all public and private stakeholders. It is therefore essential to work within a framework that is comprehensive enough to accommodate such diversity of perceptions while providing a practical mechanism for concrete implementation.

Inaction should not be the response. It is even more important today to build that common understanding and fine-tune the design of the environmentally sound management framework to the progressive accumulation of scientific and technical knowledge. Activities on the ground should help us in uncovering the reality of what environmentally sound management means in practical terms, and to better articulate the responsibilities and contribution of all stakeholders. When it comes to promoting the transfer of technology for the environmentally sound management of hazardous wastes and other wastes produced locally, international cooperation as conceived in the Convention will not suffice. The multi stakeholder approach called for in the Basel Declaration on Environmentally Sound Management is a step forward in building the appropriate interface between Governments, intergovernmental organizations and the private or public operators.

As a baseline, each country should have both the capacity and capability to manage the generation, treatment, storage and transport, disposal and recovery of hazardous wastes, and enforce its legislation in this field. However, the needs vary greatly from countries that have developed a waste management system from those that have not. For instance, countries that are vulnerable to import of hazardous wastes because of lack of adequate infrastructure, rely heavily on prohibition of both import and export. While such prohibition could be subject of inconvenience for countries with developed infrastructure and enforcement capabilities as it may hamper existing trade in wastes and hazardous wastes. Many Parties to the Basel

Convention are not in a position to manage hazardous wastes in an environmentally sound manner and are faced with a dilemma. Indeed, financial and other investments in the management of wastes or hazardous wastes generated locally may act as a disincentive for additional resources to minimize generation. Hence, the notions of differentiated but common responsibility and access to new and affordable technologies that are environmentally friendly. The establishment of the Basel Convention Regional Centres in Africa, Asia and the Pacific, Eastern and Central Europe and Latin America and the Caribbean represents a regional delivery mechanisms for facilitating the access or transfer of sound and proven technologies for the management and minimization of wastes.

A WORLD OF COMPLEXITY

The world community should build new partnerships engaging national and local authorities, the civil society, industry and business and intergovernmental bodies to face up to the complexities of waste management. Waste management and disposal often is seen as conflicting with the priorities for poverty reduction or food security. Clean up, disposal and prevention costs compete with other financial commitments taken at the national or international level for development. The introduction of cleaner production methods may entail high capital costs and the need for trained personnel that many countries cannot afford. A further complication concerns the use of sound technologies for the destruction or disposal of wastes.

There are uncertainties about how to achieve irreversible transformation of the hazardous content in a waste, the levels of destruction of the hazardous materials or the concept of destruction efficiency,the choice of favoring work on Best Available Technique and Best Environmental Practice. What sort of technologies to use when the hazardous content of a waste is low. Integrated approaches like the life-cycle management of materials are promising avenues to address such complexity. At the global level efforts are made for the multilateral environmental agreements to be mutually supportive and complementary. This is of particular relevance to the chemical – and waste – related conventions including the Basel Convention, the Stockholm Convention on Persistent Organic Pollutants and the Rotterdam Convention on the Prior Informed Consent Procedure for Certain Hazardous Chemicals and Pesticides in International Trade.

Also, the market of wastes is growing and is shifting from national to world markets. Many companies operate more and more on a regional or worldwide scale. Large quantities of waste are being subject to trade generating substantial economic benefits; a portion of this market concerns wastes destined for recovery operations, in particular non-ferrous metal wastes and precious metal wastes.

In addition, the recovery of metals from end-of-life equipment, such as electrical and electronic waste, has the effect of delocalising the dismantling and recovery operations to developing countries. Finally, wastes destined for final disposal will follow the path of least resistance if enforcement is inadequate, the national legislation is incomplete or inexistent, the national currency is low, environmental and occupational standards are low or liability regimes do not exist. These factors associated with high disposal costs in the countries generating the hazardous or other wastes have the potential to trigger illegal traffic.

A WAY FORWARD

Corporate performance should lead to improved environmental performance. In the building up of a multistakeholder partnership, industry and business could converge around a core set of environmental practices that would be illustrative of the way to implement the environmentally sound management approach. Such practices could operate at the facility level while taking into account the wider requirements for environmentally sound management at the regional, national or international level such as the needs for the transfer of know-how and capacity building. Governments will continue to build the global foundation through instruments like the Basel Convention to ensure a world level playing field. Integrated policy framework supported by the relevant legislative architecture and incentive mechanisms should enable stakeholders to operate in a way that is compatible with global standards while retaining their competiveness.

We must act collectively and in fairness and address country needs taking into account the principle of common but differentiated responsibility. Advances in waste management and recycling are an obvious priority. The development and access to environmentally sound technologies is critical. Instruments like the Basel Convention are in the forefront of environmental protection and in the design of the policies and standards that are required to meet our common objectives. Environmentally sound management de facto entails a continuous improvement of performance.

But our will to make the world better, freer of hazardous materials can only materialize through solidarity taking into consideration the immense needs of developing countries and countries with economy in transition in this domain. Globalization can either aggravate existing perversions or cure them; it is a choice of society.

THEME ONE:

GLOBAL AND INTERNATIONAL COMMITMENTS

TWO CASES OF SOLID WASTE MANAGEMENT, BOTH SUPPORTED BY EU FINANCING, FOR TWO COUNTRIES, WITH DIFFERENT STAGES OF DEVELOPMENT

A F Ferraz

M H Tavares

PROCESL - Engenharia Hidraulica e

Portugal

ABSTRACT. The paper summarizes two different situations, in relation to waste management, concerning countries at two particular stages of development. Both waste systems were and are being implemented by European funds and under European strategy and legislation. The two concrete waste management situations to be dealt with are: Valorlis, Portugal, serving a population of about 295,000 inhabitants; and a small Caribbean island, the Commonwealth of Dominica, with about 75,000 inhabitants. In order to link both cases, the paper shows how two distinct states of development lead to the establishment of different priorities and the choice of specific waste strategies, techniques and technologies.

Keywords: Waste, Management, Collection, Transport, Treatment, Technique, Technology, Legislation, Stage, Priority.

A F Ferraz, is a Director and Commercial Manager with PROCESL - Engenharia Hidraulica e, Portugal

M H Tavares, works for PROCESL - Engenharia Hidraulica e, Portugal

INTRODUCTION

Whenever general European Union principles, policy and legislation are dealt with by Member or non-Member States, Waste Management Systems always have to consider National priorities as a must.

The general environmental strategy of the European Union is to be found in the European's Programmes of Policy and Action in relation to the Environment and Sustainable Development. The strategy involves general guidelines and proposes new approaches to environmental issues, based on preventive actions and on necessary changes in the behaviour pattern of modern society.

Community legislation on waste takes, in most cases, the form of Directives, which need to be transposed into National law of the Member States of the Union. However there are some Community laws and regulations which are directly applicable at Member State level, not requiring any transposition in order to be enforced.

For non-Member States, community legislation is only applicable in particular cases where the European Union is directly financing project implementation.

Thus, this paper presents, after a brief description of the main framework of EU policy concerning waste, two particular and different waste systems, but both financed by the EU, one, in Portugal, belonging to a Member State, and the other, in the Commonwealth of Dominica, belonging to a Developing State.

MAIN FRAMEWORK OF EU POLICY CONCERNING WASTE

The policies and legislation of the EU on solid wastes - taking into account mainly Directive 75/442/EEC dated 15th. July amended by Directive 91/156/EEC dated 18th March and Resolutions 90/C 122/02 and 97/C 76/01 dated respectively 7th July 1990 and 24th February 1997 - are framed by the following main principles:

- The producer or the owner of wastes is responsible for all the management operations concerning their wastes;

- Any waste management operation without adequate authorisation is forbidden, as well as any operation carried out in a non-authorised plant or in disregard of the procedures established by law and by technical regulations;

- All waste management operations - namely production, storage, transport or delivery/reception - must have an updated registration, the waste producers being obliged to send their waste registration to the competent authorities, on a yearly basis, according to specific regulations defined by each country;

- The costs associated with waste management, as well as to their final destination, must be supported by the waste producer (or by their owner, when their producer is unknown or undefined).

The prevention of waste production is a major concern, either through the reduction of waste itself or through the separation at source of certain wastes whose characteristics - nature, hazard risk - require a specific treatment. The promotion of reuse, recycling and other types of wastes valorisation must be emphasised, in order to recover materials or energy, and/or to reduce costs of treatment facilities.

Packaging waste is focused-on in particular, with minimum objectives being settled until the end of 2005, in terms of weight percentage of valorisation and recycling (see Directive 94/62/EC dated 20th December).

The main principles relating to waste elimination and final disposal are the prohibition of discharge of wastes on dumps, and the authorisation of any treatment and final disposal operations, types of which allowed by law are defined.

EU Council issued specific legislation about incineration of urban wastes - Directives 89/369/EEC of 8th June and 89/429/EEC of 21st June - and landfilling of wastes - Directive 1999/31/CE of 26th April.

Directives about incineration of urban wastes are mainly concerned with the prevention of environmental impacts associated to air emissions, with limits on concentrations of some major pollutant gases, as well as other operational restrictions, concerning namely the solid wastes handled by these plants.

The Landfill Directive is mainly concerned with the prevention of underground water pollution and biogas control, although other important aspects are dealt with, such as procedures for the admission of wastes, as well as stability control, closure and post-closure care and monitoring of water, leachate and gas. The conditions to be ensured during the construction of the landfill facility are a major issue, several requirements being settled to guarantee adequate permeability levels, proper lining and stormwater drainage.

PRESENTATION OF TWO CASES

The Case of Valorlis, in Portugal

Valorlis Solid Waste System serves a geographic area of 2,157 km^2, with a population of approximately 300,900, in 2002. The solid waste received and treated in the Leiria sanitary landfill was about 110,534 tonnes, also in 2002.

Valorlis makes a separate collection of each fraction for six municipalities, Batalha, Leiria, Marinha Grande, Ourém, Pombal e Porto de Mós. In 2002, the fraction quantities collected were 2,227 tonnes of glass, 1,736 tonnes of paper and 416 tonnes of package.

The Valorlis Waste System is responsible for the following main sub-system/activities, where the constructions were financed by the Cohesion Fund:

• Closure of six dumping areas located in each of the six municipalities;

- Implementation of a safe treatment and final disposal system, through a sanitary landfill, for non recyclable waste;

- Implementation of a pre-treatment for leachate, including constructed wetlands. The pre-treated effluent is discharged into the municipal sewage system;

- Construction of a biogas collection system, to be converted for electricity generation;

- Construction of three transfer stations serving Batalha, Porto de Mós and Ourém e Pombal for waste transfer to the sanitary landfill;

- Installation of an effective separate package system for recycling;

- Installation of approximately 640 igloo systems (1 igloo per 462 inhabitants). In 2002, 12 underground igloos were installed, 2 in each municipality, with the aim of reducing visual impact;

- Construction of a central separation plant, near the landfill;

- Construction of a compaction unit to optimize available space;

- Separate collection and transport of the igloo systems, is dealt with by different and independent circuits, by means of specific vehicles. Control and management is performed by a digital system, based on a Geographical Information System (GIS);

- Valorlis developed an igloo management digital system, SIDIGEC, as a tool for management control and decision making. SIDIGEC's main purposes are: optimization of circuits and collection frequency, validation of collected quantities, identification of particular problems and development of a data basis and statistics as a predictive tool.

Figure 1 shows a GIS application to an igloo distribution and Figure 2 shows a representation of "SIDIGEC".

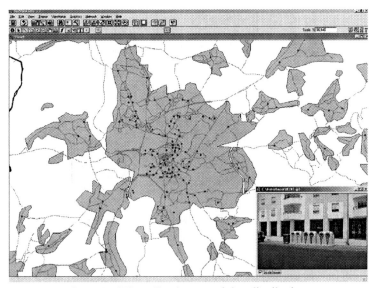

Figure 1   GIS application to an igloo distribution

Figure 2   Representation of SIDIGEC

The Case of Dominica in the Caribbean

Dominica is a small developing island State, of about 750 km², with a population of approximately 75,000 (2001). The population is primarily distributed in coastal communities, with the heaviest concentration in Rouseau, the capital, Portsmouth and Marigot.

Dominica's economy is primarily agricultural. A specific objective in Dominica is to emphasize eco-tourism, while preserving the island's natural and environmental resources. The development of eco-tourism requires increased efforts to conserve the environment.

The generated waste is estimated of about 15,500 tonnes per year, but no exact data exist as far as volume, density or composition of waste. The equivalent permanent resident of visitors is about 4,000 (2002), but it is expected to increase in the next years about 11% annually. More than 50% of the solid waste is generated in Rouseau area.

Dominica is served by two dumping areas, where all the waste is dumped: Stock Farm, near Rouseau, receiving annually around 6,500 tonnes of waste and Portsmouth in the north of the island, which accepts annually around 3,000 tonnes of waste. The collection of waste is without source separation. Only hospital and industrial hazardous waste are disposed in special sites.

The priority was to construct a new landfill with adequate standards and of sufficient capacity to allow for environmental acceptable solid waste disposal for the whole island over a period of 15 to 20 years. The restoration of the two obsolete dumping areas was the next priority to preserve the environment. The new landfill at Fond Collet, with a capacity of 170,000 m³, is presently under construction. It is located in a quarry area, north of Rouseau. It comprises a basal liner of HPDE, a leachate collection and treatment system, drainage of surface water and the control of landfill gas.

FINAL COMMENTS

After the fulfilment of the European Union and the National legislation, the priorities of Valorlis were regulated by management and development target criteria. The whole system is now on operation. Only the biogas collection and utilization system will be ready to operate in the end of the current year. The system constant improvement is a concern and a practice of the management entity.

The main concern for Dominica, considering that tourism is very important, was a safe final disposal. The landfill is to operate in the beginning of 2004.

ECOLOGICAL BALANCE OF SILICON
USING RICE HULL ASH IN AGRICULTURE

N B Prakash

S Ito

National Agricultural Research Centre

Japan

ABSTRACT. Rice hull, a major by-product of the rice milling industry, can be a significant source of energy in rice producing countries. The rice hull ash (RHA) obtained from combustion presents disposal problems. However, its silicon content offers the opportunity for recycling of RHA in agriculture. Rice is a known silicon accumulator and the plant benefits from Si nutrition, and hence an effective method of recycling of RHA is essential for the ecological balance of silicon in rice farming. Application of RHA in rice nurseries and in main fields is a common practice in many developing countries. The burning temperature influences the composition of RHA gas as well as the form of silicon, which has to be considered for its effective recycling in agriculture. A greenhouse study was conducted to evaluate the bioavailability of silicon for rice from RHA obtained at different burning temperatures. The silicon content in soil solution was higher in the RHA obtained at 400-500°C. The average soil solution Si content from planting till harvest was higher for the 400°C treatment compared to other treatments. The content of silicon in soil solution and rice yields were decreased with RHA obtained at higher temperatures. Maximum grain yield, Si content of rice straw and hull, and its uptake were also higher for the 400°C treatment These facts suggest that RHA obtained at lower temperatures is effective for recycling as a source of Si in rice farming.

Keywords: Silicon, Temperature, Rice hull ash, Ecological balance, Recycling, Rice soils.

N B Prakash is a faculty member in the Department of Soil Science and Agricultural Chemistry in the University of Agricultural Sciences at Bangalore, India and is currently working as JSPS post-doc fellow in the Department of Soils and Fertilizer, National Agricultural Research Center at Tsukuba, Japan. His research focuses mainly on recycling of plant silicon materials and interaction of silicon with other nutrient elements in rice soils.

S Ito is the chief of the Lab of Soil Management, Department of Soils and Fertilizer in the National Agricultural Research Center at Tsukuba, Japan. He is currently undertaking research into the development of low input rice cultivation utilizing local organic materials.

INTRODUCTION

Silicon is an integral and quantitatively major component of the soil-plant system that exists in nature and in agriculture. Plant residues are used as an Si source both intentionally and incidentally in agriculture all over the world. Among the plant residues, rice hull contains very high amounts of silicon and is being used as a source of fuel in many industries. Burning rice hull as fuel to generate energy results in the waste product, rice hull ash (RHA). RHA is rich in silica and can be an economically viable raw material for many industries [1,2]. The importance of RHA in several industries, such as steel manufacturing (where it is used to prevent rapid cooling of steel and to ensure uniform solidification), cement industries, ceramic bricks, sodium silicate and silicagel synthesis, and as an adsorbent of minor vegetable oil components is highly acknowledged in the literature. Although various uses for rice hull and RHA have been suggested in the literature, their disposal or utilization remains a major concern.

While the yield of rice increased substantially during the green revolution, a trend of declining yields per unit area has been reported lately despite high inputs of fertilizer and pest control agents [3]. Recent studies suggest that the depletion of plant-available silicon may be a major reason for the yield decline observed under intensive rice cultivation [4,5]. However, sustained high rice yields in Japan are most likely due to the widespread application of silicate slag at rates of around 1.5 to 2 t /ha [6,7]. Silicate slags are an expensive Si source and most rice farmers of tropical and subtropical countries cannot use them at the rates of 1 or 2 t/ha/year and hence major emphasis has to be given on recycling of plant-derived silicon materials.

Application of RHA to soil normally occurs near rice mills and/or their disposal sites. Numerous research reports suggest that black to gray RHA can be effectively used in rice nurseries and in the main rice fields for achieving healthy seedlings and increasing yields [5,8,9]. Though RHA contains high amounts of silicon, the form and type of silicon is more important with respect to recycling in agriculture. The burning temperature of rice hull is not uniform in all the industries, which can yield different types of RHA (black, gray and white), thereby influencing its silicon availability to plants.

This paper examines the feasibility of recycling RHA as a source of silicon to maintain the ecological balance of depleted silicon in rice farming.

RECYCLING OF PLANT SILICON MATERIALS
WITH SPECIAL REFERENCE TO RHA

The silicon content of rice straw and rice hull is relatively high compared to other parts of the rice plant. Although rice straw contains 4.0% Si, practically no reports are seen in the literature on its recycling as a source of Si in subtropical and tropical rice cultivation. For many reasons, proper recycling of rice straw and rice hull is not common among most rice farmers of developing countries in Asia [10]. However, the addition of rice straw and rice straw compost is a common practice among most Japanese rice farmers.

In Australia, France, Indonesia, Italy, Malaysia, Myanmar (Burma), the Philippines, Spain, Thailand, and in California (United States) rice straw is generally burned [11,12]. The practice of burning rice straw in rice fields is a common practice in northern India, which results in the loss of valuable organic matter and nutrients, besides causing environmental pollution [13]. However, the extent of silicon availability due to recycling of rice straw or its ash in soils is less well documented in literature.

Table 1   Available quantity of plant Silicon materials for recycling

PLANT MATERIAL	Si CONTENT, %	AVAILABLE QUANTITY, million t	TOTAL Si AVAILABILITY, million t
Rice straw	4	712	28.5
Rice hull	8	117	9.3

Rice hull is a major byproduct of rice farming. Assuming that harvested rice is composed of 20% rice hull, the 585 million tonnes (t) of unmilled rice produced in the world in 2001 [14] would generate around 117 million t of rice hull and about 9.3 million t of Si (Table 1).

Environmental regulations limit disposal of the hulls in landfills. The thermal energy requirements of rice processors and power generators, high transportation costs owing to the low packing density of hulls, and the lack of other significant uses for rice hull make burning the most common practice. RHA is thus a keystone material as a source of Si for many industries.

Though RHA is a derivative of rice hull, the amount of nutrients it possesses mainly depends on the burning temperature. The effect of burning on rice hull/husk was first reported in Japan where it is referred to as "carbonized rice husk" [15].

Table 2   Effect of carbonized rice husk application on the Si content of rice seedlings

TREATMENT	SiO_2 CONTENT, %
Control (0)	5.04
Burnt rice husk ashes (160°C)	5.68
Carbonized rice husk	
160 °C	6.99
320 °C	8.11
480 °C	9.21
Paddy seed bed	
With Carbonized rice husk	9.12(13)
Without Carbonized rice husk	7.40(11)
Upland seed bed	
With Carbonized rice husk	7.58(2)
Without Carbonized rice husk	5.30(1)

Figures in parenthesis indicate number of seedbeds tested

Carbonized rice husk was prepared by burning the rice husk slowly at as low a temperature as possible. The carbonized rice husk is a good Si source compared to unprocessed rice husk[15]. Application of carbonized rice husk increased greatly the Si content of rice seedlings (Table 2). Si in carbonized rice husk was taken up at a greater rate by rice seedlings than that in the ashes of rice husk burned according to the common practice. Application of carbonized rice husk significantly increased the Si content of young rice seedlings in the seed-beds of farmers.

The analysis of more than 60 samples of 10 different paddy varieties of Malaysia [16] reported that 21.33% of the rough rice comprised of rice husk, while 13% of the husk constituted rice husk ash. The nutrient content of rice husk ash was 80.26% silica, 0.38% phosphorus, 1.28% potassium, 0.21% magnesium and 0.56% calcium. Si contents were lower for Malaysian varieties while American [17] and Indonesian [18] varieties show a greater range for P, K, Mg and Ca content (Table 3).

Table 3   Range of percentage composition of nutrients in Rice Hull Ash

NUTRIENT	MALAYSIA[1]	AMERICA[2]	INDONESIA[3]
Si	71.43-88.52	91.1-97.0	86.90 - 97.30
P	0.32-0.46	0.1-1.3	0.2 – 2.85
K	0.87-1.50	0.4-2.5	0.58 - 2.50
Mg	0.12-0.30	0.1-1.2	0.12 - 1.96
Ca	0.35-0.71	0.2-1.4	0.20 - 1.50

[1]Hashim et al.,1996, [2]Houston,1972, [3]Wen-Hwei,1986

The Si application as rice hulls (9.4%) or its ash (44.4% Si) increased yield of rice on an Ultisol with very high P fixing capacity at Cavinti, Laguna, Philippines[19]. Field experiments conducted with IR 8 at Polonnaruwa in SriLanka have shown that rice hull ash application of 0.74 t/ha with the recommended level of fertilizers (69:20:18 kg NPK/ha) yielded an additional 1.0-1.4 t/ha, whereas further increase in ash additions decreased the yield [20]. Amendment by rice hull ash for saline acid soils in Thailand increased rice yield only in the first year of application [21].

The effect of RHA, rice straw and the method of N application (prilled or briquetted urea) on a transplanted rice crop was studied through field experiments in Malawi during 1995 and 1996 [22]. The RHA alone increased the rice grain yields in all three experiments up to 12%; however, the increase was not statistically significant (Table 4). The combination of rice straw and RHA along with optimum N rates (60 kg ha^{-1}) increased the rice grain yields significantly in 1996 winter season, although the increase was not significant in the other two experiments. The grain yield increase due to RHA was consistent, and ranged from 187 to 448 kg ha^{-1} in all the experiments

Table 4   Effect of RHA and rice straw on grain yield ofrice grown
in a field experiment in Malawi (1995-1996)

TREATMENT[1]	1995 WINTER	1996 SUMMER	1996 WINTER	AVERAGE, kg/ha
Control	2149	2032	1620	1933.67
A1S0N0	2128	2309	1718	2051.67
A1S1N60-PU	5045	3973	5549	4855.67
A1S1N60-UB	6299	4181	5861	5447.00
A0S1N60-PU	4048	3584	5889	4507.00
A0S1N60-UB	6112	3733	5426	5090.33

[1]A-Rice Hull Ash (applied at 1 kg m^{-2} to the nursery beds to a depth of 6cm before sowing seed)
S-Rice Straw (applied at 2.5 kg/plot by puddling 5cm length straw to a depth of 22-27 cm)
N-Nitrogen applied at 0 and 60 kg ha^{-1} as PU(Prilled Urea) or UB (Urea Briquette)
1-Applied,   0-Not applied

Review of the literature indicates that recycling of RHA has a more beneficial effect as a source of silicon in improving the crop growth and yield, as well as reducing the incidence of several diseases both at the nursery stage and in the main field (Table 5). The RHA at the rate of 1 kg m^{-2} can be effectively used in rice nurseries and region/location specific studies are necessary to quantify the amount of RHA necessary for recycling in the main fields of rice. However, RHA is already being used as a silicon source, knowingly or unknowingly, in rice seedbeds and main fields in many countries.

Table 5 Application of RHA as Si source for rice seedbed and main fields of rice

NO	SOURCE		QUANTITY	TIME	OBSERVATION	REF
1	RHA		0.74 t ha^{-1}	M	Grain yield increase due to RHA ranged from 1.0-1.4 t ha^{-1}	[20]
3	RHA		Not Available	M	Addition to saline acid soil increased rice yield in the first year of application	[21]
4	RHA		1 kg m^{-2}	S	Reduced the incidence of leaf blast in seedlings	[8]
5	RHA	&	RHA@2 kg m^{-2}	RHA-S	Reduced incidence of neck	[23]
	RS		RS @ 2t ha^{-1}	RS-M	blast in seedlings	
6	RHA	&	RHA @2 kg m^{-2}	RHA-S	Reduced incidence and	[24]
	RS		RS @ 2t ha^{-1}	RS-M	severity of leaf scald disease	
7	RHA		250 mg kg^{-1}	Pot culture-basal	Increased Si content of rice at all three stages of growth	[25]
8	RHA		0, 0.5,1.0 & 2 kg m^{-2}	S	Higher seedling dry matter. Increased uptake of P, K and Si by shoots.	[26]
9	RHA	&	RHA- 1 kg m^{-2}	RHA-S	Grain yield increase due to	[22]
	RS		RS- 2.5 kg/plot	RS-M	RHA ranged from 187-448 kg ha^{-1}	
10	RHA		2 & 4 t ha^{-1}	M	Increased grain yield besides mobilizing soil P	[27]

RHA- Rice Hull Ash,   RS-Rice Straw,   M- Main field,   S-Seedbed (Nursery)

EFFECT OF BURNING TEMPERATURE ON RICE HULL FOR RECYCLING OF RHA IN AGRICULTURE

With a gross calorific value of 3000 kcal/kg, rice hull is capable of highly efficient combustion, and can serve as a very good source of fuel in many industries including mini power plants of 1 to 2 MW capacity in developing countries. The burning temperature of rice hull influences the composition of RHA and the form in which silicon is present, which has to be considered for its effective recycling in agriculture. Rice hull normally contains 7.1% of silicon [28] and the content of silicon in rice hull ash ranges from 32-40% [5]. The burning temperature of rice hull when used as source of fuel is not necessarily uniform and may result in black to gray RHA which may contain varied amounts of plant-available silicon. A small study was conducted to evaluate the bioavailability of silicon for rice from RHA obtained at different burning temperature.

Rice Hull Ash Preparation

A known weight of rice hull sample was taken in the porcelain dish and held in an electric muffle furnace(KM-160 ADVANTEC) heated to a specific temperature for a period of three hours. The rice hull samples were heated from 300 to 900°C and ashes were named after their respective combustion temperatures (RHA-300, RHA-400, RHA-500, RHA-600, RHA-700, RHA-800 and RHA-900).

Greenhouse experiment

Rice Hull and RHA obtained at different temperature (300 to 900°C) of fifteen g each was mixed with a known quantity (7kg) of sand in a plastic pot. A treatment without these sources was used as the control. One g each of N, P_2O_5 and K_2O were added as polyolefine coated urea, super phosphate and potassium chloride respectively, along with dolomite(1g/pot) and surface soil (50g/pot). A ceramic filter tube of length 10cm connected with a plastic tube was laid into the center of the sand filled pot. The system was submerged by the addition of deionized water. Two seedlings of rice were transplanted in each pot. The pots were arranged in a completely randomized design on a greenhouse table. Three replications were used in the study. Deionized water was used for watering daily to keep it submerged during the study. Each time about 10-15 mL of soil solution was collected by using syringe fixing to the plastic tube attached to the ceramic filter kept in the soil. Si content in the soil solution was analyzed by ICPEAS. After the harvest of the crop, the dry weights of shoot and unhulled rice were recorded. Si content in rice straw and hull were analyzed after digesting with H_2SO_4 in presence of H_2O_2 and the precipitated Si was estimated by using gravimetric method.

Higher grain yield was noticed in the plants treated with RHA-400 followed by RHA-300 when compared to other treatments, while maximum straw yield was noticed in RHA-300 treated plants (Table 6). The Si content in straw and hull was found to be significantly higher in RHA-400 treated plants and decreased with those RHA obtained at higher temperatures. The total uptake of Si per pot was also found to be higher in RHA-400 treated pots. These facts indicate that RHA obtained at lower temperature is effective in its dissolution and release of Si, thereby increasing plant growth and yield of rice.

Table 6   Effect of Rice Hull and RHA obtained at different temperature on rice

TREATMENTS	SiO_2, % IN STRAW	SiO_2, % IN HULL	GRAIN YIELD, g/plot	STRAW YIELD, g/plot	TOTAL SiO_2 UPTAKE, g/plot
Control	1.87	2.21	34.08	58.80	1.88
Rice Hull	3.01	3.92	36.08	64.83	3.32
RHA-300	3.22	5.63	38.01	68.93	3.87
RHA-400	4.36	5.83	41.50	58.63	4.85
RHA-500	3.37	4.72	36.14	56.80	3.47
RHA-600	2.78	3.32	32.01	54.50	2.62
RHA-700	2.18	2.83	33.16	59.90	2.22
RHA-800	1.71	2.31	29.32	65.36	1.75
RHA-900	1.60	2.05	34.98	54.96	1.59
Mean	2.64	3.75	35.01	60.7	2.96
S.E(d.f.= 16)	0.455	0.703	2.496	6.37	0.441

Analysis of soil solutions collected at different intervals indicated that there was significant difference in the Si content among the treatments throughout the growing season of rice (Figure 1a, 1b). The content of Si in soil solution was higher in the pots treated with RHA-400 compared to other treatments on the day of planting and subsequently in RHA-300 for a period of two weeks after planting. In general, the soil solution Si was higher in RHA-300 and RHA-400 treatments followed by RHA-500 and RHA-600 treatments (Figure 1a). However, average soil solution Si content from planting till harvest was higher in RHA-400 treatment compared to other treatments (Figure 2). The average soil solution Si content and Si concentration of soil solution at different growth stages of RHA-700, RHA-800 and RHA-900 treated pots were on par with that of the control indicating less release of Si from RHA obtained at higher temperatures (Figure 1b, 2).

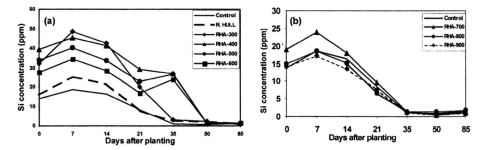

Figure 1   Silicon in soil solution at different growth stages of rice crop (2002)

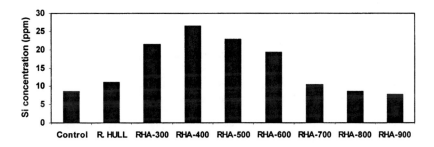

Figure 2   Average Silicon content in soil solution (planting till harvest)

ECOLOGICAL BALANCE OF SILICON

Rice is a silicon accumulator plant. In order to produce 5 t/ha of grain yield, normally 230-470 kg of Si is removed per crop from the soil and intensive rice cultivation without silicon fertilization can lead to a declining or stagnating rice yield in many countries. Silicon depletion has been noticed with intensive cultivation of high yielding varieties in traditional rice growing soils of many countries, if farmers are not replacing the Si removed by rice. Savant et al.,[5] have conducted an extensive review of the literature documenting the positive impact of Si applications on rice yields under subtropical to tropical conditions, while Lian [29]and Elawad and Green [30] have done so for temperate countries of the world. Many soils in Africa, Asia, and in Latin America are highly weathered and desilicated and Si management may be important for increasing and sustaining rice productivity.

It is well known that the uptake of nutrients by the crop and the nutrients supplied to the crop, are in balance with each other: the so-called "mineral balance". What this means is that "what comes out of the soil has to be returned to the soil" to maintain the ecological balance.

Rice straw and hull has several off-farm uses leading to a lack of recycling in rice farming in many countries. A significant proportion of RHA is being used as a source of silicon in several industries rather than in agriculture. However, recycling of RHA as a source of Si needs to be done in such a way that will ensure minimum damage to the environment for maintaining silicon in rice field soils.

CONCLUSIONS

RHA is already being used as a source of silicon in rice seedbeds and main rice fields in many countries. Effective recycling of RHA mainly depends on the burning temperature of rice hull, which influences the composition of RHA and the form the silicon takes. The present study suggests that RHA obtained at lower temperature is effective for recycling as a source of Si to maintain the ecological balance of silicon in rice farming.

ACKNOWLEDGMENTS

We would like to acknowledge the support of Japan Society for the Promotion of Science for carrying out this work.

REFERENCES

1.   KAMATH,S.R., PROCTOR,A., Silica gel from rice hull ash: preparation and characterization. Cereal Chemistry, Vol.75, 1998, pp484-487.

2.   KALAPATHI,U., PROCTOR,A., SHULTZ,J., A simple method for production of pure silica from rice hull ash. Bioresource Technology, Vol.73, 2000, pp257-262.

3.   CASSMAN,R.G., DEDATTA,S.K., OLD,D.C., ALEANTARA,J., SAMSON,M., DESCALSOTA,J., DIZON,M., Yield decline and nitrogen economy of long-term experiments continues on irrigated rice systems in the tropics', In R.LAL, and B.A.STEWART, eds, Soil Management: Experimental Basis for Sustainable Quality, CRC Press Inc, Boca Raton, FL, 1995, pp181-222

4.   SAVANT,N.K.,.DATNOFF,L.E., SNYDER,G.H., Depletion of plant available silicon in soils: a possible cause of declining rice yields. Communication in Soil Science and Plant Analysis, Vol.28, 1997a, pp1245-1252.

5.   SAVANT,N.K., SNYDER,G.H., DATNOFF,L.E., Silicon management and sustainable rice production. Advances in Agronomy, Vol. 58, 1997b, pp151-199.

6.   KONO,M., Effectiveness of silicate fertilizer to Japonica varieties. Tropical Agriculture Research Series, Vol.3, 1969, pp241-247.

7.   TAKAHASHI,E., MA,J.F., MIYAKE,Y., The possibility of silicon as an essential element for higher plants. Comments in Agricultural Food Chemistry, Vol.2, 1990. 99-122.

8. KUMBHAR,C.T., NEVASE,A.G., SAVANT,N.K., Rice Hull Ash applied to soil reduces leaf blast incidence. International Rice Research Notes, Vol.20, 1995, pp23-24.

9. SAWANT,AS., PATIL,V.H., SAVANT,N.K., Rice hull ash applied to seedbed reduces deadhearts in transplanted rice. International Rice Research Notes, 1994, pp21-22.

10. PONNAMPERUMA, Straw as source of nutrients for wetland rice. In "Organic matter and Rice" International Rice Research Institute, Los Banos, Laguan, Philippines, 1984, pp117-136.

11. TANAKA,A., Methods of handling rice straw in various counties. International Rice Communication Newsletter. Vol. 22, 1973, pp 1-20.
12. TANAKA,A., Role of organic matter. In "Soils and Rice, International" International Rice Research Institute, Los Banos, Laguan, Philippines, 1978, pp605-620.

13. MISHRA,B.,P.K., BRONSON,K.F, Decomposition of rice straw and meneralization of carbon, nitrogen , phosphorus and potassium in wheat field soil in western Uttar Pradesh. Journal of Indian Society of Soil Science, Vol.49(3): 2001, pp419-424.

14. FAO World Agricultural Center, FAOSTAT Agricultural Statistic Data Base Gateway, 2002

15. ISHIBASHI,, H., Effect of silica contained in carbonized rice husk on the growth of rice seedlings. Bulletin Faculty of Agriculture, Yamaguchi University, Vol. 7, 1956, pp333-340.

16. HASHIM,A.B., AMINUDDIN,H., SIVA,K.B., Nutrient content in Rice Husk Ash of some Malaysian Rice Varieties, Pertaanika. Journal of Tropical Agricultural Science, Vol.19(1), 1996, pp77-80.

17. HOUSTON,D.F., Rice Chemistry and Technology. St.Paul, Minnesota, American Association of Cereal Chemists. 1972.

18. WEN HWEI, H., Rice Hulls Rice: Production and Utilization, AVI Publishing Company Inc., West port Connection, California, USA, 1986.

19. GARRITY,D.P., MAMARIL,C.P., SOEPARDI,G. Phosphorus requirements and management in upland rice-based cropping systems. In "Proceedings of The Symposium on Phosphorus Requirements for Sustainable Agriculture in Asia and Oceania," March 6-10, 989,. International Rice Research Institute, Los Basons, Laguna, Philippines. 1990 pp.333-347

20. AMARASIRI,S.L.. Organic recycling in Asia-Sri Lanka. FAO Soils BullEtin, Vol. 36, 1978, pp119-133.

21. DITSATHAPORN,C., RUNGSANGECHAN,P., YUVANIYOM,A., SUTTAVAS,A., ARUNIN, S., Effect of rice hull ash and plowing depths on saline-acid soil improvement. Research Report, Agriculture Development Research Center, Bangkok, Thailand, 1989, pp184-194

22. SISTNANI,K.R., REDDY,K.C., KANYIKA,W., SAVANT,N.K., Integration of Rice Crop Residue into Sustainable Rice production System, Journal of Plant Nutrition, Vol.21(9), 1998, pp1855-1866.

23. KUMBHAR,C.T., SAVANT,N.K., In "Silicon in Agriculture" (DATNOFF,L.E., SNYDER,G.H., KORNDORFER,G.H., Eds.). Elsevier Science,The Netherlands, 2001a, pp382.

24. KUMBHAR,C.T., SAVANT,N.K., In "Silicon in Agriculture" (DATNOFF,L.E., SNYDER,G.H., KORNDORFER,G.H., Eds.) Elsevier Science, The Netherlands, 2001b, pp382.

25. TALASHILKAR,S.C., CHAVAN,A.S., Effect of Rice Hull Ash on Yield and Uptake of Silicon and Phosphorus by rice cultivars at different growth stages, Journal of Indian Society of Soil Science, Vol.44, 1996, pp340-342.

26. SISTANI,K.R., SAVANT,N.K., REDDY,K.C., Effect of Rice Hull Ash silicon on rice seedling growth. Journal of Plant Nutrition, Vol.20, 1997, pp195-201.

27. PRAKASH,N.B., NAGARAJ,H., VASUKI,N., SIDDARAMAPPA,R., ITOH,S., Effect of recycling of plant silicon for sustainable rice farming in south India, 17th World Congress of Soil Science, Bangkok, Vol.2, 2002, pp698.

28. YOSHIDA,S., The physiology of silicon in rice. Technical Bulletin No.25. Food Fertilizer Technological Center, Taipei,Taiwan.1975

29. LIAN,S., Silica fertilization of rice. In " The Fertility of Paddy Soils and Fertilizer Applications for Rice," Food Fertilizer Technological Center, Taipei, Taiwan,1976, pp97-220.

30. ELAWAD,S.H., GREEN,V.E.,Jr., Silicon and the rice plant environment: A review of recent research Il Riso, Vol,28, 1979, pp235-253.

THEME TWO:

EUROPEAN WASTE DIRECTIVES AND PRIORITIES

EUROPEAN WASTE MANAGEMENT DRIVERS TOWARDS A SUSTAINABLE EUROPE AND HOW SCOTLAND CAN INFLUENCE THE AGENDA

K Collins

Scottish Environment Protection Agency

United Kingdom

ABSTRACT. This paper describes the key European drivers towards the sustainable management of waste resources. The paper aims to provide a comprehensive outline of these drivers as a resource for delegates interested in specific aspects of integrated waste resource management. Current and pending directives are briefly described. Other papers deal with some of these Directives in more detail. Proposed Directives are listed, with further details provided as an annexe. The relevance of the European 6th Environment Action Programme is described. The means by which Scotland interacts with and influences the European agenda is discussed and conclusions provided as to how this could be improved in the future; this will form the basis of the presentation.

Keywords: Directives, Hierarchy, Drivers, Waste, 6th Environment Action Plan, Europe, Radar, Policy.

K Collins is Chairman of the Scottish Environment Protection Agency (SEPA). He is also a Board Member of the Institute for European Environment Policy, a Member of the Management Board of the European Environment Agency and Vice-President of the National Society for Clean Air.

INTRODUCTION

The basis of the move towards sustainable waste management in Europe is Council Directive 75/442/EEC as amended by Directive 91/156/EEC (the framework directive). The framework directive sets out the basic principles of waste management which are fundamental to the issue of sustainable development. These are that:

- Waste must be prevented or its production reduced and its harmfulness must be reduced. These objectives are to be achieved through the development of clean technologies which reduce the use of natural resources and by technical developments and marketing of products to ensure the smallest contribution by their manufacture, use or final disposal to increasing the amount of harmfulness of waste and pollution hazards;

- If waste is produced it should be recovered by means of recycling, re-use or reclamation or used as a source of energy and residues which have no value finally disposed of safely.

Where recovery or disposal takes place, the framework directive requires that the activity takes place without endangering human health and without using processes or methods which could harm the environment. Particular regard has to be paid to the need to undertake the activities without risk to water, air, soil and plants and animals, without causing a nuisance through noise or odours and without affecting the countryside or places of special interest.

In addition, these core principles are supported through the related principles of proximity whereby wastes should be managed as close to their point of arising as possible; self sufficiency for each member state in its ability to manage its wastes; the waste hierarchy showing the options that are most sustainable as the preferred management options; use of best available technology (BAT) and producer responsibility. Together they underpin most of the existing and forthcoming EC environmental legislation such as, for example the landfill Directive, the pollution prevention and control Directive, the waste incineration Directive and the draft Directive on composting biodegradable waste. In addition the 6th Environment Action Programme provides a framework for integrated environmental activity aimed at improving the environment of member states and contributing to global environmental improvement. Aspects of this programme impact directly on waste resource management. The wide range of specific drivers are discussed in more detail in the following sections.

SUMMARY OF KEY EC POLICY DRIVERS

INCINERATION AND THE WASTE INCINERATION DIRECTIVE (WID) 2000/76/EC

The Waste Incineration Directive (WID) 2000/76/EC came into force on 28 December 2000. Member States were required to transpose it into national law by 28 December 2002. The aim of the WID is to prevent or to limit as far as practicable negative effects on the environment, in particular pollution by emissions into air, soil, surface water and groundwater, and the resulting risks to human health, from the incineration and co-incineration of waste.

Key Provisions Of The Directive

- The European Parliament and the Council of the European Union considered that a single text on the incineration of waste would improve legal clarity and enforceability;
- Covers thermal treatment of solid and liquid wastes, both hazardous and non-hazardous;
- Covers both incineration and co-incineration plants;
- Brings into force tighter emissions standards and wider monitoring regime;
- New incineration and co-incineration plants need to meet the requirements of the WID straight away;
- Existing plants needing to meet the requirements of the WID by December 2005;
- It will supersede previous Directives on incineration of municipal and hazardous waste (89/369/EEC, 89/429/EEC, 94/67/EC).

Current Activity

2% of current municipal waste arisings in Scotland were converted to energy in incineration plants in 2002 [1]. There are two municipal waste incinerators operating in Scotland; one in Dundee and the other on the Shetland Isles.

The Dundee facility is authorised to accept 105,000 tonnes of municipal solid waste and 15,000 tonnes of commercial and industrial wastes per year. Energy in the form of electricity is produced by the plant and is sold into the National Grid.

The Shetland facility is authorised to deal with 26,000 tonnes of municipal solid waste per year and the waste heat is used in a district-heating scheme.

Implications For Waste And Resource Management In Scotland

- Improved environmental practices;
- Thermal treatment likely to increase as consequence of other European legislation e.g. Landfill Directive diversion targets;
- Potential need to introduce greater source segregation of waste;
- Recently there have been a number of European Court judgements on the role and place of thermal treatment of municipal waste. The full implications of these have yet to be considered, but they appear to back the idea that municipal waste incineration is inherently a disposal operation, and also clarifying the conditions under which waste incineration should be regarded under EU law as a recovery or disposal operation.

Barriers To Thermal Treatment

- Negative public perception and opposition;
- High capital costs;
- Lack of bankability for emerging technologies;
- Relatively cheap costs for other waste management options;
- Low landfill tax rate;
- Economies of scale, especially in remote areas.

Opportunities For Business

- Investment opportunities as biodegradable waste targets in Landfill Directive need to be complied with;
- Increasing stability in market from knowledge of new standards and strategies e.g. National Waste Plan;
- New employment opportunities.

Environmental Benefits

- Tighter emission standards;
- Wider monitoring requirements;
- More sustainable treatment of waste than landfill;
- If accompanied with source segregation then increase in volumes of materials that can be reused, recycled, composted etc;
- Gaining energy from waste;
- Less landfill void space needed.

Tighter requirements in the WID will also contribute to improvements in air quality under the Air Quality Daughter Directives and the targets in the National Emissions Ceiling Directive.

THE WASTE ELECTRICAL AND ELECTRONIC EQUIPMENT (WEEE) DIRECTIVE 2002/96/EC

The Waste Electrical and Electronic Equipment (WEEE) Directive 2002/96/EC came into force on 13th February 2003 and Member States have 18 months to transpose it into National Law i.e. until August 2004. It is an extended producer responsibility directive aimed at reducing the quantity of WEEE and increasing its re-use, recovery and recycling. It also aims to improve environmental performance of all operators involved in the life cycle of electrical and electronic equipment (EEE).

Key Provisions Of The Directive

- Separate collection of WEEE sets at a target of 4kg per head of population per annum. This implies free take-back of WEEE from private households, including retailer take-back (collection in-store or by third parties) on delivery of a new appliance;
- High standards for treatment of WEEE;
- Producers to meet recovery and re-use targets (ranging from 50% to 80%);
- Producers to provide for the financing of the collection (from a central point), treatment and recovery/recycling of WEEE from private households as well as providing financial guarantees for the management of future WEEE;
- Provision of relevant information to:
  - Users - regarding their role in contributing to the collection of WEEE including locality and types of facilities available;
  - Authorised Treatment Facilities - identifying components and materials, including location of hazardous substances in WEEE;
  - EU - on amount of EEE put on the market and the relevant levels of recycling achieved.

Current Activity

Of the 900,000 [2] tonnes of WEEE arising per annum in the UK (42,000 tonnes for Scotland), only 50% enters the recycling chain, with even smaller amounts recovered. The rest goes to landfill which may cause heavy metal contamination as well as loss of potentially valuable and otherwise finite natural resources.

Implications For Waste And Resource Management In Scotland

- Improved environmental practices;
- New investments in the recovery infrastructure;
- Need for integrated approach to meet these high standards and targets by forming strategic partnerships through the whole life cycle chain of WEEE.

Barriers To Recycling

- Economies of scale, especially in remote areas (e.g. Highlands and Islands);
- Current attitude that reprocessed or reused goods are inferior;
- Lack of markets for some of the recycled material (e.g. plastics).

Opportunities For Business

- Reduced raw material and waste management costs;
- New treatment technologies emerging through R&D;
- New employment opportunities.

Environmental Benefits

- Reductions in amount of landfill (estimates show an avoided impact of 133,000 to 340,000 tonnes per annum for the UK);
- Improvements to air and water quality;
- Contribution to reduction to local disamenity;
- Increased awareness amongst consumers and businesses in general on waste issues.

The Directive will also contribute to Scotland's sustainable development objectives on waste resources as set out in the National Waste Strategy: Scotland 1999 and the National Waste Plan 2003.

THE LANDFILL DIRECTIVE

The Landfill Directive 1991/13/EC came into force on 26th July 1999. It sets out certain technical requirements for the construction and operation of landfill sites aimed at preventing escape of emissions such as landfill gas, leachate, noise and smells etc. It also encourages waste minimisation and recovery by applying tight targets on the reduction of biodegradable municipal and waste which can be sent to landfill, by banning certain waste from landfills and by requiring that waste be treated before landfill.

Key Provisions Of The Directive Relate To

- Site location to ensure that the site does not pose a serious risk to the environment or human health and well-being;
- To protection of soil and water through requiring measures to be taken to control the ingress of rainwater and surface waste and by the establishment of standards for a liner and geological barrier to prevent liquid escaping from the site;
- Landfill classification into 3 categories: "Hazardous" which may only take hazardous waste; "non-hazardous" which may take only non hazardous or inert waste; and "inert" which may take only inert waste;
- After care provisions which require that landfill emissions must be monitored for as long after its closure as may be required by the competent authority which regulates the site;
- Reduction targets for biodegradable municipal waste sent to landfill, based on the quantity landfilled in 1995, of 25% by 2010, 50% by 2013 and 65% by 2020;
- A ban on landfilling liquid waste, explosive waste, certain infectious clinical waste, tyres and waste which does not meet the criteria for acceptance at the various classes of landfill;
- Treatment of waste prior to landfill to change its characteristics in order to reduce its volume or hazardous nature, facilitate its handling or enhance recovery.

Current Activity

About 12 million tonnes of controlled waste is landfilled in Scotland each year. Of this some 3 million tonnes is waste collected by local authorities from households and other sources. Recycling rates for local authorities' waste vary from authority to authority but average out at about 5%. The rate will, however, increase sharply in response to targets set by the Scottish Executive and as progress is made towards meeting the landfill directive targets. The plans prepared by the landfill industry in Scotland to show how existing landfill will be modified to comply with the directive have been assessed by SEPA and it is clear that many sites will close before the final compliance date is reached.

Implications For Waste And Resource Management In Scotland

- Improved landfill standards;
- New investment in the industry infrastructure;
- Improved recovery of resources from the waste streams.

Barriers To Recycling

- Development of appropriate recycling facilities;
- Market Development;
- Public attitudes;
- Lack of integrated product policy for the lifetime management of products.

Opportunities For Business

- Development of new technologies to take advantage of recovered material;
- Development of new technologies to reduce waste at source;
- Development of new landfill to meet new standards will lead to larger more profitable landfills for waste which can only be landfilled.

Environmental Benefits

- Reduction in methane discharges;
- Increased value recovery from waste resources;
- Recovery of energy from a renewable source;
- Enhanced environmental protection;
- Reduction in use of raw material and resources used to obtain it;
- Reduced disamenity;
- More sustainable waste management.

THE END OF LIFE VEHICLES DIRECTIVE

Motor vehicles have been around for just over 100 years. They have brought great social and economic benefits, but at significant environmental cost. The End-of-Life Vehicles (ELV) Directive (2000/53/EC) introduces measures which are aimed at reducing waste production and promoting recycling and recovery of waste materials.

Key Provisions Of The Directive

- To restrict the use of certain heavy metals in the manufacture of new vehicles;
- The establishment of adequate systems for the collection of ELVs;
- Introduction of 'Certificates of Destruction';
- Vehicle manufacturers or importers to pay 'all or a significant part' of the costs of take back for complete ELVs;
- Minimum technical requirements for sites where ELV are stored and where treatment is to take place;
- De-pollution requirements (e.g. draining of fluids, removal of components identified as containing mercury);
- Treatment operations to promote recycling.

Current Activity

It is estimated that the number of motor vehicles reaching the end of their natural life – end of life vehicles (ELVs) – in Europe, will rise from 8 million in 1997 to more than 12 million in 2015. Scotland generates about 140,000 ELVs annually, which represents at least 140,000 tonnes of waste.

Implications For Waste And Resource Management In Scotland

Although a vehicle dismantling system exists in Scotland it will need to be adapted to ensure that both the requirements of the Directive can be met and that there are adequate facilities for vehicle users resident in Scotland's Highlands and Islands.

Barriers To Recycling

- Economies of scale in remote areas (e.g. Highlands and Islands);
- Under developed markets for some of the recyclates (e.g. plastics).

Opportunities For Business

Although the implementation of the ELV Directive may in some cases signal the demise of small local vehicle dismantlers, it will provide new business opportunities for those who work in partnership with vehicle manufacturers. De-pollution requirements, which are likely to be quite labour intensive, could also result in employment opportunities.

Environmental Benefits

Reduced waste production;
Increased value recovery from waste resources;
Reduction on the reliance of virgin natural resources;
- Reducing and controlling hazardous substance use;
- Reduction in potentially detrimental emissions to water and air.

BIOLOGICAL TREATMENT FOR ORGANIC WASTE

While at present there is no EU based composting directive, an EU-initiative to improve the present situation for biodegradable waste (bio-waste) management and to help meet the targets of the Landfill Directive 1999/31/EC has been proposed based on Articles 95 and 175 EC Treaty.

Key Provisions Of The Directive

- To promote the biological treatment of bio-waste by harmonising the national measures concerning its management in order to prevent or reduce any negative impact on the environment, thus providing a high level of environmental protection;
- To protect the soil and ensure that the use of treated and untreated bio-waste results in benefit to agriculture or ecological improvement;
- To ensure that human, as well as animal and plant, health is not affected by the use of treated or untreated bio-waste;
- To ensure the functioning of the internal market and to avoid obstacles to trade and distortion and restriction of competition within the Community.

Current Activity

In Scotland, around 30,000 tonnes of compost is being produced annually, some (the minority) is of high quality and sold to consumers as soil conditioners and mulches. Most of the compost currently produced is part of a pre-treatment process prior to landfill.

Implications For Waste And Resource Management In Scotland

The launch in February 2003 of the National Waste Strategy Scotland committed the composting industry to build and operate, by 2010, sufficient composting plants to handle approximately 300,000 tonnes of organic waste per year.

To meet the requirements of the Landfill Directive (1999/31/EC) it is essential that most of the compost produced will be of a sufficiently high quality to be used as a product rather than as pre-treatment prior to landfill. To aid the industry, the British Standards Institute in cooperation with The Composting Association, WRAP (Waste Resource Action Plan) has launched a composting standard the BS PAS (Publicly Available Specification) 100 quality standards for composted bio-waste.

Quality product is seen as vital for the protection of the environment and with the introduction in May 2003 of the EU Animal By-Products Order (Directive EC1774/2002), most composts being produced for the market place should be of the highest quality and fit for purpose.

Barriers To Composting

- Moving away from a mixed waste collection culture to a source segregated collection system;
- Finding sufficient markets.

Opportunities For Business

- Marketing opportunities;
- Composting equipment sales & design;
- New technology opportunities;
- Plant operations;
- Training.

Environmental Benefits

- Reductions in amount of landfill (estimated at over 500,000 tonnes in Scotland by 2020);
- Improvements to soil fertility;
- Restoration of contaminated land;
- Reduction in emissions of Methane from landfill;
- Increased awareness of waste issues.

THE PACKAGING WASTE DIRECTIVE

The Packaging and Packaging Waste Directive was adopted in 1994 as a means of addressing the environmental impacts associated with some 58 million tonnes of packaging waste believed to arise in the EU each year.

Key Provisions Of The Directive

- To minimise the amount of packaging placed on the market;
- To ensure that packaging is capable of being recovered or recycled;
- To reduce the content of certain heavy metals in packaging;
- To require a minimum of 50% recovery and recycling of packaging waste from 2001.

Implementation Of The Directive

Although this Directive is not a 'producer responsibility' Directive, most Member States, including the UK, have transposed the relevant requirements using domestic producer responsibility provisions. In the UK, the relevant legislation is the Producer Responsibility Obligations (Packaging Waste) Regulations 1997 and the Packaging (Essential Requirements) Regulations 1998.

Current Activity

Ten million tonnes of packaging waste is produced in the UK each year of which around one million tonnes arises in Scotland. In order to meet the mandatory recovery and recycling targets, the UK government has imposed packaging waste recovery obligations on 'producers' of packaging. Recovery of packaging waste is carried out by over 400 recovery and recycling companies that have been accredited by the UK environment agencies. These companies generate packaging waste recovery notes (PRNs) for every tonne of packaging waste that is recovered and sell them to producers of packaging. The UK is reported to have increased its recovery of packaging waste from 38% in 1999 to 48% in 2001 at a cost of over £150 million to producers.

Implications For Resource And Waste Management In Scotland

Although the Regulations have allowed the UK to increase recovery of packaging waste, there is no evidence that there has been any reduction in resource use, or that there has been any net reduction in the amount of overall waste that is subject to final disposal. The increase in recovery of packaging waste may merely have displaced other waste to landfill instead. In addition, the Regulations have, to some extent, driven a preference for recycling rather than reuse. Much stronger measures are needed to encourage minimisation of new packaging and reuse of used packaging. Although the overall capacity for packaging waste recovery is reasonable, there is a need for expansion of the collection infrastructure in Scotland and the UK and further development of certain material recovery sectors, in particular plastics.

Barriers To Success Of The Regulations

- Poor collection infrastructure;
- Limited recovery capacity in certain material sectors, in particular plastics;
- Subdued growth of domestic recovery infrastructure within some material sectors as a result of increased export markets;
- Limited markets for recyclate.

Opportunities For Business

- Producers can reduce obligations by reducing resource use and reusing packaging.
- Reprocessors of packaging waste can receive additional funding for their activities by becoming accredited.
- Reprocessors can potentially form partnerships with Compliance Schemes to develop new recovery technologies or expand on existing infrastructure.

OTHER PENDING DIRECTIVES, MEASURES OR COMMUNICATIONS

Further details of pending directives are provided in Annex 1.

Communication From The Commission On Environmental Indicators
Communication On Sustainable Use Of Resources
Communication On Integrated Product Policy (IPP)
Measure On Environmental Product Declarations
Measure On Sustainable Production And Consumption
Communication On An Action Plan On Environment And Health

THE 6TH ENVIRONMENT ACTION PROGRAMME (EAP)

Adopted in March 2002, the 6th EAP ('Environment 2010: Our future, Our Choice') sets out the European Union's priorities and objectives for sustainable development up to 2010 in the following four priority areas:

1) Climate change;
2) Nature and biodiversity;
3) Environment and health;
4) Sustainable use of natural resources and management of waste.

The strategy for development of the 'sustainable use of natural resources and management of wastes' priority is still under evaluation, but is likely to include the actions detailed in Table 1.

Table 1   6th EAP: Key Actions on sustainable Use of Natural Resources

KEY ACTIONS
• Estimation of the materials and waste streams in the Community, including imports and exports using for example the instrument of material flow analysis;
• Review of the efficiency of policy measures, including economic instruments, and the impact of subsidies relating to natural resources and waste policy instruments with the aim of conserving resources and encouraging the uptake of resource-efficient technologies, products and services;
• Decoupling resource use and waste generation from economic growth and significantly reducing waste through improved waste prevention initiatives, better resource use efficiency and a shift to more sustainable consumption patterns;
• Development and implementation of a broad range of instruments including research, technology transfer, market based and economic instruments, programmes of best practice and indicators of resource efficiency;
• Establishment of goals and targets for resource efficiency and the diminished use of resources, decoupling the link between growth and adverse environmental impact;
• Integration of resource efficiency into Integrated Product Policy (IPP), eco-labelling schemes, green procurement policies and environmental reporting

The 6th EAP strategy on sustainable resource use and waste management provides a useful European context for the further development of a more resource efficient Europe.

INFLUENCING THE EUROPEAN AGENDA

As the preceding discussion illustrates, the agenda which informs the European waste strategy is focused on a number of key principles: waste hierarchy; self sufficiency; proximity; BAT; and producer responsibility. In turn, these principles underpin the evolving body of legislation which collectively represents the key driving force in shaping the European waste strategy's contribution towards a sustainable Europe.

In turning to consider SEPA's capacity to influence the agenda which informs the European waste strategy, it is worth making a distinction between the formulation and implementation stages of the European policy process.

The progressively more dominant role of the European Union in driving domestic environmental policy agendas through legislative developments increasingly makes its institutions – the Commission, the Council of Ministers and the Parliament – a key focus for the representation of often competing interests (for example, sectors of the business community or non-governmental organisations) intent on aligning the policy agenda with their own objectives. This representation of competing interests is at its most intense at the policy formulation stage. This is due to the potentially crucial implications which the formulated policy (in the form of EC legislation) may have for these interests during the implementation phase. Consequently, many interest groups recognise the need to try and shape that process by lobbying the various institutions directly or indirectly (via national governments) during the policy formulation phase.

SEPA's engagement with the policy process at the formulation stage is indirect. As a non-departmental public body, SEPA is accountable to the Scottish Executive and ultimately to the Scottish Parliament. As such, it is precluded from performing a lobbying function similar to that undertaken by conventional interest groups seeking to promote their own, often narrowly defined, sectoral objectives. Instead, SEPA works closely with both the Scottish Executive and the Department of Environment, Fisheries and Rural Affairs to assess the domestic policy implications of proposed EC legislation. This 'bottom-up' role, based on technical assessments of proposed legislation, provides an important Scottish perspective on the efficacy of such legislation. In turn, this is fed into the European decision-making process via the Scottish Executive's EU Office in Brussels. The Scottish Executive's EU office works closely with UKREP, the group of civil servants drawn from a range of UK Government Departments to represent UK interests in the European Union.

SEPA's legislative horizon scanning 'forward radar' system provides an important mechanism which enables the agency to contribute to the policy process at the formulation stage. It also enables SEPA to identify areas where further research may be needed to inform future European policy developments. Arguably, SEPA has the potential to play an even more significant role in this context through further development of the agency's networking capacity at both the UK and EU level. Effective influence is made possible by the sharing and exchange of forward intelligence on external policy and legislative developments. This enables the respective external liaison priorities to be discussed, and where possible, joint forward planning to take place with government and other stakeholders.

SEPA's capacity to exert influence in the formulation phase of the policy process is necessarily indirect. However, the agency is able to exert significant influence in promoting the European policy agenda during the implementation phase of the policy process. There are particular reasons for this.

SEPA works within a statutory framework to fulfil its main aim of providing an efficient and integrated environmental protection system for Scotland which will improve the environment and contribute to the Scottish Ministers' goal of sustainable development. This framework embraces the legislative policy drivers that largely shape the implementation of the European Waste Strategy following their transposition into domestic law. The enforcing of laws and regulations emanating from EC Directives remains a fundamental tool with which SEPA exerts influence on the process of implementing the European agenda. In addition, in Scotland, SEPA has a fundamental role by working in partnership with the Scottish Executive to shape the National Waste Strategy for Scotland which provides it with a strategic as well as a policy and regulatory role. This high level of integration of these various tools in Scotland should see a rapid and significant reorientation of how we manage the complex issues of sustainable waste resource use in the future.

Moreover, SEPA's influence on the implementation phase of the strategy is further reinforced through the provision of focused advice; an example of which includes the production of guidance in relation to the Waste Incineration Directive which sets out a clear context for the development of thermal recovery from waste in Scotland as a means of addressing significant public concern on the use of incineration as a waste management technique. SEPA also seeks to influence the implementation of the European Waste Strategy through tools which include the provision of environmental information within the context of the National Waste Strategy. This in includes the annual publication of a Waste Data digest.

In addition to the above, SEPA also seeks to influence the implementation of the European agenda outwith the confines of regulatory policy drivers. In this context, tools such as the provision of focused advice and environmental awareness raising are deployed in relation to the development of waste management projects for co-funding through the European Structural Funds.

CONCLUDING REMARKS

Europe, and indeed Scotland, is at a threshold of change in the management of waste resource materials. This is driven by the growing awareness of the finite nature of many of the key resources we depend on and on the environmental impacts of our production and consumption patterns; our lifestyles.

It is crucial that we respond to these pressures in a timely, integrated and focused manner. The development of the policy and statutory drivers emanating from Europe are both a threat and an opportunity. They are a threat if we are not prepared and planning ahead. The consequences of this are inadequate and poorly thought through implementation measures, poor public perception and additional costs including, significantly, those arising from infraction proceedings. The opportunities relate to the benefits of taking greater responsibility for the lifestyles we lead in terms of a cleaner, safer and more sustainable environment and economy.

To maximise the benefits of the drivers that are emanating from Europe at a significant rate we must ensure that we have the necessary 'forward radar' in place to maximise the planning time to implement often complex measures. We need to be integrated in our communications and activities, be they policy, regulatory or strategic, with all of the key players relevant to our various nations. We also need to recognise the challenge of addressing the tri-partite nature of our societies where government, economy and society merge and often conflict. This needs a greater maturity in our ability to develop more effective partnerships across these sectors. In Scotland such a framework is provided by the National Waste Strategy and the National Waste Plan.

Finally we must recognise that in terms of the global environment there are no islands. We as the human race and the environment we depend on are intimately connected. Addressing this reality is the great challenge of the 21st century.

ANNEX 1 – FUTURE PROPOSALS

Proposal to Amend Directive 94/62/EC on Packaging and Packaging Waste

This would amend Directive 94/62/EC on packaging and packaging waste and would fix new targets to be achieved by 2008. Recycling targets by weight would be: overall recycling 55-80% (up from 25-45%), glass 60%, paper/board 60%, metals 50%, plastics recycled back into plastic 22.5% (up from 15% for all materials), and wood 15%. Feedstock recycling for plastics would count towards overall recycling target, but not the specific 22.5% plastics recycling target. Would include guidelines on what is and is not packaging. Expected to be adopted 2004.

Proposal to Amend Directive 2002/96/EC on Waste Electrical & Electronic Equipment

This would amend Directive 2002/96/EC on waste electrical and electronic equipment (WEEE). From 13/8/05, producers of new non-household electrical and electronic products would take-back old products (put on the market after 13/8/05) that are equivalent or fulfil the same function. If no new products are made to replace historical products, users (other than private households) would be responsible for managing historical non-household WEEE. Expected to be adopted 2003.

Expected Proposal For a Directive on Mining Waste

Would regulate the construction, operation, closure and post-closure phase of a disposal facility. Would require mine and quarry operators to have a waste management plan, and an internal emergency plan for measures to be taken inside a mine or quarry. Operators would have to hold a permit before mining could begin. Would provide for disposal facilities to be classified according to dimension and potential risk. Would oblige operators to monitor a mine after closure and minimise any environmental impact, and to guarantee that they could cover the potential costs of the waste facility and post-closure requirements, based on the facility's "likely environmental impact". Expected to be adopted 2005.

Expected Communication on Environmental Issues Of PVC

To examine the environmental issues related to PVC throughout its lifecycle. Would concentrate mainly on the use of certain PVC additives, including phthalates, lead, and cadmium, as well as the treatment of PVC waste. May suggest measures with a view to making a legislative proposal later, or propose voluntary means to deal with PVC. Commission is discussing voluntary commitments made by industry to phase out lead stabilizers in PVC by 2015 and increase recycling of PVC by 200,000 tonnes by 2010.

Expected Proposal For a Directive on Sewage Sludge

This would amend Directive 86/278/EEC on sewage sludge use in agriculture. Would aim to promote the use of sludge in areas such as silviculture and land reclamation. Would extend the definition of sewage sludge to cover sludge from urban wastewater treatment plants, septic tanks, domestic wastewaters from dwellings, and certain industrial sectors including food and drink (dairy, baking and confectionery, beverages), fibre and paper, and the leather industry. Would impose stricter limit values for heavy metals in sludge and soil and extend the list of heavy metals covered (cadmium, copper, nickel, lead, zinc and mercury) to include chromium. May introduce limit values for a number of organic compounds, including dioxins. Would require prior biological, chemical or heat treatment of sludge, according to the specific use to be made of land (e.g. sludge to be spread on land where the public has access would require a higher degree of treatment). Expected to be adopted 2005.

Expected Proposal For a Directive on Batteries and Accumulators

This would repeal Directives 91/157/EEC, 98/101/EC and 93/86/EEC on batteries and accumulators and extend their scope to cover all types of batteries and accumulators. Producers would finance collection, treatment, recovery, and safe disposal of waste batteries. Batteries/accumulators would have to be easily and safely removable from appliances and accompanied with instructions. Batteries/accumulators and appliances containing them would be marked for separate collection and to indicate which heavy metals they contain. Producer responsibility for managing waste batteries would be implemented either by means of a legislative requirement, or on the basis of a voluntary agreement. Would set: collection targets in the range of 30-40% or 60-70%or 70-80%for consumer batteries, 70-80% or 80-90%or 90-100% for car batteries, and 60-70% or 70-80% or 80-90% for batteries containing cadmium; and recycling targets of 45-55% or 55-65% or 65-75% for consumer batteries, 50-60% or 60-70% or 70-80% for car batteries, and 50-60% or 60-70%or 70-80% for batteries containing cadmium. A recovery target for cadmium from batteries would be set. Ni-Cd batteries would be banned where commercially viable substitutes are available. Expected to be adopted 2005

Expected Proposal For a Commission Decision on Classification of Waste-to-Energy Processes

To establish criteria that incineration should fulfil to be regarded as energy recovery. To amend entries D10 (incineration on land as a disposal operation) and R1 (use principally as a fuel or other means to generate energy as recovery operation) of the Waste Framework Directive 75/442/EEC. The European Commission's (CEC) position on cases pending in the European Court of Justice (ECJ) is that burning waste in industrial processes (e.g. cement kilns, power plants) and municipal incinerators should be considered recovery. The European Standardisation Committee is discussing criteria for classifying waste-to-energy processes. CEC may propose either a legislative amendment, or a Communication to clarify waste-to-energy classifications. The ECJ delivered rulings on two cases on 13/2/03. It states that waste burnt "principally" as a fuel with the genuine primary objective of generating energy should be classified as for recovery. The energy generated and recovered should be greater than the energy consumed and the surplus is used to produce heat or electricity, as in industrial processes [3]. Waste that is burned in municipal incinerators with the primary objective of eliminating it should be classified as for disposal, even if energy is generated and recovered [4]. The rulings also say that the waste's calorific value, the amount of harmful substances contained in incinerated waste, or whether the waste has been mixed or not does not need to be considered if these conditions have been met. CEC wants to talk to Member States about the possibility of drafting a legislative amendment, to bring waste management rules into line with European case-law.

Expected Proposal For a <u>Directive</u> on Biodegradable Waste

This would aim to promote separate collection and treatment of biodegradable waste to help meet targets in the Landfill Directive (99/31/EC) for reducing the amount of waste sent to landfill: by 2006, reduction to 75% of total biodegradable municipal waste produced in 1995; reduction to 50% by 2009; reduction to 35% by 2016. Would set out options in the following order: 1) prevention/reduction; 2) reuse; 3) recovery/recycling; 4) composting of waste recovered separately; 5) mechanical/biological treatment of unsorted biodegradable waste; and 6) use as a source of energy. Member States would establish biodegradable waste collection schemes within 3 years in urban areas for a population above 100,000 and within 5 years in agglomerations over 2,000. Would introduce permit requirements for treatment plants, with possible exemptions for plants with a total annual output of 100 tonnes of compost. Would introduce EU-wide compost quality standards in 3 classes. Less stringent standards would apply to a 3rd class lower-quality compost, to be known as sabilised bio-waste. Sets out emission limit values for biogas generated by bio-waste treatment. Would address handling and safe disposal of catering waste. Commission has asked its Scientific Committee to give an opinion on a study on heavy metals and organic compounds from waste used as an organic fertiliser. It would use this to propose limits for pollutants in compost. Expected to be adopted 2006

Expected Proposal For a <u>Measure or Guidance</u> Document on Definition Of Waste

To clarify the waste definition under the Waste Framework Directive 75/442/EEC. Would determine under which circumstances and at what stage a material should be considered a waste and when a waste becomes a raw material or product again (e.g. after treatment or by fulfilling conditions relating to composition, etc.). Would suggest a definition of "secondary raw material". The European Court of Justice (ECJ) has ruled that a substance is waste if: 1) it is subject to recovery; 2) it is a residue from the manufacturing process of another substance; 3) the only possible use to be given to the substance is elimination; 4) its composition is not appropriate to the use; 5) it requires precautionary measures to protect the environment [5]. CEC is awaiting a final ECJ ruling on a case concerning steel packaging. A preliminary opinion states that scrap steel is to be classified as waste until it has been melted and turned into ingots, sheet or coil, when it can be considered recycled (and, therefore, no longer waste). Once the ECJ delivers a ruling on this case, and based on others dealing with disposal/recovery classifications, CEC will consider how the waste definition could be clarified/updated. CEC is discussing what a "secondary raw material" is, which would contribute to a discussion on when waste is no longer waste. CEC is discussing whether the waste definition should be addressed in a legislative framework, or if it should form part of the debate on recycling. An ECJ Advocate General has issued an opinion (10/4/03) that reiterates an earlier case and confirms that rock left over from mineral extraction operations, which is stored indefinitely awaiting further use, should be classified as waste.

Expected Communication on Waste Recycling

This discussion document on a "Thematic Strategy" for waste recycling, would set out actions to identify priority wastes for recycling. Selection would be based on criteria linked to the results of analyses that identify where recycling produces an obvious net environmental benefit. The ease and cost of recycling the wastes would also have a bearing. Would set out indicative recycling targets. Would establish monitoring systems to track and compare progress by Member States to ensure that the collection and recycling of these priority waste streams occurs. Would identify policies and instruments to encourage the creation of markets for recycled materials. Would ensure separation of wastes at source, further development of producer responsibility and more environmentally sound waste recycling and treatment technology. May tackle the debate on reuse over recycling in CEC waste management hierarchy. Would review the definition of recycling. Would highlight resources that may be obtained from waste, such as energy recovery. Expected to be presented 2003

Expected Proposal For a Measure on Waste Contaminated With Persistent Organic Pollutants

This would establish a strategy for identifying waste consisting of, containing, or contaminated with persistent organic pollutants (POPs). Would provide for contaminated waste to be handled, collected, transported, and stored in an environmentally sound manner. When disposing of contaminated waste, the POP content would have to be destroyed or POP characteristics irreversibly transformed. The Commission will probably launch a study in preparation for a later proposal to set thresholds, and propose more general provisions now, based on the Stockholm Convention on POPs. CEC is getting ready to issue a single legislative proposal to regulate all aspects of POPs, including waste aspects.

Expected Proposal For a Directive on Waste Oils

This would update Directive 75/439/EEC on the disposal of waste oils (amended by Directive 87/101/EEC) to take account of technological progress and bring provisions into line with other Community waste legislation. May revise the threshold (50 parts per million) for polychlorinated biphenyls (PCBs) content in waste oils at which they become subject to the requirements of Directive 96/59/EC on the disposal of polychlorinated biphenyls and polychlorinated terphenyls (PCBs/PCTs). A study (2/02) commissioned by CEC on the regeneration and incineration of waste oils finds that in 1999, 47% of collected waste oils were incinerated to create energy, 24% were regenerated, 28% were burnt illegally, and 1% were discarded [6]. The Waste Oils Directive gives priority to regeneration over incineration, but Member States are widely using waste oils as fuel in industrial plants. The study finds that, taking a full life-cycle analysis into account, regeneration does not always have environmental advantages that far outweigh those of incineration. Commission might revisit current rules requiring Member States to give priority to regenerating, rather than incinerating, waste oils, once the outcome of infringement cases against 12 Member States is known.

Expected Proposal For a Commission Decision on Classification of Disposal & Recovery Operations

To revise and update classifications of operations as "disposal" or "recovery" of waste (Annexes IIA and IIB of Framework Directive 75/442/EEC on waste). CEC says that the current lists of recovery and disposal operations need to be updated. It says that new operations should be added, ambiguities clarified, and it wants to check if recovery outputs meet environmental concerns. The European Court of Justice has delivered a further ruling stating that waste which is to undergo more than one treatment should be classified according to the first treatment or operation. Waste's calorific value is not relevant for determining if an operation is recovery or disposal [7]. Expected to be adopted 2004.

Expected Proposal For a Regulation on Shipments of Waste

This would amend Regulation (EEC) 259/93 on the supervision and control of shipments of waste within, into and out of the EC. Exporters would require explicit prior written, rather than tacit, consent from importers before shipments of hazardous, semi-hazardous, and mixtures of waste could proceed. Exporters would send notification of a shipment to the authorities of the exporting country, who would then notify authorities of importing and transit countries. National authorities could object to a shipment based on national regulations or if the shipment's classification is in dispute (whether the waste is being shipped for disposal or to be incinerated and the resulting energy recovered). Would introduce new models for documentation. Expected to be adopted 2005.

Communciation From the Commission on Environmental Indicators

To establish a set of six headline environmental indicators to monitor the implementation of the specific objectives outlined in the EU strategy for sustainable development, covering: climate change; natural resource management; protection of public health; and land use and transport. These indicators will measure total greenhouse gas emissions, urban air quality, municipal waste generation, the energy intensity of the economy, volumes of transport (tonnes and passenger km) relative to GDP, shifts in the modal split of transport ,and the share of renewables in electricity consumption. Would form part of an open list of 20 indicators, from which six would be chosen each year by CEC for a synthesis report on progress towards sustainability, economic and social objectives. Outlines a second list for which indicators may be developed: chemical consumption, biodiversity, resource productivity, hazardous waste generation, recycling rates of selected materials, life expectancy air quality, water quality, municipal waste volumes and waste prevention. CEC released updated environmental indicators. Greenhouse gas emissions in the EU have not changed, but the energy intensity of the economy has decreased. The amount of municipal waste collected and landfilled remained the same, but the amount incinerated went up slightly.

Expected Communication on Sustainable Use of Resources

To launch a debate on a thematic Strategy on the Sustainable Use of Resources. This would outline why action is required and which resources are considered a priority. May propose: economic instruments to encourage the uptake of resource-efficient technologies, products and services; promotion of research and development into less resource-intensive products and product processes; best-practice programmes for business; removing subsidies that encourage overuse of resources; integration of resource efficiency considerations into other policies, and environmental reporting. May also: estimate waste streams; review the efficiency of policy measures and the impact of subsidies relating to natural resources and waste; establish goals and targets for resource efficiency; promote extraction and production methods to encourage eco-efficiency; and develop best practice programmes and indicators of resource efficiency. A recent study on resource use produced for CEC says that the EU's total material requirements have been decoupling relatively from economic growth, implying that resource efficient production is already favoured in the EU [8]. However, this has not led to a decrease in the EU's overall resource consumption. The report states that the EU cannot rely on a "business as usual" situation to lead a real cut in resource use, and that policies will have to be designed to influence resource use. The report states that it is not scientifically possible to prioritise environmental impacts of resource use, and that a political decision will have to be taken on which resources should be the target of policies in the short and long term.

Expected Communication On Integrated Product Policy (IPP)

This would set out a strategy for integrating environmental criteria into the manufacture and design of products (integrated product policy-IPP), detailing objectives and the tools to implement them. Some of the potential tools preferred by CEC to implement IPP are standardisation, green public and private procurement policy, environmental taxation, life-cycle analyses, voluntary environmental agreements, environmental product declarations, environmental taxes, environmental management systems. Would develop a method to identify the most environmentally damaging products. Would start "pilot projects" to look at the effects on the environment of certain products in the food, construction products, automotive and electronic sectors. References to stimulating eco-design have been removed, and it no longer recommends the use of environmental product declarations to stimulate green private procurement.

Possible Proposal For a Measure on Environmental Product Declarations

This would aim to set a harmonised standard for environmental product declarations (EPD).This could be based on life-cycle analyses undertaken by manufacturers which would provide information about the environmental impact of a product. Commission says it may use EPDs to create a demand for greener products by providing consumers with information on a product environmental performance. To do this CEC would first have to establish an EU-wide database of life-cycle analysis (LCA) information before proposing legislation. In a recent report for CEC by external consultants, EPDs were evaluated to see how they could be used as part of an Integrated Product Policy. The report concludes that CEC is well placed to play a role in developing EPDs. It says that CEC could harmonise EPDs by using European or international standards, stimulate the supply of EPDs by providing LCA data on products, and stimulate demand for EPDs through public procurement criteria. CEC stated in its draft Communication on integrated product policy (IPP) that it will decide whether to propose a measure on EPDs at the end of 2003.

Expected Measure on Sustainable Production and Consumption

This would look at how to make production and consumption sustainable. Since the World Summit on Sustainable Development in Johannesburg in September 2002 there has been an ongoing discussion on how to make the EU's production and consumption sustainable; this would be a contribution to that debate. Direct Material Consumption, for the EU as a whole, grew from 15.0 tonnes per capita in 1983 up to 16.7 tonnes per capita in 1989. After a slight decrease in the early 1990s, it stayed fairly constant at around 15.5 tonnes per capita in the second half of the 1990s (varying greatly between different Member States).

Expected Communication on an Action Plan on Environment and Health

This would aim to deal with the growing problem of illness and disease caused by environmental pollution. Would focus on research into this area, and would set up a pan European health and environment monitoring system. Would set objectives to reduce exposure to pollution and prevent environment related diseases. May set up pilot projects to look at PCBs and Dioxins in the Baltic area, heavy metals (cadmium, nickel etc) in air and soil, and exposure to endocrine disrupting substances. Commission says that research has shown that 20% of diseases in the EU may be linked to the environment, and that more research is need into this area. CEC says that diseases such as respiratory illnesses, development disorders are increasing and can be linked to environmental factors [9].

REFERENCES

1.   NATIONAL WASTE PLAN 2003, SEPA, February 2003.

2.   ICER UK Status report on Waste from Electrical and Electronic Equipment 2000.

3.   Commission of the European Communities vs Federal Republic of Germany, Articles 45-46, Case C-228/00, 13 February 2003.

4.   Commission of the European Communities vs Grand Duchy of Luxemburg, Articles 40-42, Case C-458/00, 13 February 2003.

5.   Criminal Proceedings against Euro Tombesi and Adino Tombesi (Case C-304/94), Roberto Santella (C-224/95), Giovanni Muzi and others (C-342/94) and Anselmo Savini (C-224/95), 25 June 1997.

6.   V MONIER, E LABOUZE; "Critical Review of existing studies and Life-Cycle analysis on the Regeneration and Incineration of Waste Oils", Taylor Nelson Sofres S.A, December 2001, 208 pp.

7.   SITA EcoServiceBV vs Minister van Volksuisvesting, Ruimtelijke Ordening en Milieubeheer, Case C-116/01), 3 April 2003.

8.   S MOLL, S BRINEZU and H SCHUTZ; "Resource Use in European Countries", European Topic Centre on Waste and Material Flows, Copenhagen, March 2003, 91 pp.

9.   COMMISSION OF THE EUROPEAN COMMUNITIES, "Protecting our children's health by cutting back pollution", Press Release IP/03/502, Brussels, 4 April 2003.

EFFECTS OF EU WASTE DIRECTIVES ON THE UK CONSTRUCTION INDUSTRY

M Sjogren Leong
Building Research Establishment
United Kingdom

ABSTRACT. The growth of waste in the EU has driven the need to increase levels of effective waste minimisation and management. Current and future legislation will be a key driver for resource efficiency within the construction industry. It will challenge the industry to manage their resources effectively and efficiently. There are obvious advantages and opportunities for the waste management industry too, with clients and main contractors requiring material waste management strategies for particular types of materials and sites. This paper gives an introduction to the key legislations and policies that may have an impact on the construction and demolition industry and their waste management strategies.

Keywords: Construction and Demolition Waste, Landfill Directive, European Waste Catalogue, Packaging Waste Directive.

Michelle Sjogren Leong is currently doing research at BRE on waste minimisation and recycling of wastes from construction and other industries. Her work includes monitoring and auditing of waste streams using SMARTWaste and conducting market surveys and industrial case studies for WRAP and DTI.

INTRODUCTION

The latest figures from the Symonds Report (Symonds 2001) show that the construction and demolition (C&D) sector produce 94 million tonnes of inert waste, of which 38.02 million tonnes are recycled into aggregates. However, we have to bear in mind that these figures only include inert materials, like concrete, bricks and blocks, cement and soil etc. The Centre for Resource Management (CRM) at BRE estimates the true figure of C&D waste production to be circa 140 million tonnes. This figure includes all waste produced in the C&D process taking into consideration other waste that is not commonly included, for example, timber, metals, plastics, packaging, furniture and many more. The BRE SMARTWaste auditing tool uses twelve waste groups and a list of 500 waste products with the potential for an infinite list. The waste groups are currently being adapted to the European Waste Catalogue. With such staggering amounts of waste involved, current and future legislation will be a key driver for resource efficiency within the construction industry. It will challenge the industry to manage their resources effectively and efficiently. There are obvious advantages and opportunities for the waste management industry too, with clients and main contractors requiring material waste management strategies for particular types of materials and sites. This paper gives an introduction to the key legislations and policies that may have an impact on the construction and demolition industry and their waste management strategies but will focus mainly on the Landfill Directive, the European Waste Catalogue and the Packaging Waste Directive.

COMMON EU STRATEGY: 6TH ENVIRONMENTAL ACTION PROGRAMME

The 6th Environmental Action Programme is a European Community framework that has been established to limit the environmental and health impacts arising from the use of natural resources. This includes taking measures to improve the resource efficiency of energy use, and implementing the sustainable use of water and of soil. However, for non-renewable resources, although affected indirectly by many different policies, there lacks a coherent Community policy focused on the overall de-coupling of resource use (resulting in environmental impacts and degradation) from economic growth. Therefore, an objective of the 6th Environmental Action Programme, is to outline the priorities for action on the environment for the next 5 to 10 years, in order to,

"... ensure the consumption of renewable and non-renewable resources and the associated impacts do not exceed the carrying capacity of the environment and to achieve a decoupling of resource use from economic growth through significantly improved resource efficiency, dematerialisation of the economy, and waste prevention."

One of the key objectives related to resource efficiency is to de-couple the generation of waste from economic growth and achieve a significant overall reduction in the volumes of waste generated through improved waste prevention initiatives, better resource efficiency and a shift to more sustainable consumption patterns. Also, for wastes that are still generated the aim is to achieve a situation where:

- The waste is non-hazardous or at least presents a minimal risk to the environment and human health.
- The majority of the waste is incorporated back into the economic cycle, especially through recycling, or returned to the environment in a useful (e.g. compost) or harmless form.

- The waste that still has to go for final disposal is reduced to an absolute minimum and is safely destroyed or disposed of.
- Waste should be treated as closely as possible to its source.

From these objectives a number of targets to be achieved within the lifetime of the Community programme have been set, so as to observe a significant reduction in the volume of waste going for final disposal and in the volumes of hazardous waste generated. To keep with the general strategy of waste prevention and increased recycling, the following aims have been decided:

- To reduce the quantity of waste going for final disposal by around 20% by 2010 compared to 2000 and 50% by 2050.
- To reduce the volume of hazardous wastes generated by the same target as above.

KEY EU LEGISLATION AND REGULATION

Working Group on Sustainable Construction

As one of the fourteen priority actions for improving competitiveness within construction, a Working Group on Sustainable Construction was established in 1999 which included three Task Groups, one of which was TG3 on C&D waste management. The main function of TG3 was to provide a document of recommendations on how to improve C&D waste management through improved planning, prevention and reclamation.

The scope of the document focused on the whole construction process including design, pre-construction, construction, demolition, reuse, recycling, final disposal, research and education. The output was to make recommendations to three core sectors of construction including Industry, Member States and their public authorities, and the European Commission and incorporated other requirements of industry and member states. For example, member states are requested to provide waste management plans to facilitate self-sufficiency, reduce movements of waste materials and establish inspections of disposal and reclamation. Reports to the Commission by individual States is to be submitted every three years, for agglomeration into a single report.

Project Group on C&D Waste

In addition to TG3, the European Commission set up the "Priority Waste Streams Programme" in 1992 and, six priority waste streams programmes were initiated including:

- Used Tyres
- End-of-Life Vehicles
- Chlorinated Solvents
- Healthcare Waste
- Construction and Demolition Waste
- Waste from Electrical and Electronic Equipment

Project groups were set up for each of the six waste streams to discuss and recommend ways that member states could improve methods of waste management of these.

In April 2000 a working document produced by the C&D waste project group described the measurement of the C&D waste stream in member states, and detailed the aims and instruments that are likely to improve C&D waste management. The document also includes a selection of recommendations which member states need to consider when developing their own waste management policies.

Currently, the European Commission wishes to introduce a recommendation (a non-binding measure) for C&D waste with the aim of improving the management of the C&D waste stream by following the waste hierarchy, giving preference to prevention over reuse, material recycling, energy extraction and lastly disposal. It will aim to reduce the impact of C&D waste on the environment whilst better utilising natural resources.

The recommendation will also encourage the substitution of hazardous substances in new buildings and make sure that waste from construction (bricks, glass, wood etc) is sorted at the point of generation. It would also include proposals for recycling targets set initially at 50% to 70% by 2005 and an increase in landfill charges. It is thought that the European Parliament would prefer binding legislation rather than just a recommendation. So far progress has been very slow. If adopted, this recommendation/legislation will have a significant impact on the demolition industry which supplies the construction sector.

European Waste Catalogue

In providing a European position on waste, it has also been necessary to provide a common reference system for waste. The European Waste Catalogue (EWC) came into force in January 2002 through an amendment to the Duty of Care regulations. It applies to all wastes in Europe whether for disposal or reclamation, and is a harmonised, non-exhaustive list using common terminology across the Community. However the inclusion of a material in the EWC does not mean that it is a waste, only when the relevant definition is satisfied is it considered waste. The EWC identifies 20 broad categories of waste and over 800 waste types based on the process, giving rise to the waste. It took effect in January 2003 in the UK under the Landfill Directive regulations where the classifications will replace the simple waste identification system on the Duty of Care transfer note.

An expert working group drawn from Member States, DGXI and Eurostat is currently reviewing the EWC. According to a Symonds' report (Symonds 1999) to the EU, Member States are interpreting the EWC in different ways. The "Top down" classification adopted by Germany and the Netherlands involves finding the most appropriate top level classification and recording the waste there, whether or not it can be assigned to one of the sub-categories. The UK uses the "Bottom up" classification by finding the most appropriate sub-category and calculating the totals of the top level classifications by adding values for each sub-category. Both approaches used should ideally lead to a common result but it does not. In another words, the "top down" approach would categorise e.g. concrete, bricks, tiles and ceramics as 17 01 00 (concrete, brick, tiles, ceramics), whether mixed together or not, while the "bottom up" approach would classify any mixtures of these waste as 17 07 00 (mixed C&D waste).

Table 1   Main waste codes most relevant to C&D waste production

WASTE CLASSIFICATIONS	TYPE OF WASTE
15	Packaging; absorbents, wiping cloths, filter materials and protective clothing
17	Construction and demolition waste (including excavated soil from contaminated sites)
20	Municipal wastes (household waste and similar commercial, industrial and institutional wastes) including separately collected fractions

Table 2   Waste Packaging

CLASSIFICATIONS	TYPE
15 01	Packaging (including separately collected municipal wastes)
15 01 01	Paper and cardboard packaging
15 01 02	Plastic packaging
15 01 03	Wooden packaging
15 01 04	Metallic packaging
15 01 05	Composite packaging
15 01 06	Mixed packaging
15 01 07	Glass packaging
15 01 09	Textile packaging
15 01 10 (Hazardous)	Packaging containing residues of or contaminated by dangerous substances
15 01 11 (Hazardous)	Metallic packaging containing a dangerous solid porous matrix (for example, asbestos), including empty pressure containers

Table 3   Construction and Demolition Waste

CLASSIFICATIONS	TYPE
17 01	Concrete, bricks, tiles, ceramics
17 02	Wood, glass and plastic
17 03	Bituminous mixtures, coal tar and tarred products
17 04	Metals (including their alloys)
17 05	Soil (including excavated soil from contaminated sites), stones and dredging spoil
17 06	Insulation materials and asbestos-containing construction materials
17 07	Mixed C&D waste
17 08	Gypsum-based construction material
17 09	Other construction and demolition wastes

Table 4   Municipal Waste (relevant classifications)

CLASSIFICATIONS	TYPE
20 01	Separately Collected Fraction
20 01 01	Paper and cardboard
20 01 02	Glass
20 01 08	Biodegradable kitchen and canteen waste
20 01 11	Textiles
20 01 13 (Hazardous)	Solvents
20 01 14 (Hazardous)	Acids
20 01 15 (Hazardous)	Alkalines
20 01 21 (Hazardous)	Fluorescent tubes and other mercury-containing wastes
20 01 23 (Hazardous)	Discarded equipment containing chloroflurocarbons
20 01 27 (Hazardous)	Paint, inks, adhesives and resins containing dangerous substances
20 01 28	Paint, inks, adhesives and resins other than those mentioned in 20 01 27
20 01 29 (Hazardous)	Detergents containing dangerous substances
20 01 30	Detergents other than those mentioned in 20 01 29
20 01 35 (Hazardous)	Discarded electrical and electronic equipment other than those mentioned in 20 01 21 and 20 01 23 containing hazardous substances
20 01 36 (Hazardous)	Discarded electrical and electronic equipment other than those mentioned in 20 01 21 and 20 01 23
20 01 37 (Hazardous)	Wood containing dangerous substances
20 01 38	Wood other than that mentioned in 20 01 37
20 01 39	Plastics
20 01 40	Metals
20 01 99	Other fractions not otherwise specified
20 03	Other municipal wastes
20 03 07	Bulky waste

Hazardous Wastes 91/689/EEC & 94/904/EC

Member states are required to implement controlled management of hazardous waste. These indicate the appropriate means necessary to collect, transport, store and manage hazardous wastes. These are defined in Annexes covering generic types of hazardous waste including pigments, paints, resins, and plasticisers, and properties of waste which render them hazardous including oxidising, harmful, carcinogenic and corrosive substances, as well as substances that yield damaging leachate or ecotoxic risks. In 2001, the hazardous waste list created by EC Decision 94/904/EC was incorporated into the EWC.

UK Special Waste Regulations

The current Special Waste Regulations are undergoing a major review, which will simplify some procedures. However, a re-designation of 'special waste' to 'hazardous waste' will reveal more types of waste covered under the new regulations. It is expected that the new regulations will take effect in 2003 as the Hazardous Waste Regulations. The two

Regulations identify all hazardous wastes with those items not currently designated special waste such as fluorescent tubes and timber creosote sleepers becoming hazardous wastes when the new regulations take effect.

Under the Landfill Directive, pre-treatment will be necessary for most hazardous wastes which are to be landfilled. Pre-treatment can reduce the hazardousness of waste or, in some cases, render it non-hazardous. Certain treatments may be used which do not alter the hazardous properties of a waste but significantly reduce the probability of the hazard having an effect. A wide variety of pre-treatment techniques are available, with biological, thermal and some physico-chemical treatments most suitable for organic wastes and physico-chemical treatment most suitable for inorganic wastes.

A number of waste streams arising from C&D waste may be deemed hazardous waste depending on the concentration of contaminants e.g. contaminated soils, asbestos, tar and tar products, treated timber, paint and varnish. Paint waste is generated by construction activities and some paints are hazardous wastes. Waste paint can be reused either by the construction company for the next job or through community repaint schemes. Between 30,000 and 40,000 tonnes of waste paint was generated in 1997/8 and 1998/9, although it is not possible to say how much of this came from the construction sector.

All wastes containing greater than 0.1% asbestos are classified as hazardous waste. Cement or bonded asbestos is a major hazardous waste stream. Re-use and recycling are not suitable options for asbestos wastes, as asbestos is banned for use in virtually all new applications. Treatment of asbestos wastes is difficult and expensive, although technologies to destroy asbestos using heat treatment or acid digestion are being explored. In the meantime the only practical option for asbestos wastes is disposal to landfill. Asbestos wastes must not be allowed to enter the inert waste stream destined for recycling as secondary aggregates. Crushing of asbestos wastes, or their use in roadways or as hardcore will result in the wastes breaking up and could release fibres into the atmosphere. There is a shortage of facilities accepting asbestos cement wastes in some areas. When co-disposal stops in 2004, asbestos can be landfilled in a non-hazardous landfill site in separate cells according to the Landfill Directive Regulatory Guidance Note 11 (EA 2002).

A large amount of contaminated waste is consigned to landfill in England and Wales. Research and development of soil treatment and cleaning technologies are allowing soils to be recycled for beneficial use, break down contaminants or extract them for further treatment and/or disposal. It is expected that there will be increasing use of such technologies to recover soils and remove contaminants both on and off site. The increasing use of treatment and recovery of contaminated soils may lead to a reduction in the quantities of waste soils being consigned as hazardous waste. However, a number of factors such as increased redevelopment of brownfield sites and the implementation of the new contaminated land regulation, may result in the redevelopment and remediation of more contaminated sites. This could lead to an increase in the quantity of waste contaminated soils being generated.

Packaging and Packaging Waste Directive 94/62/EC

The Packaging and Packaging Waste Directive introduced in 1994 sets targets to be achieved for both the recovery and recycling of packaging waste, which are revised every five years. In December 2001 the European Commission released a proposal for a revision of the

Packaging and Packaging Waste Directive which focuses on the new five year targets to be met by June 2006 (30 June 2009 for Greece, Ireland and Portugal). It also clarifies some of the definitions of the original text.

The European Parliament voted on the proposal and an amended version was then passed to the European Council of Ministers for review in October 2002. The targets were again altered and the deadline date was changed to 31 December 2008 (2012 for Greece, Ireland and Portugal). The targets put forward by the Council were:

- Overall recovery (recycling plus energy recovery), minimum 60% with no maximum.
- Overall recycling 55% minimum (80% maximum).
- Minimum material specific recycling targets:
  o Glass 60%
  o Paper and board 60%
  o Metal 50%
  o Plastics (mechanical and chemical recycling) 22.5%
  o Wood 15%

Currently, the next stage for the EU Environment Council is to decide on a common position between the Parliament and Council's view. Their proposal will then be sent back to the Parliament for a second reading. It is thought that a compromise of 2007 may be chosen for the deadline.

UK Producer Responsibility Obligation (Packaging Waste) Regulations

The Packaging Waste Regulations implements the recovery and recycling targets set out in the EC Directive on Packaging and Packaging Waste. The UK recovery and recycling targets for 2003 are 59% for overall recovery and 19% for material specific recycling. Shared producer responsibility means that all parts of the UK packaging chain contribute towards meeting the recycling and recovery targets. The percentage obligation depends on the type of activity undertaken by the company on the packaging. The share of responsibility is: 6% for raw material manufacturers, 9% for converters, 37% for packer/fillers, and 48% for sellers. Also, there is 100% obligation on transit packaging around imported goods. All businesses with an annual turnover of £2 million or more that handle 50 tonnes or more of packaging each year must comply with the regulations.

The construction sector is obligated to the packaging waste regulations but is currently unregulated. However, the Environment Agency is beginning to show an interest in this sector as the amount of packaging generated is relatively large. Packaging plays an important role, it protects products, and is used to promote and identify the products. In terms of its use in the construction process, almost every product used on a site is delivered wrapped in, sat on or held together by packaging of some kind

There is a plethora of information on the recovery and recycling of packaging waste from households, but very little concerning packaging waste from construction sites. The majority of studies relating to waste on construction sites cover the topic of material waste management, which does not pay particular attention to the levels of packaging on sites. Little is known regarding the amount, type and cleanliness of packaging material being disposed of on construction sites. Previous BRE studies however, have shown that packaging materials can constitute as much as 50% of the volume of waste leaving a construction site.

The recyclability of the packaging products found in a BRE waste audit of nine sites is presented in Figure 1. As much as 34% of the packaging could have been recycled without the need for cleaning. More than 9% was reusable referring mainly to timber pallets that can be used again. Reusable packaging that is broken also consisted of timber materials (broken timber pallets or broken timber packaging) which can be recycled with the rest of the timber materials. About 27% of the packaging was recyclable but contaminated mainly by soil that could be recycled after some cleaning. Only 7% of the packaging was not recyclable or reusable, in other words, 93% of packaging waste can be diverted from landfill. And if packaging waste makes up approximately 26% of overall waste in a construction project, the amount of waste, if segregated, can be very significant.

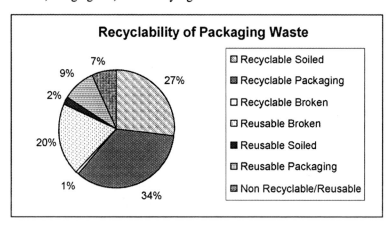

Figure 1   Recyclability of Packaging Products

The Landfill Directive

The Landfill Directive was introduced into UK law in 2002 through the Landfill (England and Wales) Regulations 2002. The Directive defines three classes of landfill depending on the type of waste they accept: hazardous, non-hazardous and inert waste. The co-disposal of waste is currently common practice in the UK. However, the Directive states that hazardous waste may only be landfilled in a hazardous waste site or if the hazardous waste is stable and non-reactive, landfilled in individual cells in a non hazardous site. Therefore the co-disposal of hazardous waste must cease by 2004. The following wastes are or will be banned from landfill by 2004:

- Explosive, oxidising or flammable wastes.
- Infectious clinical waste.
- Liquid wastes, except those suitable for disposal at an inert waste site.
- Tyres (whether whole or shredded).

The aim of the Directive is to provide measures, procedures and guidance to prevent or reduce negative effects to the global environment and all its cycles from landfilling of waste during the whole lifecycle of the landfill site. All waste, except for inert waste, is to be pre-treated before landfilling to reduce its volume or hazardous nature, or to aid recovery. Pre-treatment may involve mechanical waste separation and sorting procedures followed by composting, anaerobic digestion, thermal treatment and other processes. It is expected that by mid 2007 at the latest, no waste will be able to be collected and taken to landfill without some weight reduction being applied, through source segregation or sorting at a waste transfer station. Treatment applies to new landfills from July 2001, landfills classed to receive hazardous waste from July 2004 and all others by July 2009. Existing sites can be classed as hazardous until July 2004, then reclassify as non-hazardous, when co-disposal ends.

The Landfill Directive also impacts on the type of waste accepted at landfills. Inert waste can be accepted for which treatment is not technically feasible. Landfills will be divided into hazardous, non-hazardous and inert. For inert landfills, operators can only accept waste that:

• Does not undergo any significant physical, chemical or biological transformations.
• Does not dissolve, burn or otherwise physically or chemically react, biodegrade or adversely affect other matter with which it comes into contact in a way likely to give rise to environmental pollution or harm to human health;.
• Its total leachability and pollutant content and the eco-toxicity of its leachate are insignificant and, in particular, do not endanger the quality of any surface water or groundwater.

Table 5   Waste accepted at landfills for inert waste

EUROPEAN WASTE CATALOGUE	DESCRIPTION	EXCLUSIONS
15 01 07	Glass (Packaging)	
17 01 07	Concrete	
17 01 02	Bricks	
17 01 03	Tiles and Ceramics	
17 02 02	Glass (C&D)	
17 05 04	Soil and stones	Excluding topsoil and peat
20 01 02	Glass (Municipal)	
20 02 02	Soil and stones	Excluding topsoil and peat

The Government proposed that appropriate measures for water control and leachate management should be determined by the Environment Agency on a site specific basis for inert sites. The tight definition of inert waste for the Landfill Directive is different to the definition used for the purposes of the landfill tax. The Government has confirmed that these definitions are for wholly different purposes and waste, which is defined as inert for the landfill tax, will remain inert for those purposes. The Government does not consider that the landfill tax regulations need to be amended to extend the definition to include the Directive definition of inert. The general requirements of the Directive for landfill engineering for the protection of soil and water will increase demand for inert waste in the construction, redevelopment and restoration of landfills. The demand for future inert waste landfill capacity is therefore likely to be reduced significantly, with a consequent reduction in future landfill numbers and licensing requirements.

Waste materials that would most concern the C&D industry, with regards to the Landfill Directive, will mainly be gypsum, composites and any biodegradable materials like timber. These materials will have to be diverted from landfill in one way or another. Certain preserved/treated timbers will be particularly difficult after 2004 when co-disposal is prohibited, pre-treatment will be required and there will only be 37 hazardous sites nationwide of which, only 10 to 15 will be accessible to commercial uses. Gypsum, although classified as hazardous waste, can be landfilled in a non hazardous landfill site in separate cells according to the Landfill Directive Regulatory Guidance Note 11 (EA 2002).

Landfill Tax

The landfill tax scheme was introduced in 1996 and applies to waste disposed of in licensed landfills. The aim of the tax is to ensure that the price of landfill fully reflects the impact that the landfilling of waste has upon the environment. It provides an incentive to reduce the quantity of waste sent to landfill and increase the proportion of waste managed by processes higher up the waste hierarchy. The revenue raised from the landfill tax is used to encourage the use of more sustainable waste management practices and technologies.

There are two rates of tax, a standard rate of £14 (2003) per tonne and a lower rate of £2 per tonne. The higher rate for mixed (non-hazardous) waste will increase by £1 every year until it reaches a rate of £15 per tonne in 2004 and the standard rate of landfill tax will subsequently be increased by £3 in 2005-06 to £18 per tonne, and by at least £3 per tonne in the years thereafter, on the way to a medium- to long-term rate of £35 per tonne. The categories of waste to which the lower rate of tax apply – generally inert waste – are set out in the Landfill tax (Qualifying Materials) Order 1996 (SI No 1528).

Mixed loads may be subject to the inert rate if there is no potential for pollution. For example stone and concrete with small amounts of plaster and wood attached may fall into the inert category. Similar active waste mixed with inactive waste will attract the higher rate. Waste materials are also exempt if temporarily disposed of, with certain conditions attached. Exemptions also apply to inert materials that are used for landfill restoration or filling former quarries.

A principle concern following the introduction of the landfill tax was the reduction in quantity of inert materials, principally construction wastes, arriving at landfill sites. Currently the majority of waste on construction (not demolition) sites are not segregated and attracts the higher rate of tax. Landfill tax has led to a significant increase in recycling and beneficial reuse of inert C&D waste especially from the demolition sector. Instead of sending excavation spoil to landfill and buying in new filling material, companies are starting to use their own wastes as a resource. Active waste to landfill has actually increased slightly and there is no clear evidence that the landfill tax is having any effect on behaviour in the construction industry. This can be attributed to the cost of landfill remaining very low even with the escalator. There has been much discussion recently over the cost of landfill tax and future increases. The tax is designed to make landfilling more expensive but is still very low in the UK currently at £14 per tonne, compared with £34 in Denmark and £45 in the Netherlands. Research has called for a doubling or tripling of the landfill tax in the UK. In the latest budget, it was announced that the tax would increase by £3 per tonne per year from 2005-6. The Government's aim is for the Landfill tax to reach £35 per tonne. The recent changes were lower than the waste industry had been expecting, with expectations that the tax should have been increased by £5 per year. As the increase was not as high as originally expected, it may not have had the immediate effect of discouraging landfilling of waste.

Table 6   Categories of waste described as inert for landfill tax purposes

GROUP	DESCRIPTION OF MATERIALS	QUALIFYING MATERIALS	
Group 1	Rocks and soils (naturally occurring)	Clay Sand Gravel Sandstone Limestone Crushed stone China clay	Stone from the demolition of buildings or structures Slate Topsoil Peat Silt Dredgings
Group 2	Ceramic or concrete materials	Glass (includes fritted enamel) Ceramics (includes bricks, bricks and mortar, tiles, clay ware, pottery, china and refractories) Concrete (includes reinforced concrete, concrete blocks, breeze blocks and aircrete blocks but excludes concrete plant washings)	
Group 3	Minerals (processed or prepared, not used)	Moulding sands Clays (includes moulding clays and clay absorbents) Mineral absorbents Man-made mineral fibres (includes glass fibres) Silica Mica Mineral abrasives	
Group 4	Furnace slags	Vitirified wastes and residues from thermal processing of minerals where the residue is both fused and soluble Slag from waste incineration	
Group 5	Ash	Bottom ash and fly ash from wood, coal or waste combustion	
Group 6	Low activity inorganic compounds	Titanium dioxide Calcium carbonate Magnesium carbonate Magnesium hydroxide Iron oxide	Ferric hydroxide Aluminium oxide Aluminium hydroxide Zirconium dioxide
Group 7	Calcium sulphate (disposed or either at a site not licensed to take putrescible waste or in contaminant cell which takes only calcium sulphate)	Gypsum and calcium sulphate plasters*	
Group 8	Calcium hydroxide and brine (deposited in brine cavity)		
Group 9	Water (containing other qualifying material in suspension)		

* Plasterboard is excluded

UK Waste Strategies

The UK waste strategies do not put too much emphasis on C&D waste, with the exception of Scotland. The Scottish National Waste Strategy identifies four priority waste stream projects, one of which is C&D waste requiring the development of a C&D Waste Action Plan that will reflect three key objectives and tasks:

- Comprehensive review of volume and location of C&D waste
- Levels of C&D waste reclamation, key players and barriers to reuse
- Future management and market development of C&D waste

The Scottish National Waste Strategy is the only Strategy in UK that has focused attention on C&D waste and identified it as a priority area. This will be watched closely by other Nations to see how its successes can be translated into their own Strategies.

Sustainable Construction Strategy

The need to reduce waste at all stages of construction was central to the message of "Rethinking Construction", the 1998 report of Sir John Egan's Construction Task Force on improving the quality and efficiency of UK construction. Improving the efficiency of the construction industry is a key objective for the Government, as set out in its strategy for more sustainable construction 'Building a Better Quality of Life', which identifies priority areas for action, and suggests indicators and targets to measure progress. It sets out action that the Government has already taken, further initiatives that are planned, and highlighted what others can do. The Government will use the strategy as a framework to guide its policies towards construction, and will encourage people involved in construction to do the same.

The Sustainable Construction strategy emphasises the importance of reducing waste at all stages of construction by focusing on the need to consider long term impacts of design, construction and disposal decisions so that materials and other resource use is optimised. The strategy encourages the industry (including clients) to consider refurbishment or renovation as an alternative to new buildings and structures. It highlights the need to avoid over-specification in materials and the scope for standardisation of components. The Office of Government Commerce published the government's own construction sustainability action plan, Achieving Sustainability in Construction Procurement in July 2000. Sixteen government departments and other bodies are now implementing this action plan. The Construction Industry's main clients take the view that the UK Construction Industry does not deliver consistent quality and value for money. Too often the performance of the industry is unreliable; projects run neither to time nor budget and too much effort and resource is invested in making good defects, carrying out premature repair and replacement, and in litigation. The central message of Rethinking Construction is that through the application of best practices, the industry and its clients can collectively act to improve their performance. The 'Rethinking Construction' report identified targets for improvement in construction productivity, profits, defect and accident reduction. As a follow up to the Egan report, the construction industry is currently looking at the future implementation of the Egan principles. In consultation with the wider public, the Strategic Forum for Construction has produced the report 'Accelerating Change', which focuses on the three improvement agendas of client leadership, integrated supply teams and people issues to deliver widespread improvement to construction processes on the ground.

CONCLUSIONS

The last decade has seen a very significant evolution of legislation, regulation and policy in relation to waste management, which is increasingly reflected in improved waste management practice. This process is ongoing. Representatives of industry, Federations, Associations and Trade Bodies working with BRE on various government funded projects have indicated that further legislation and regulations are required if we are to achieve the ambitious targets for sustainable waste management. However, this needs to be targeted at specific issues that are both economically viable and beneficial to the growth of the industry itself.

The Landfill Directive when implemented in 2004 will have a significant impact on co-disposal and pre-treatment, this together with the rising landfill tax will make site waste management more economically palatable. With the Landfill Directive and more attention being fixed on packaging waste on construction sites, a rise in the recycling rate will inevitably be the outcome.

As EU directives on waste management and the rising cost of landfill put increasing pressure on the industry, workable solutions for the management of C&D wastes need to be found. We can see it as both an opportunity and a challenge for the industry and also for those already involved in waste management and those who may diversify into C&D waste recycling.

REFERENCES

1. ANDERSON, M., CONROY, A & C, TSIOKOU. "Construction Site Packaging Wastes: A Market Position Report". BRE Information Paper, IP 8/02. BRE. Garston. 2000.

2. BUILDING RESEARCH ESTABLISHMENT (BRE). "Final Report on EU Directives". DTI Construction Industry Directorate. London. 2003.

3. ENVIRONMENT AGENCY (EA). "Landfill Directive. Regulatory Guidance Note 11" Environment Agency. Bristol. 2002.

4. EUROPEAN COMMISSION (EC). "Environment 2010: Our future, Our choice. The Sixth Environment Action Programme. European Commission. Brussels. 2001.

5. SYMONDS GROUP. "Survey of Arising and Use of Construction and Demolition Waste in England and Wales 2001", Environment Agency. Bristol. 2002.

6. SYMONDS GROUP. "Construction and Demolition Waste Management Practices, And Their Economic Impacts". DGXI European Commission. Brussels. 1999.

THE LANDFILL DIRECTIVE – IMPLEMENTATION ISSUES (THE MISSING LINK)

C Gray
Scottish Environment Protection Agency
United Kingdom

ABSTRACT. The concept behind the Landfill Directive is environmentally sound and, where definitions are clear (as with the technical standards for landfill basal liners), the industry knows what is expected and can invest accordingly. Where timescales and definitions are not clear due to Directive overlaps (IPPC versus Landfill Directive), Directive terminology ("as soon as possible") or domestic interpretations (hazardous versus special waste), regulators and waste managers are uncertain about requirements, and industry cannot be expected to invest in major capital works in an atmosphere of uncertainty.

Keywords: Environmental regulators, Landfill Directive, Scotland, Waste management.

Mr C Gray works for the Scottish Environment Protection Agency (SEPA).

INTRODUCTION

The Case

The overall aim of the Directive is, by way of stringent operational and technical requirements on waste and landfills, to provide for measures, procedures and guidance to prevent or reduce as far as possible negative effects on the environment, in particular the pollution of surface water, groundwater, soil and air, the greenhouse effect, and risk to human health during the whole life cycle of a landfill.

Of particular note is the stated intention to:

1. reduce the amount of biodegradable waste going to landfill (with targets and deadlines);
2. ban certain wastes from landfill;
3. ensure that the costs of landfill are reflected in gate prices;
4. require pre-treatment of wastes prior to landfilling; and
5. utilise landfill gas where possible and flare the gas where utilisation is not possible.

Landfill Directive Timescales

Landfill Directive 1991/31/EC	adopted 16 July 1999
Domestic legislation to implement required by:	16 July 2001
Hazardous landfill sites to be classified and prohibited wastes banned by:	16 July 2002
Site conditioning plans to be received from all operational sites by:	16 July 2002
Co-disposal of hazardous and non-hazardous wastes to cease:	16 July 2004
Existing sites to comply with Landfill Directive requirements "as soon as possible" and by 16 July 2009 at latest:	16 July 2009

Scottish Timetable as Implemented

First consultation:	February 2001
Direction to SEPA requiring hazardous landfill classification and prohibition of hazardous waste disposal in non-hazardous landfill sites with submission of site conditioning plans by 16 July 2002:	June 2002
251 licences modified and 15 sites classified as hazardous:	July 2002
Second Consultation:	August 2002
Landfill Regulations take effect:	April 2003

Key Issues

Overlap between Directives:
Principal Directives applicable to landfills:

a.	Landfill Directive.	(99/31/EC)
b.	Waste Framework Directive.	(75/442/EEC)
c.	IPPC Directive.	(96/61/EC)
d.	Hazardous Waste Directive.	(91/689/EEC)
e.	Groundwater Directive.	(80/68/EEC)
f.	Water Framework Directive	(2000/60/EC)

The overlap with the IPPC Directive is particularly problematic as this has three major implications in Scotland:

a.   A PPC permit is required (not a stand alone landfill permit).
b.   BAT is deemed to be met if the technical standards in the Landfill Directive are met.
c.   The timescale for IPPC compliance runs to 2007 (not 2009).

In effect most landfills therefore have to comply with the Landfill Directive by 2007 and not 2009 with confusion over the format of a Landfill PPC permit.

Definition of terms
There are a number of issues with the definition of terms either contained in the Directive or as presently applied in domestic Scottish legislation e.g.

a.   Hazardous waste/special waste.
b.   Pre-treatment.
c.   Restoration/disposal.

Timetable for implementation - the Missing Link
In effect, due to both the overlap with the IPPC Directive and the domestic implementation of the Landfill Directive, the timescale for improving the standard of landfill, and as a consequence reducing the number of landfills, has been compressed from a 10 year implementation period (1999 - 2009) to a 4 year implementation period (2003-2007). This situation presents all of us with a range of challenges related to building and commissioning new waste management facilities in time to handle the large quantities of waste which will be diverted from landfills.

Theoretical advances in waste management and recycling cannot be converted into real improvements on the ground without an appropriate regulatory framework in place.

CONCLUSIONS

By the end of 2007, licensed landfills in Scotland will be reduced from 260 in 2001 to around 100 with approximately 45 non-hazardous sites and 55 inert landfills. There are presently plans in place for 2 hazardous waste landfills.

The overall effect of implementing the Landfill Directive will be a reduction in numbers of landfill sites and a massive increase in costs. The reduction in numbers is particularly important in a Scottish context given the large haul distances involved to those landfill sites remaining or to alternative waste management facilities. The relatively small volumes of waste generated in large parts of Scotland act as a disincentive to invest in alternative facilities.

Delay and confusion in the application of the Directive has compressed the implementation timescale with the likely effect being a shortfall in alternative waste management facilities for the short to medium term.

This compression of the implementation period is particularly important in relation to planning requirements for new facilities and the present run in period for such facilities. The Missing Link in this process is the linkage between removing landfill as a waste disposal option replacing it with other more advanced waste management techniques based locally.

The overall objectives of the Landfill Directive are likely to be achieved in Scotland in due course but there will be a painful transition with likely increases in fly-tipping and disposal "deserts" in large parts of Scotland for a period of time requiring extensive road haulage with its associated environmental disbenefits.

These problems could have been minimised by greater clarity in the Directive, the prevention of duplication in Directive application (why include landfilling as an IPPC activity?) and the application of the intended Landfill Directive timescale.

Potential Solutions

1. Review the planning system to prevent undue delays in determining applications (particularly notable in relation to waste management facilities).

2. Put the necessary resources into resolving definitions and timescales to clarify requirements and to give the industry the certainty needed to invest.

3. Simplify the regulatory regime domestically (eg a non PPC permit for landfills could still meet the requirements of the IPPC Directive.)

4. Provide the necessary support for local authorities to provide alternative waste management facilities where uneconomic for the private sector to invest.

5. Take steps to ensure that penalties for fly tipping are realistic as the increased costs of transport and landfill will act as an incentive for waste producers to fly tip waste on a routine basis.

CEMENT AND MICRO SILICA BASED IMMOBILISATION Of DIFFERENT FLUE GAS CLEANING RESIDUES FROM MSW INCINERATION

D Geysen

C Vandecasteele

KU Leuven

M Jaspers

G Wauters

INDAVER

Belgium

ABSTRACT. Flue gas cleaning residues from MSW incineration cannot be easily reused so that landfilling is at the moment the only management option. The wastes are generally treated prior to landfilling to prevent leaching of heavy metals so that no intensive effluent treatment is needed. Usually, cement is used for immobilisation, but for some wastes high amounts of cement are needed in order to comply with landfill leaching limit values. In this study different flue gas cleaning residues are immobilised with micro silica and with cement. Leachate concentrations of the flue gas cleaning residues immobilised with micro silica comply with German, Flemish and future European landfill criteria. Compliance with these leaching limits is only just attainable using cement alone.For both treatment methods, next to the formation of C-S-H, pH remains an important leaching controlling factor. With micro silica lower leachate pH values can be reached than with cement. The compressive strengths of the micro silica based product is however lower than for cement based products.

Keywords: Immobilisation with cement, Immobilisation with micro silica, Flue gas cleaning residues, Heavy metals, Leaching, Landfill acceptance criteria.

D Geysen is PhD student at the Laboratory of Environmental Technology, Department of Chemical Engineering, KULeuven. He studies immobilisation techniques for flue gas cleaning residues from MSW incineration.

C Vandecasteele is Professor, chairman of the Department of Chemical Engineering and head of the Laboratory of Environmental Technology of KULeuven. His main fields of research are environmental problems related to solid wastes containing heavy metals, membrane processes for waste water treatment and solvent purification.

M Jaspers is PhD and member of staff of the Quality, Environment, Safety and Health (QESH) Department of INDAVER. INDAVER is a major waste treatment company in Belgium also active in other European countries.

G Wauters is PhD and director of the QESH Department of INDAVER.

INTRODUCTION

Incineration of municipal solid waste produces ca. 200 kg of bottom ash and 40 kg of air pollution control residues per tonne of incinerated waste. Reuse options are available for bottom ash, but for air pollution control residues these are scarce and are in most countries not considered as good waste management practice. Landfilling is typically the final option. Although multi-layers of PE-liners and clay prevent contamination of the groundwater around the landfill, it is still necessary to prevent leaching of contaminants from the waste by immobilisation. This not only reduces the need for intensive effluent treatment, but also reduces the risks for groundwater contamination if containment would fail. Therefore leaching limit values are used to decide whether a waste can be accepted on a landfill or not. From 16.07.2005 EC landfill leaching limit values should be applied in The European Community [1]. If for certain elements, the concentrations in the leachate of a waste sample exceed these leaching limit values, the waste has to be treated to reduce this leaching.

Cement is most commonly used for immobilising wastes. The leachate pH of cement based solidified/stabilized alkaline wastes is however high due to high concentrations of $Ca(OH)_2$ in cement and in some wastes [2]. Amphoteric metals such as Pb and Zn have an increased solubility at elevated pH. Therefore it would be interesting if the pH could be decreased to minimise leaching of amphoteric metals [3].

In this study the effect of adding micro-silica on the leachability of heavy metals from flue gas cleaning residues is studied and compared with cement treated wastes.

LEGISLATION

The EC directive 1999/31 "on the landfill of waste" is already effective from 16.07.1999 and is implemented in national legislations [4]. The council decision 2003/33 with EC landfill acceptance criteria should be implemented in national legislations at the latest on 16.07.2005. In Table 1, leaching limit values for the acceptance of wastes on the different landfills described in Flemish and German legislation are compared with the new European leaching limit values. In Flanders, until now, only one set of leaching limit values was applied for the three different landfill categories; landfill for inert waste, for non-hazardous waste and for hazardous waste. In Germany, different leaching limit values are given for each type of landfill. The EC council decision also distinguishes between three different levels for the three different landfill categories.

The Flemish leaching limit values for the acceptance of wastes on landfills are comparable to the German values for "wastes not derived from human settlements" [5, 6]. The German Z_3 and Z_4 leaching limit values are for wastes derived from human settlements. The implementation of the European leaching limit values for inert and non hazardous waste landfills makes that leaching limit values become more severe in Flanders and in Germany. For Ba, Sb, Mo and Se no standards are in use until now. The leaching limit values for landfilling of hazardous wastes on hazardous waste landfills will become less severe (e.g. Pb leaching limit values for hazardous wastes will increase from 2 mg/l to 5 mg/l). The national authorities and landfill operators have to indicate which landfill will become a landfill for hazardous waste and which landfill will become a landfill for non-hazardous wastes in the context of the EC directive. The choice will depend upon the protection measures applied at the landfill but also economic interests do play a roll. The EC directive accepts the storage of treated hazardous wastes on non-hazardous waste landfills if, after treatment, they are not reactive and comply with the limits for non-hazardous waste landfills.

Table 1   Flemish, German and European landfill acceptance criteria

mg/l	FLEMISH	GERMAN Z₃	GERMAN Z₄	GERMAN Z₅	EC INERT	EC NON-HAZARDOUS	EC HAZARDOUS
As	1	0.2	0.5	1	0.05	0.2	2.5
Ba	-	-	-	-	2	10	30
Cd	0.5	0.05	0.1	0.5	0.004	0.1	0.5
Cr	0.5ᵃ	0.05ᵃ	0.1ᵃ	0.5ᵃ	0.05ᵇ	1ᵇ	7.0ᵇ
Cu	10	1	5	10	0.2	5	10
Hg	0.1	0.005	0.02	0.1	0.001	0.02	0.2
Mo	-	-	-	-	0.05	1	3
Ni	2	0.2	1	2	0.04	1	4
Pb	2	0.2	1	2	0.05	1	5
Sb	-	-	-	-	0.006	0.07	0.5
Se	-	-	-	-	0.01	0.05	0.7
Zn	10	2	5	10	0.4	5	20
Cl⁻	1000ᵈ	-	-	10000	80	1500	2500
CN⁻	1	0.1	0.5	1	-	-	-
F⁻	50	5	25	50	1	15	50
NH₄⁺	1000	4ᶜ	25ᶜ	1000	-	-	-
NO₃⁻	30	-	-	30	-	-	-
SO₄²⁻	1000ᵈ	-	-	5000	100	2000	5000
TDS	< 10 weight %			< 10 weight %	400	6000	10000
Conduct. µS/cm		5000	to be indicated	100000			
pH	4-13		to be indicated	4-13			

a: limit values for Cr(VI), b: limit values for total Cr (easily soluble fraction), c: NH₄-N, d: no limit for hazardous waste landfills

For some hazardous wastes it is very difficult or expensive to treat them in a way to attain the non-hazardous waste leaching limit values.

CHARACTERISTICS OF FLUE GAS CLEANING RESIDUES

An example of hazardous wastes for which treatment prior to landfilling is necessary are flue gas cleaning residues from municipal waste incineration. In this work 6 different flue gas cleaning residues are described: wastes 1 to 3 are semi-dry Ca(OH)₂-scrubber residues; waste 4 is a semi dry NaHCO₃-scrubber residue; waste 5 is an electrofilter fly ash; and waste 6 is a boiler ash. Table 2 gives total concentrations for the different residues. Pb and Zn are the heavy metals with the highest concentration.

Wastes 5 and 6 contain higher concentration of Ba in comparison to the scrubber residues. The concentrations of Ni, Cr, Cd, Mo, Se and As are lower than commonly applied soil reclamation standards.

Table 2   Total content of major elements and heavy metals
from different air pollution control residues.

CONCENTRATION, mg/g	WASTE 1	WASTE 2	WASTE 3	WASTE 4	WASTE 5	WASTE 6
Al	4.5	5.5	8.63	1.7	50.7	52.5
As	0.02	0.01	0.02	< 0.01	0.05	0.04
Ba	0.33	0.32	1.06	0.05	2.1	2.2
Ca	348.8	287.3	262.7	90.8	156.6	161.8
Cd	0.12	0.2	0.17	0.1	0.24	0.03
Cr	0.08	0.1	0.12	< 0.1	0.13	0.18
Cu	0.55	0.5	0.43	0.15	0.84	0.42
K	25.3	-	22	17.3	36.6	8
Mg	5	-	5.4	1.3	12.8	13.2
Mo	< 0.01	< 0.01	0.02	-	0.04	0.02
Na	19.2	24.6	21.8	296.4	31	10.7
Ni	0.05	0.01	0.06	< 0.01	0.1	0.15
Pb	3.05	4.6	4.84	3.65	6.22	1.4
Sb	0.74	0.28	0.34	-	0.51	0.4
Se	0.05	0.01	< 0.02	< 0.01	0.02	0.04
Zn	6.49	7.76	12.28	3.65	11.45	4.55

Table 3 gives leaching values for the different residues. Leaching is performed with a batch test at L/S = 10 and shaking during 24 hours. The particles of the waste are smaller than 4 mm

The Pb leachate concentrations for all these residues are above the leaching limit values for landfilling on hazardous waste landfills (Germany and Flanders: 2 mg/l; EC: 5 mg/l). Only waste 4 has also a Zn leachate concentration (71.3 mg/l) above the EC leaching limit value of 20 mg/l. In addition, waste 2 has a Zn concentration (18.0 mg/l) in the leachate above the German and Flemish leaching limit values for hazardous waste landfills (10 mg/l). The EC hazardous waste leaching limit value for Se is exceeded by waste 1 and waste 4. Non of the wastes is in accordance with the Se leaching limit value for non hazardous wastes. The EC hazardous waste leaching limit value for Sb is exceeded by waste 4.

In the leachate of waste 5, 22.7 mg/l Ba is measured which is close to the EC limit of 30 mg/l for hazardous waste and above the EC limit of 10 mg/l for non-hazardous wastes.

Next to leaching of heavy metals, also the high level of TDS (total dissolvable solids) and the high Cl leaching of the scrubber residues necessitate treatment prior to landfilling (Table 3). The TDS for the $Ca(OH)_2$-scrubber residues is approximately 20 %. For the $NaHCO_3$ residu, more than 70 % dissolved under these test conditions. The EC limit for hazardous wastes, (10000 mg/l) corresponds with 10 % TDS. Boiler ash and fly ash comply with this limit, but exceed the limit for non-hazardous waste landfills.

Reducing the leachability of Pb and to a lesser extent of Zn from these wastes is an important goal. The solubility of both metals is strongly influenced by pH as indicted in Figure 1 for waste 1. The solubility goes through a minimum and at a pH around 12.5 the Pb leachate concentration of the untreated waste is comparable to the concentration at low pH. The Zn leachate concentration at a pH of 12.5 is still significantly lower than the leachate concentration at low pH.

Two treatment methods to reduce the leaching of heavy metals (in this case Pb and Zn) are discussed and compared here. The first treatment method consists of adding cement to the waste, for the second method micro silica is used.

Table 3   DIN 38414-S4 leaching values for major elements, heavy metals and pH from different air pollution control residues.

CONCENTRATION, mg/l	WASTE 1	WASTE 2	WASTE 3	WASTE 4	WASTE 5	WASTE 6
Al	0.1	< 0.1	1.4	15.5	< 0.1	0.2
As	0.14	0.05	0.04	0.26	0.02	0.01
Ba	7.2	2.2	3.8	0.15	22.7	0.31
Ca	8420	10228	8262	52.1	2723	1635
Cd	< 0.1	< 0.1	< 0.1	< 0.1	< 0.1	< 0.1
Cr	0.1	0.3	0.1	0.2	0.6	0.4
Cu	0.3	0.2	0.2	0.4	0.1	0.2
K	2284	2104	2758	1680	3014	613
Mg	1.2	0.2	1.7	8.1	1.6	0.7
Mo	0.39	0.16	0.63	0.25	0.86	0.33
Na	2532	2410	2994	4530	2322	794
Ni	0.1	0.1	< 0.1	0.1	0.1	0.2
Pb	103.1	159.1	97.8	143.8	172.7	5.51
Sb	0.01	< 0.01	0.01	1.24	< 0.01	< 0.01
Se	0.98	0.19	0.34	1.00	0.31	0.09
Zn	5.6	18	5.5	71.3	3.4	3.67
Cl⁻	16700	20920	19250	31410	1490	2200
TDS %	18.8	22.9	21.2	72.5	9.2	2.0
pH	12.49	12.53	12.25	13.2	12.48	12.43

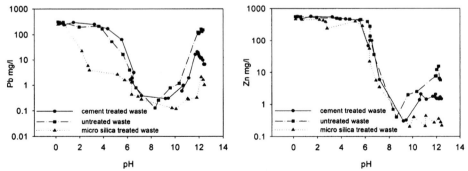

Figure 1   Pb (left) and Zn (right) concentrations in the leachate of waste 1, untreated, treated with cement and treated with micro silica.

TREATMENT OF FLUE GAS CLEANING RESIDUES WITH CEMENT

Cement is most commonly used to immobilise wastes containing heavy metals. Next to the physical entrapment of pollutants, heavy metals react with the hydrated cement matrix which consists mainly of calcium silicate hydrate (C-S-H).

Leaching of the different above-described residues treated with cement is studied here. The treatment consists of mixing waste with cement and water in such an amount that a homogeneous paste is obtained. For each waste, different cement to waste ratios are applied. Leaching is done at L/S = 10 with samples crushed to below 4 mm. The pastes are hardened for one or five weeks prior to leaching. A CEM I cement is used because previous studies showed that with this type of cement, the lowest Pb leachate concentrations could be obtained after one week of hardening time [7].

Figure 2 shows Pb leaching results after 1 week of hardening for the different wastes. A clear decrease in lead leachate concentration is observed for the four scrubber residues when more cement is applied in the immobilisation recipe. For the fly-ash (waste 5) and for the boiler ash (waste 6), a maximum in Pb leachate concentration is observed. The initial increase in leachate concentration for these 2 wastes (fly ash and boiler ash) is attributed to the increase in pH shown in the graph at the right, which is more pronounced than for the other residues. Only the leachate of cement treated waste 6 (boiler ash), which contains the lowest total Pb concentration of all these residues, complies with the leaching limit values. For none of the scrubber residues nor for the fly ash the German and Flemish Pb leaching limit value of 2 mg/l is attained, even after 5 weeks of hardening (Figure 3).

The EC Pb leaching limit value of 5 mg/l is attained for all wastes considered (except waste 4), when 0.5 g cement / g waste is applied in the recipe and leached after 5 weeks hardening. The pH of the cement treated waste leachates remains high or even increases at increased addition of cement. Also an increase in Ba leaching is observed with a maximum of 15 mg/l for waste 1 treated with a CEM I type of cement (not shown here). Analysis of cement and leaching of cement showed that Ba can leach from cement and particularly from the types of cement containing high concentrations of clinker. No decrease in Cl leaching could be attained (not shown here).

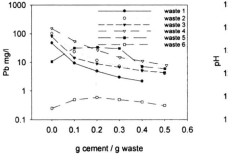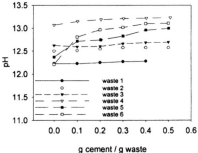

Figure 2   Pb leachate concentrations (left) and pH (right) of the leachate of cement treated wastes leached after 1 week of hardening time.

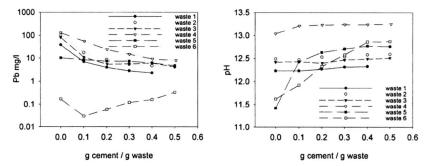

Figure 3   Pb leachate concentrations (left) and pH (right) of the leachate of cement treated wastes leached after 5 weeks of hardening time.

TREATMENT OF FLUE GAS CLEANING RESIDUES WITH MICRO SILICA

In a previous work it is indicated that different siliceous materials can be applied for treating hazardous wastes containing heavy metals [7].   Micro silica produces a reduction in the heavy metal leachability of alkaline flue gas cleaning residues.   Micro silica, also called condensed silica fume, is a residue of the production of ferro silicon alloys in an electric arc furnace. The off gases of the electric arc furnace contain SiO, which, in contact with oxygen, is oxidised to SiO_2 and condenses into spherical particles of pure and amorphous SiO_2 (90 - 94 % by weight of SiO_2, average particle size from 0.1 to 0.5 µm, specific surface area between 20 and 30 m^2/g).   The silica dissolves in alkaline conditions and forms in the presence of Ca^{2+} or at decreased pH a gel or a precipitate.   Reaction products are formed, similar to the cement hydration products in which also heavy metals can be incorporated. Boiler ash, fly ash and scrubber residue can act as source for Ca.   In this way, in contrast to cement addition, no alkaline additive is used and the residual alkalinity of the waste is partly consumed for the dissolution/precipitation of silica.

Figure 4 shows Pb leachate concentrations of the different micro silica treated wastes leached after 1 week of hardening.   After one week of hardening, lower Pb leachate concentrations are attained than for cement treated wastes. Pb leachate values of waste 4, waste 5 and waste 6 comply with leaching limit values for landfilling on non-hazardous waste landfills even at a relatively low micro silica dosage.   For waste 4 and waste 6 a significant decrease in pH is observed whereas the pH of the scrubber residues and of the fly ash remains approximately constant. For the scrubber residues, high micro silica dosages have to be applied to attain the leaching limit values for landfilling on hazardous waste landfills.

After 5 weeks of hardening (Figure 5) Pb leachate concentrations clearly decreased in comparison to 1 week of hardening. This is partly due to a decrease in pH. But even when no decrease in pH is observed (waste 1), the Pb leachate concentration decreases. That the decrease in Pb leaching is not merely a consequence of the pH decrease upon addition of micro silica, appears clearly from Figure 1. For a given pH around 12 the Pb concentration in the leachate is much lower for the micro silica treated waste than for the untreated and for the cement treated waste. The EC hazardous waste Pb landfill leaching limit value of 5 mg/l can be attained and even attaining the EC non-hazardous waste landfill leaching limit value for Pb of 1 mg/l is possible.

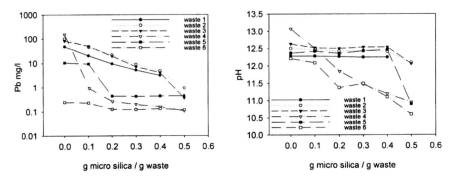

Figure 4: Pb leachate concentrations (left) and pH (right) of the leachate of micro silica treated wastes leached after 1 week of hardening time.

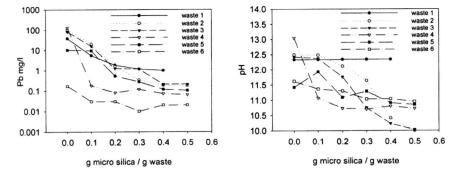

Figure 5: Pb leachate concentrations (left) and pH (right) of the leachate of micro silica treated wastes leached after 5 weeks of hardening time.

COMPRESSIVE STRENGTH MEASUREMENTS

The strength development was studied for three different wastes (waste 1, 2 and 6). Samples were prepared by combining micro silica and cement with waste at a dosage of 0.3 g additive per g waste. In civil engineering, strength development is testes on samples stored in humid (100 % RH) conditions. Here, next to these humid conditioned samples also samples stored in lab environment were tested.

Results of samples stored in lab environment are shown at the left side of Figure 6. The result at the right side of Figure 6 is for samples stored in humid conditions. A maximum strength of 25 Mpa was observed after 30 days of hardening in lab environment for waste 1 treated with cement. With micro silica, a maximum strength of 13 Mpa was observed for waste 1 after 90 days of hardening.

Lower strengths are developed for samples stored in humid air (right side of Figure 6). With micro silica, a measurable strength after 90 days of curing was only observed for the boiler ash (waste 6). For both other wastes, swelling due to uptake of water resulted in the formation of cracks (Figure 7, left). After attaining a maximum in strength for waste 1, the formation of cracks reduced the strength to almost zero. The faster uptake of water for waste 2 made it not possible to measure strength already after 7 days.

Cement treated samples of waste 1 and 2 stored in humid air can resist against swelling and the formation of cracks. Salts are deposited on the surface of samples of waste 1 and waste 2 (Figure 8, right).

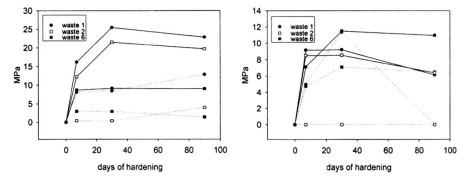

Figure 6 Strength development of some wastes treated with cement (solid lines) and with micro silica (dotted lines), stored in laboratory environment (left) and in 100 % relative air humidity (right).

Figure 7 Crack formation in waste 2 treated with micro silica and stored in humid air conditions (left). Salt deposition on waste 2 treated with cement and stored in humid air conditions (right).

CONCLUSIONS

With regard to the EC landfill leaching limit values of the Council decision 2003/33; Pb, Zn, Cl, the high % TDS and to a lesser extend Se and Sb can be of concern when landfilling flue gas cleaning residues from MSW incineration.

When treating different MSW flue gas cleaning residues with cement, 0.5 gram of cement per gram of waste is necessary to attain the EC hazardous wastes landfill leachate limit value for Pb. Only the studied boiler ash complied with this limit. When increasing the amount of cement in the recipe, an increased leachate concentration of Ba is observed but the EC hazardous waste landfill limit is not exceeded. With micro silica lower Pb leachate concentrations are attained for all residues and even the EC non-hazardous landfill leaching limit value for Pb can be attained.

The Cl leachate concentrations of scrubber residues are almost 10 times higher than the hazardous waste landfill limits. Also the % TDS exceeds the standard value. For these reasons it is generally necessary to landfill the solidified waste in "salt cell conditions". Washing the waste before solidification might be an alternative.

Compressive strength measurements show relatively high values for cement treated samples stored in laboratory environment. With micro silica lower values are attained but they are still acceptable for landfilling. In 100 % relative humid conditions, the developed strengths after 90 days curing are significantly lower. Samples of scrubber residues treated with micro silica and stored in humid air absorb water due to the presence of $CaCl_2$. Cracks are formed and the compressive strength becomes too low for measurement.

ACKNOWLEDGMENTS

This research work was supported by INDAVER N.V.

REFERENCES

1. EC, Official Journal of the European Communities, 16.01.2003, Council Decision 19 December 2002, Establishing criteria and procedures for the acceptance of waste at landfills pursuant to Article 16 of and Annex II to Directive 1999/31/EC, Brussels, 2003.

2. BOBROWSKI A, GAWLICKI M, MALOLEPSZY J, Analytical Evaluation of Immobilization of Heavy Metals in Cement Matrices. Environmental Science and Technology, 1997, 31, 745-749.

3. CHONG YOON RHA, SEONG KEUN KANG, CHANG EUN KIM, Investigation of the stability of hardened slag paste for the stabilization/solidification of wastes containing heavy metal ions, Journal of Hazardous Materials, 2000, 73, 255-267.

4. EC, Official Journal of the European Communities, 16.07, 1999, Council Directive 1999/31/EC, on the landfill of waste, Brussels, 1999.

5. AMINAL 1995, Besluit van de vlaamse regering van 1 juni 1995 houdende algemene en sectorale bepalingen inzake milieuhygiëne, (VLAREM II), versie 2002.

6. BMU, TA-Abfall, Technische Anleitung zur Lagerung, chemisch/physikalischen, biologischen Behandlung, Verbrennung und Ablagerung von besonderes überwachungsbedürftigen Abfällen, 1991.

7. GEYSEN D, SCHROOTEN L, IMBRECHTS K, VANDECASTEELE C, VAN GERVEN T, JASPERS M, BROUWERS E, WAUTERS G, Immobilisation of lead in flue gas cleaning residues from MSW-incinerators of Flanders (Belgium), in Proceedings of the Eighth International Waste Management and Landfill Symposium, 647- 656, 1-5 October 2001, S. Margherita di Pula, Cagliari, Italy.

THE CHALLENGE FOR GLASS RECYCLING

R Cocking
British Glass Manufacturers' Confederation
United Kingdom

ABSTRACT. Glass recycling in the UK has been underway for over 25 years but has only grown to a level of 35 per cent. The UK's unique PRN system, introduced in 1997, has created a low cost collection infrastructure, but not yet delivered significant increases in glass recycling rates. With new European legislation set to massively increase the UK's glass recycling targets, this briefing paper looks at the developments that the glass packaging industry believes will need to be implemented in order to achieve a higher recycling rate.

Keywords: Revision to the EU Packaging and Packaging Waste Directive (PPWD), Packaging Recovery Notes (PRN), Kerbside collection.

Rebecca Cocking is Recycling Manager for the British Glass Manufacturers' Confederation and is responsible for liaising with local and national government on glass recycling issues for all types of glass.

INTRODUCTION

Glass recycling in the UK has been underway for over 25 years but has only grown to a level of 35 per cent. The UK's unique PRN system, introduced in 1997, has created a low cost collection infrastructure, but not yet delivered significant increases in glass recycling rates. With new European legislation set to massively increase the UK's glass recycling targets, this briefing paper looks at the developments that the glass packaging industry believes will need to be implemented in order to achieve a higher recycling rate.

British Glass

The British Glass Manufacturers' Confederation is the organisation that represents the UK's glass industry. From our offices in Sheffield we act as the Industry's focal point, playing the principal role in communicating the concerns and aspirations of our 100 plus members to the Government, the European Union and other external interest groups and trade bodies. With more than 40 years experience of serving our members, we are also instrumental in promoting glass as the first choice material for containers, flat glass, domestic glass, scientific glass and glass fibre applications. Our technical arm, Glass Technology Services, is a world-renowned centre of excellence, helping companies around the globe to solve technical and production challenges.

Industry Sectors Represented

British Glass represents all sectors of the UK glass industry, these include:

- container – producers of bottles and jars
- flat glass – producers of glazing and architectural glass
- domestic glass - makers of glasses and cookware
- decorative glass – producers of bowls, vases and other decorative items
- scientific glass – makers of test tubes, beakers and a vast range of scientific glass
- glass fibre applications – producers of insulation, exhaust muffles and electronic components
- glass manipulators – makers of car windscreens

What We Do

In addition to representing the industry and promoting glass we provide our members with a range of services and benefits including:

- Promoting the use of glass as a material essential to modern life
- Representing the industry to government, EU and trade bodies
- Informing members of legislative proposals and changes
- Keeping members up-to-date with technological advances
- Providing members with legal advice
- Offering guidance on environmental issues
- Promoting and developing glass recycling
- Organising industry training and seminars
- Providing members with technical advice and consultancy through GTS
- Working with other trade bodies with shared interests
- Encouraging members to share information for their mutual benefit

International Glass Recycling

Although overall the national recycling rate has increased compared to the whole of Europe, the United Kingdom is still lagging behind. Only Greece and Turkey are less active than the UK when it comes to recycling, see Table 1. There are a number of reasons why other European countries are better at recycling than the UK. Whilst the UK PRN structure is a 'cheapest cost' system the downside is that investment in the infrastructure is proportionally lower and so our infrastructure is not as advanced as other European countries. In addition the public's attitude to recycling in the UK is different. Many continental European nations see recycling as a way of life and as such public participation is high, the same is not the case in the UK and greater efforts are needed to educate the public.

Table 1   European Glass Recycling 2001

COUNTRY	TONNES COLLECTED [1]	NATIONAL RECYCLING RATE
Switzerland	294,000	92%
Netherlands	400,000	78%
Austria	200,000	83%
Sweden	144,000	84%
Norway	44,000	88%
Germany	2,666,000	87%
Finland	46,000	91 %
Denmark	120,000 [2]	65% [2]
France	1,950,000	55%
Belgium	279,000	88%
Portugal	122,000	34%
Italy	1,100,000	55%
Ireland	46,000 [2]	40% [2]
Spain	5706,000	33%
United Kingdom	736,000	34%
Greece	44,000	27%
Turkey	73,000	24%
TOTAL	**8,775,000**	

[1] From the general public and from bottlers [2] Estimate

However, the main difference is the number of bottle banks per head of population. The countries with the highest recycling rates have a bottle bank density of one to 1,500 people. The UK has about 2,700 people on average per site. If there were more bottle bank sites in the UK, there would be a greater opportunity for the public to use them. There would also be fewer incidences of them being full. British Glass believes that doubling the number of bottle banks in the UK is essential to hitting the recycling targets for glass.

The key issues to increasing glass recycling in the future are making glass recycling easy and convenient for the public, ensuring the public are aware of the benefits of their recycling activity and educating the consumers of the future. Our European partners have provided a blueprint for success. Everywhere you turn there are recycling banks, recycling is widely promoted, including on TV and recycling awareness and participation is a part of everyday school activity.

Glass Recycling In The UK

In the UK around 50 per cent of the glass recycled is green – mainly imported wine and beer bottles – however, only about 16 per cent of the glass our customers ask us to make is green. This colour imbalance has always created a challenge for glass recycling.   We have worked hard to increase the amount of green glass we recycle and currently all the green bottles we make in the UK contain around 85 per cent recycled green glass.

UK Colour Splits

In the future as we collect more glass we will be able to recycle a little more green but much more clear and brown glass. However, there will be more green than the container industry alone can use, see Figure 2.   Alternative markets for glass are growing, such as export to European container manufacturers, water filtration, shot blasting, additives to bricks or ceramics and aggregates. We believe that in the future these alternatives will provide a market for ALL glass that cannot be used for closed-loop container manufacture.

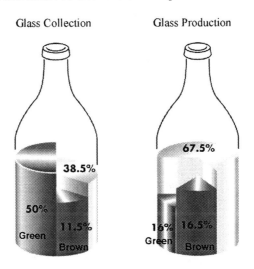

Figure 1 Colour split of glass collection and glass manufacture

UK Production vs Recycling 2001 (35%)

Glass recycling is a fantastic example of sustainability. Because glass can be recycled endlessly without any loss of quality, we can recycle the same glass over and over again to make new bottles and jars, see Figure 2. This process:

- Saves energy
- Cuts quarrying
- Reduces landfill
- Creates employment.

'Closed-loop' recycling delivers the highest environmental benefits, again and again. So we believe that as much recycled glass as possible should be used to make new bottles and jars.

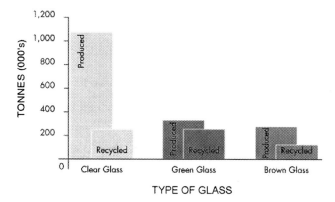

Figure 2   UK production vs recycling (35%)

UK Production vs Recycling 2006 (65%)

However, there are other uses for recycled glass, which deliver some environmental benefits and are better than throwing the glass away:

- Water filtration – diversion from landfill and reduction in quarrying
- Additives to bricks, cement or ceramics – some energy saving and a reduction in landfill and quarrying
- Shot blasting – diversion from landfill and reduction in quarrying
- Aggregates – diversion from landfill and reduction in quarrying

Our calculations show that recycling glass into new bottles and jars delivers an environmental benefit that is 50 times greater than recycling glass into aggregates, even when transport is taken into account, see Figure 3.

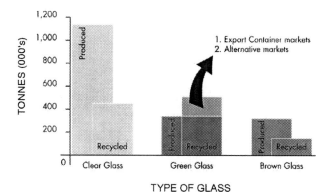

Figure 3   UK Production vs Recycling 2006 (65%)

Aggregates v Closed Loop Recycling

Journey to CONTAINER RECYCLING		*Journey to AGGREGATES RECYCLING*	
Distance	203 miles	Distance	20miles
Payload	25 tonnes	Payload	25 tonnes
Fuel consumption	7 mpg	Fuel consumption	7 mpg
Energy used per tonne	1.2 gallon (57 kWh)	Energy used per tonne	0.1 gallon (6 kWh)
Energy saving per tonne	7.0 gallon (345 kWh)	Energy saving per tonne	0.0 (0.0 kWh)
Energy Saving	5.9 gallon (288 kWh)	Energy Saving	-0.1 gallon (-6 kWh)

With nearly 1.5 million tonnes of glass being thrown away every year, landfill remains the biggest end market for glass. This is a major lost opportunity. For the last 25 years there has been slow but steady growth in glass recycling. In that time 7.7 million tonnes of glass have been through the bottle bank system, delivering massive energy savings, reduced emissions and significant reductions in quarrying and landfill. The UK has more than 50,000 bottle banks, about half the European average, just 10 per cent of homes have kerbside collection of glass and only 20 per cent of pubs and clubs have commercial collections. With this collection infrastructure we only recycle 30 per cent of our bottles and jars. As a nation we still throw away nearly 1.5 million tonnes of glass every year. The European Union's 'Packaging & Packaging Waste Directive' means we will have to achieve a much higher recycling rate for glass - probably 65 per cent by 2006/2008.

Effect of Proposed 65% Recycling Rate

Clearly we have along way to go. The glass packaging industry currently has capacity for at least 1.1 million tonnes of recycled glass. Combined with alternative uses, this is enough to meet these high targets, Figure 4, if the glass can be collected.

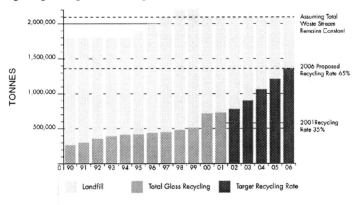

Figure 4 Glass recycling targets

We recognise that this can only be achieved by all those involved in the collection system, including the glass packaging industry, working together to make glass recycling even more accessible to the public.

PRODUCER RESPONSIBILITY IN THE EC

K Riddick

Scottish Environment Protection Agency

United Kingdom

ABSTRACT. "Producer Responsibility" is becoming increasingly popular in European Waste Management. It is designed to encourage manufacturers and retailers to reduce resource use, extend product life span, reduce waste disposal and increase recovery and recycling of discarded products. This presentation summarises current use of the producer responsibility as it is used to deal with packaging, electrical equipment and vehicles within the EC.

Keywords: Producer responsibility, Packaging, End of life vehicles (ELVs), Waste electrical and electronic equipment (WEEE)

Karen Riddick has been employed by the Scottish Environment Protection Agency since 1997. She is currently a Senior Consultant and is responsible for the management of SEPA's Producer Responsibility Unit. Prior to her position with SEPA, Karen spent 6 years in Local Authority Trading Standards and Consumer Protection. She holds the Institute of Trading Standards Diploma in Consumer Affairs, a Degree (BSc (Hons)) in Environmental Science from Manchester University and a Masters (LLM) in Environmental Law from DeMontfort University, Leicester.

INTRODUCTION

'Producer Responsibility' is the principle that requires 'producers' (including manufacturers, retailers and importers i.e. those who are responsible for placing products onto the market) to take responsibility for the environmental impacts that their products may have. Producer Responsibility has evolved from the 'Polluter Pays Principle'. It adopts the basic premise that those responsible for environmental damage must bear the costs of dealing with that damage. However, unlike the traditional Polluter Pays Principle, 'producer responsibility' is more focused on prevention of pollution and incorporates responsibility for wider lifecycle environmental impacts as opposed to just 'end of pipe' pollution.

The definition and application of 'producer responsibility' varies throughout the world. The Organisation for Economic Co-operation and Development (OECD) provide a useful definition of the principle as "an environmental policy approach in which a producer's responsibility, physical and/or financial, for a product is extended to the post consumer stage of a product's life cycle" [1]   The OECD also identifies two related features of producer responsibility application,

- the shifting of the financial and physical responsibility for management of waste products away from local governments and back to the producers of the products.

- the associated incentives which encourage producers to consider environmental objectives at the design stage of product manufacture i.e. imposing responsibility for post consumer management of waste onto producers can encourage design changes which make collection, dismantling and recycling easier to carry out. For example, assembly of products using fewer screws can speed up dismantling, a reduction in number of material types used in products can increase 'recyclability' and components can be designed with reuse in mind.

Hence, Producer Responsibility can address cradle to grave environmental problems by encouraging minimisation of resource use, extension of product life span and recovery/recycling of products once they have been disposed of as waste. A crucial objective of Producer Responsibility is that the real environmental costs of product manufacture, use and disposal should be incorporated into the price of that product.

The mechanisms that are used to apply the producer responsibility principle in a particular country often depend on whether the core priority of a government is focused on sustainable development, product policy or waste management.

- Where sustainable development is a priority, the policy mechanism may be focused more on the resources consumed in production.

- If product policy is the priority then the policy mechanism may be focused more on the environmental impact of the consumption and post consumption stage.

- If waste management is the priority then the policy mechanism will tend to focus on reducing waste that is subject to final disposal and will commonly include recycling and diversion targets.

PRACTICAL APPLICATION OF THE PRODUCER RESPONSIBILITY PRINCIPLE IN EC DIRECTIVES

The EC has adopted the use of the producer responsibility principle primarily as a waste management tool. Indeed, it forms a central concept to recent waste management Directives including those relating to end of life vehicles and waste electrical/electronic equipment. The focus of these Directives is placed on mandatory recovery, recycling and reuse targets designed to reduce the amount of waste being subject to final disposal. However, there are often other persuasive provisions which relate to wider environmental impacts of products.

The use of the producer responsibility principle in these Directives is also influenced by the EC's desire that responsibilities should be shared amongst all those in the product supply chain rather than placed on a single point in the chain. This recognises that the 'polluter' label cannot always be placed on one person in the product supply chain but that it should indirectly be applied to, for example, those persons who create a demand for the product in the first place, or those who dictate the product specifications.

The EC has identified a number of priority waste stream projects on which to focus the producer responsibility principle. These include packaging, vehicles, electrical/electronic equipment, tyres and batteries. A summary of the Directive provisions relating to end of life vehicles, waste electrical and electronic equipment and packaging is given below.

END OF LIFE VEHICLES DIRECTIVE

Around 12 million vehicles are discarded every year in the EC. Between 70% - 75% by weight of an average vehicle is recovered or reused (mostly metal recycling and some reuse of spare parts). However, the remaining 25% is disposed of. Directive 2000/53 on end of life vehicles (ELVs) was adopted in October 2000 and is currently being transposed into domestic law. The ELV Directive aims to reduce the amount of waste disposal from vehicles by imposing mandatory reuse, recovery and recycling targets. It also aims to improve the environmental standards of all operators involved in the vehicle life cycle, particularly with respect to end of life treatment. The main requirements of the Directive are summarised below.

- Economic operators (this term includes producers, dismantlers and shredders among others) to establish adequate systems for the collection of ELVs.

- Last owners must be able to return their vehicles into an authorised collection system free of charge from January 2007.

- Member states must ensure that ELVs can only be scrapped ('treated') by authorised dismantlers or shredders who meet tightened environmental treatment standards – these standards are outlined in Article 6 and Annex 1 of the Directive.

- Member States to set rising reuse, recycling and recovery targets which must be met by economic operators by January 2006 and 2015. These are;

  - A minimum of 85% reuse and recovery and minimum 80% reuse and recycling per average weight per vehicle and year to be achieved by January 2006

  - A minimum of 95% reuse and recovery and minimum 85% reuse and recycling per average weight of vehicle and year to be achieved by January 2015.

- Producers (vehicle manufacturers or importers) to pay 'all or a significant part' of the costs of vehicle take-back and treatment from January 2007.

- Member States to restrict the use of heavy metals (lead, mercury, cadmium and hexavalent chromium) in new vehicles from July 2003.

WASTE ELECTRICAL AND ELECTRONIC EQUIPMENT DIRECTIVE

It is estimated that around 6 million tonnes of WEEE arises in the EC each year, of which 1 million tonnes is generated in the UK. WEEE is one of the fastest growing waste streams and is expected to double in the next 15 years. Directive 2002/96/EC on Waste Electrical and Electronic Equipment (WEEE) was adopted in February 2002 and is currently being transposed into domestic law. The objective of the Directive is to reduce the amount of WEEE produced by Member States and to promote reuse, recovery and recycling of equipment. It also aims to improve the environmental performance of all operators involved in equipment lifecycle, particularly those who carry out end of life treatment. The main provisions of the Directive are shown below.

- A collection target of 4 kg per head of population per annum to be achieved by 31 December 2005

- Economic operators (this term includes producers, dismantlers and shredders among others) to establish adequate systems for the collection of WEEE.

- Last owners must be able to return their waste equipment to specified collection points free of charge.

- WEEE must only be scrapped ('treated') by authorised dismantlers or shredders who meet tightened environmental treatment standards – these standards are outlined in Article 6 and Annex II and III of the Directive.

- Reuse, recycling and recovery targets must be met by economic operators by December 2006. These targets vary within the 10 identified categories of WEEE and range from 50% - 80% recovery.

- Producers are to finance the collection of household WEEE from collection points and the subsequent treatment of equipment.

- In order to increase the collection and recycling of WEEE, new equipment must be labelled advising consumers not to discard WEEE in their household waste but to use special collection schemes. Information relating to the location and type of material used in manufacture must also appear in new equipment. This is to assist in treatment and recycling of equipment.

In addition to the provisions of the WEEE Directive, the Directive on Restriction of Certain Hazardous Substances in Electrical and Electronic Equipment (RoHS) requires that lead, mercury, cadmium, hexavalent chromium, PBB and PBDE flame retardants are substituted in equipment (note that some exemptions to these provisions will apply).

PRODUCER RESPONSIBILITY FOR ELVS AND WEEE

The Directives on ELVs and WEEE demonstrate a shared approach to achieving objectives. 'Producers' i.e. manufacturers, brand owners, retailers and importers must bear financial and physical responsibility for meeting the mandatory targets but other operators in the vehicle lifecycle must also improve their environmental performance.

In terms of the two defined objectives of producer responsibility (as identified in the introduction and background), it is clear that, within these Directives, the physical and financial responsibility for end of life products is shifted away from government and back to producers. However, until the Directives are transposed and implemented it is not possible to assess the potential success of incentives for producers to incorporate environmental considerations at the design stage. Unlike the Directive's treatment and recovery/recycling requirements, product design requirements which aim to facilitate ease of dismantling and recovery/recycling are 'encouraged' rather mandatory.

It is generally assumed that the mandatory reuse, recovery and recycling targets will indirectly encourage producers to consider the design of their products i.e. producers may minimise resource use, extend product life span and design products for ease of recycling and reuse since this will reduce their obligations and the associated costs of compliance. Whether this indirect approach can be successful depends on whether there is a net benefit to making design changes; in other words, producers will likely compare the costs of compliance associated with making design changes as opposed to those associated with making no design changes. In addition, the indirect incentive to incorporate design changes during manufacture also depends on whether individual producer responsibility is adopted i.e. whether producers are responsible for their own products or whether they can meet obligations by reusing, recovering or recycling waste products manufactured and sold by other producers.

PACKAGING AND PACKAGING WASTE DIRECTIVE

It is estimated that some 58 million tonnes of packaging waste is produced in the EC each year. EC Directive 94/62 on Packaging and Packaging Waste was adopted in 1994 with the overall objective of reducing the environmental impact of packaging and packaging waste. Contrary to common belief, the Packaging and Packaging Waste Directive is not currently a 'producer responsibility' Directive since it does not require 'producers' to bear the physical and financial responsibility for meeting the required environmental objectives (note that 'producer responsibility' may be introduced into the Directive in future revisions) . However, most member states have already transposed the Directive using producer responsibility measures. The main provisions of the Directive are;

- To minimise the amount of packaging placed on the market
- To restrict the content of certain hazardous substances
- To ensure that adequate collection systems are established
- To ensure that certain recovery and recycling targets (a minimum of 50% recovery and 25% recycling of packaging waste) are achieved from 2001 onwards.

Implementation of the Packaging and Packaging Waste Directive by Member States presents an opportunity to observe the advantages and disadvantages of various applications of the 'producer responsibility principle'. In the UK, the application of producer responsibility (through the Producer Responsibility Obligations (Packaging Waste) Regulations 1997) has resulted in an increase in packaging waste recovery and recycling from 34% in 1998 to 54% in 2002. During that time, producers have paid an estimated £250 million towards recovery and recycling. This increase in packaging waste recovery and recycling suggests that producer responsibility has been a successful policy mechanism to adopt. However, since the introduction of the Regulations, various other trends have developed that highlight some weaknesses and more negative consequences of producer responsibility.

Firstly, there has been no notable reduction in the amount of packaging placed on the market and there are few examples of producers that have perceived enough of an incentive to redesign their packaging. Secondly, there has been a potential shift away from the principles of the waste hierarchy in that recycling has taken preference over reuse i.e. in the UK, producers can only meet their obligations by recovering and recycling packaging waste, hence existing reuse activities may be curtailed in favour of recycling. Thirdly, in terms of overall environmental benefit it is possible that, although packaging waste disposal has decreased, the disposal of non packaging waste may have increased as a result of the 'subsidy' paid by producers to those who recover and recycle packaging waste i.e. non packaging waste that was previously recycled may now have been displaced to landfill in favour of packaging waste that is more economically viable to recycle. In addition to affecting overall environmental benefit, the favouring of one waste material over another can potentially distort the market. Finally, producer responsibility for packaging waste in the UK has excluded consumer involvement i.e. there are no provisions placed on householders and consumers to collect and sort their packaging waste. Given that considerable proportions of packaging waste arise in the household waste stream, the absence of links between consumers and producers is now regarded as a major problem in terms of collecting enough waste to enable targets to be met.

These observations suggest that, whilst the basic nature of the producer responsibility principle is sound, practical application can have other adverse effects or may be of limited success if all parties in the lifecycle are not included. These factors may potentially influence success of future producer responsibility initiatives.

FUTURE PRODUCER RESPONSIBILITY IN THE EC

The Producer Responsibility Principle is still in relative infancy but is becoming increasingly popular throughout the world. It is likely to be applied to further products in the EC. Experience of producer responsibility in the EC to date has been based on voluntary individual use of the principle within member states rather than an overall application of the principle within a specific Directive. To date, it can be argued that the producer responsibility principle is a logical approach to environmental protection but that application methods need further refining. Further experience of the producer responsibility principle at EC level will be gained once the recently adopted Directives relating to WEEE and ELVs are transposed and implemented by member states.

REFERENCES

1. OECD, Extended Producer Responsibility – A Guidance Manual for Governments, 2001

THEME THREE:
NATIONAL GOVERNMENT POLICY

PRESENT WASTE MANAGEMENT IN THE JAPANESE CONSTRUCTION INDUSTRY

T Kondo

Institute of Technologists

Japan

ABSTRACT Nowadays, mankind understands the need to maintain the earth's environment and protect human health and safety. They are urgent objects that must be realized for a sustainable society. Although construction activities create useful and pleasant living spaces, they cannot avoid some destruction of the earth's environment. Against the background of the global problems, activities for protection of the environment, including waste management are being positively promoted in Japan's construction industry field. In this paper, present measures to reduce waste in this sector are described. They include Japanese government policy regarding waste management, the present construction waste condition, strategies of Japan's private construction sector and development of new construction methods that take waste reduction into consideration.

Keywords: Construction industry, Waste management, Waste reduction, Construction methods, Renewal construction methods, Zero emissions

Dr Teruo Kondo is a Professor of the Institute of Technologists in Japan. The Institute of Technologists was established in April 2001 primarily for education and training. Previously he worked for one of the largest general constructors in Japan where he did extensive research in building durability and maintenance. Recently he has investigated environmental management regarding the construction industry and regarding indoor air quality. He holds executive positions in the Architectural Institute of Japan and Japan Society of Finishing Technology. Dr. Kondo is also involved in discussions with Japan's Ministry of Land, Infrastructure and Transport regarding construction specifications and guidelines.

INTRODUCTION

Our material form of civilization has brought rapid urbanization to the world and a good quality of life has been realized for many people. On other hand, warming of the earth, ecological destruction, overuse of resources, waste and many other problems have made it clear that the entire environment supporting human life is threatened. It is a well-known fact that the earth's environmental problems are attributed to our construction activities.

Against this background, recent waste management in Japan's construction industry is described in this paper.

THE GOVERNMENT OF JAPAN'S RECYCLING PROGRAM FOR DEMOLISHED BUILDING WASTE

Background of enactment

The Ministry of Land, Infrastructure and Transport published "Recycling program for demolished building waste" in October 1999. The policy statement consisted of the following four points:

1. Strategy for new construction that accelerates building longevity: Control demolished building waste from the design and planning stage
2. Strategy to improve separation of demolished building waste: Discussion of an inspection system for the demolition process
3. Strategy for accelerating recycling of demolished building waste: Support for improvement of recycling equipment
4. Strategy for formation of a recycling market: Increase the use of recycled materials in public projects and formation of an information exchange system

Basis of demolished building waste in Japan

Of the resources put back into use in Japan, 50% of it reappears in the construction industry, and the waste volume from the construction industry is 20% of Japan's industrial total. The volume of final scrap from construction is 40% of the whole. Ten years ago, 90% of Japan's illegal dumping volume was from construction, and it is 60% now.

It is clear that construction activities have harmful effects on the global environment in terms of both of resources used and wasted. All parties concerned with construction projects must understand the matter and push on with waste control and recycling.

Public works projects have caused the production of great quantities of concrete and asphalt concrete. The recycling of these materials has increased, and renewal projects in Japan now represent an important example of accelerated recycling. However, a general understanding of the cost of demolition and recycling is still needed to justify limiting waste to the smallest quantity.

Therefore, waste volume control and enhanced recycling systems are currently important themes for Japan's construction industry.

The present condition of construction waste

Based on a survey by the Ministry of Health and Welfare, Japan's available space for the residual volume of final scrap was estimated to last just 3.1 years as of the end of March 1997, based on the amount of final scrap being generated at that time.

However, gypsum board and other materials that had been accepted as stable wastes suitable for final scrap, have been reclassified as controlled wastes under the Waste Scrap and Cleaning Law of December 1997. Furthermore, stable wastes that mix or stick to other materials are also now classified as controlled waste.

A special law related to dioxin reduction was established in 1997. It declared that burning waste is not recycling because most dioxins are produced through combustion. Wood from the demolition of buildings is low quality and was being burned to reduce the volume of waste. However, the government has mandated that the combustion volume be decreased. The essential reduction of wood waste through recycling is now an important area of concern.

A rapid increase in the volume of construction project waste is anticipated in the near future because of the number of buildings erected in Japan since the 1960s which are now in need of renewal. Therefore, there is a keen demand to drawn-up and achieve measures to reduce the volume of waste and to accelerate recycling in the construction industry.

ARCHITECTURAL CHAPTER FOR A GLOBAL ENVIRONMENT

The background of the Chapter

The "Architectural Chapter for a Global Environment" was established by the Architectural Institute of Japan, several other architectural organizations and contractors. The purpose of the Chapter is to accelerate construction activities that give careful consideration to the global environment. The nature of the chapter is described below.

"Recognizing the relationship between today's global environmental issues and architecture, we, representing the five following architectural organizations in Japan, hereby enact the "Architectural Chapter for a Global Environment", and herewith declare to tackle, with each other, realizing a sustainable cyclic society.

June 1st, 2000

Architectural Institute of Japan
Japan Federation of Architects & Building Engineers Associations
Japan Association of Architectural Firms
Japan Institute of Architects
Building Contractors' Society

The Twentieth Century saw a tremendous development of a materialized civilian and rapid urbanization all over the world, including Japan. These brought us a convenient and comfortable lifestyle, positioning human beings at the core of nature. As a result, one such consequence, unfortunately, has been the emergence of various global problems, such as global warming, destruction of the earth's ecosystem, overuse of natural resources, and the accumulation of wastes. They are now threatening the global environment, the basic background for all life. And it is now obvious that our architectural activities play considerably important roles in causing such phenomena.

In this regard, we now recognize the urgent need to realize a sustainable society, by preserving the global environment whilst maintaining the health and safety of human beings at the same time. Architecture itself is not self-defining, and must be viewed within the context of its region, as well as the global environment. Therefore we hereby declare, as our basic objective of the 21st Century, that we will endeavour to create architecture with the following characteristics, in cooperation with all the concerned people.

1. Architecture shall be planned, designed, built, operated and maintained as a long-term social property, with a life that will span multiple generations. <Longevity>

2. Architecture shall constitute an element of a sound social environment which is in harmony with the natural environment and which co-exists with the diversity of life on the earth. <Symbiosis>

3. Architecture shall minimize consumption of energy throughout its lifetime, and maximize the use of natural and unused energy sources. <Energy Conservation>

4. Architecture shall incorporate reusable and recyclable resources and materials of minimum environmental loads, and minimize consumption of natural resources throughout its lifetime. <Resource Conservation and Cycling>

5. Architecture shall be created as a cultural component, respecting the local history and identity, relating to "genius loci", and succeeding for future generations as a good incubator. <Succession>"

Furthermore, the Japan Society of Finishing Technology and 23 other finishing technology organizations published the "Environment Declaration" in March 2002. It is believed finishing materials and application methods in building technologies have the greatest influence on the global environment. The Environmental Declaration consists of these five characters:

1. High Durability and Longevity
2. Symbiosis
3. Resource Conservation and Cycling
4. Health and Safety
5. Landscape Preservation

ACTIVITIES FOR ZERO EMISSIONS

Zero Emissions in the Japanese Construction Industry

The goal of "zero emissions" is the creation of a construction industry where all waste from production activities is eliminated and a sustainable cyclic society is realized. Even if something is waste in a certain industry, it has the possibility of being a resource in another sector. The center of this vision is a connected circle between plural industry fields and the creation of a protected environment in a society without waste. It is worth nothing that perfect zero emissions within only one industry field is impossible in the strict sense. Figure 1 indicates the concept of zero emissions in the construction industry.

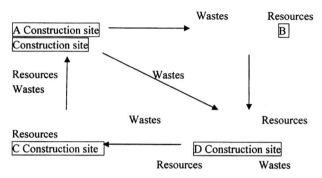

Figure1   The concept of zero emissions in the construction industry

Also, a single construction site cannot achieve perfect recycling. Furthermore, the recycling circle cannot be completed by a single construction company, and some of the waste from one construction site must be used as a resource for a different construction site. This is the current zero emission system in the Japanese construction industry.

Perfect recycling of waste includes "reuse" and "recycle". "Reuse" means to use again for the same purposes, and the "recycle" means to reuse a materials in a way that has no connection to its original purpose. "Recycle" includes "material recycle" which directly reuses as the same materials, and "chemical recycle", which produces raw materials for other materials by chemical process.

Utilization of the wastes from other industry

The use of the wastes from other industries as construction materials is not directly part of the zero emissions concept. However, it is an important consideration for society on a national level.

Furnace cement, fly ash, furnace slag aggregate, slag for roads and so on had been used in large quantities before recycling and environmental problems were considered. Nowadays, completely recycled concrete is in use and the use of recycled aggregate has been increasing. Electric furnace steel is used for most rebar products. It is a well-known fact that gypsum which is collected in sulfur removal apparatus from exhaust gases, is used as a material for gypsum board in large quantities. Asphalt roofing sheets, which have a long history in membrane waterproofing materials, are produced from old newspapers that are impregnated with liquid.

Enforcement of the "Green purchase law"

The "Green Purchase Law" has been enforced since January 2001. This law accelerates the government's purchase of environmental protection goods. The purchase of many recycled and reused materials has increased because of the law.

ACTUAL ACTIVITIES IN THE PRIVATE SECTORS

Considerations in planning and design stages

Construction of a building does not consist of only the construction stage. The most important issue in waste management is to be conscious of the building's life cycle in the planning and design stage so that materials applied to the building do not become waste. In addition to physical durability of structural materials and building elements, the building fixtures and finishes, must have the ability to be easily renewed. Flexibility in building use and enough space to accommodate changes must be considered. Therefore, the open-building concept that clearly separates support/infill has attracted the attention of people all over the world and must be incorporated from the planning and design stage.

Waste management in the construction stage

The consideration of environmental preservation has been indispensable in recent construction projects. This always includes both reduction of waste and the above-mentioned reuse and recycling. It is referred to as the 3R activities: "reduction, reuse and recycling". The following are typical 3R activities in the construction stage.

1. Minimizing packing material for building materials brought to the construction site.
2. Reduction of waste through on-site manufacturing of material.
3. Reusing of scaffolding materials instead of scrapping them.

More thorough separation of collected waste is contributing to reuse and recycling. Furthermore, some contractors are now adopting zero emissions activities during the actual construction stage.

Research on recycling building systems

Generally speaking, when construction problems are discussed, not only materials but also construction systems must be taken into account. The research on recycling systems has so far been limited. However, based on previously reported research results, the following points are made, and the direction of the technological developments in the future is indicated. For connecting and dismantling, engagement and fastening systems are considered recyclable, whereas welding and adhesion are not. The proactive introduction of engagement and fastening systems is being encouraged.

Systems for reducing waste in repair and modernization technologies

Recently, developments of modernization in construction methods have accelerated. It was once customary for deteriorated building finishing materials to be removed and discarded as wastes. However, recently developed technologies have taken the global environment into consideration by avoiding the removal of existing finishing materials. These technologies utilize an application of anchoring pins to prevent the delamination of finishing layers. Some representative methods are as follows:

1. A new renewal method for external walls by net overlaying, as indicated in Figure 2.
2. A new renewal method for roof waterproofing by FRP panel with anchoring, as indicated in Figure 3.
3. A new renewal method for roof waterproofing by galvanized steel sheet with anchoring, as indicated in Figure 4.
4. A new renewal method for external metallic walls using plastic film (Figure5).

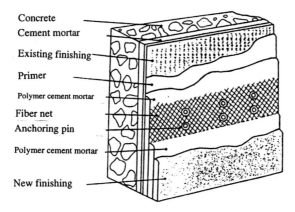

Concrete
Cement mortar
Existing finishing
Primer
Polymer cement mortar
Fiber net
Anchoring pin
Polymer cement mortar
New finishing

Figure 2   A new renewal method for external walls by net overlaying

Figure 3   A new renewal method for roof waterproofing by FRP panel with anchoring

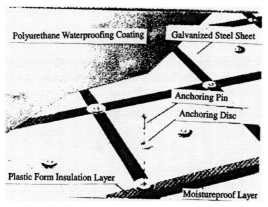

Polyurethane Waterproofing Coating   Galvanized Steel Sheet
Anchoring Pin
Anchoring Disc
Plastic Form Insulation Layer
Moistureproof Layer

Figure 4   A new renewal method for roof waterproofing by
            galvanized steel sheet with anchoring

Figure 5   A new renewal method for external metallic walls using plastic film

CONCLUSIONS

Recent examples of waste management in the Japanese construction industry field are described in this paper. Specific laws concerning environment preservation and waste management have been established.

Reusing materials, recycling and the reduction of waste have all accelerated in Japan's construction industry. Furthermore, attempts at zero emission have increased in all Japanese industries.

Recently technologies for repair and modernization have taken the global environment into consideration by avoiding the removal of existing finishing materials. These technologies utilize an application of anchoring pins to prevent the delamination of finishing layers.

However, waste management in the Japanese construction industry has not been enough. The problem cannot be resolved by material technology alone. With more consideration at the planning and design stages, significant progress should be achieved in the near future.

REFERENCES

1. THE MINISTRY OF CONSTRUCTION JAPAN: "Acceleration of separated collection and recycling of building demolition waste for establishment of resources circulating society", Building Letter, 1999, No.399, pp.1

2. YASHIRO T., "Zero emissions", Seko, Jan. 2001, pp.22

3. KINKI BRANCH ARCHITECTURAL INSTITUTE OF JAPAN: "Text of present and future of zero emissions in construction production", 1999

4. CONSTRUCTION PRICE ASSOCIATION: "Handbook of recycled materials for construction use 99", 1999

5. ARCHITECTURAL INSTITUTE OF JAPAN et. al.: "Architectural Chapter for a Global Environment", Jun. 2000

6. INDUSTRY TIMES CO.: "Recycling industry Plan 2001version", 2000

7. MANABE T. et al.: "Basic research on recycling system part 1", Proceedings of Annual meeting Architectural Institute of Japan, Tokai, 1994, pp.883

8. MANABE T. et al.: "Basic research on recycling system part 2", Proceedings of Annual meeting Architectural Institute of Japan, Hokkaido, 1995, pp.565

9. KONDO T. : "Possibility of recycling for exterior wall systems", Proceedings of CIB W70 International Symposium on Management, Maintenance & Modernization of Building Facilities, Singapore, 1998, pp. 263

10. KONDO T. et al.: "A new renewal method for external walls by net overlaying", Proceedings of 7th International Conference on Durability of Building Materials and Components, Stockholm, 1996, pp. 1292

11. KONDO T. et al.: "A new renewal method for roof waterproofing by FRP panel with anchoring", Proceedings of CIB W70 International Symposium on Management, Maintenance & Modernization of Building Facilities, Brisbane, 2000, pp. 263

WASTE POLICY IN WALES: THE CASE OF CONSTRUCTION AND DEMOLITION

C Cohen

A Flynn

J Ryder

Cardiff University

United Kingdom

ABSTRACT. The challenges of dealing with the total amount of waste that is produced, and finding environmentally and socially acceptable disposal routes for waste is a pressing public policy issue. The creation of devolved administrations in Wales and Scotland and regional bodies in England has helped to create a patchwork of responsibilities and approaches to waste management that has potentially further complicated an already complex policy area. In this paper we argue that Wales and England have similar approaches to their waste management strategies. However there are differences in waste composition between the two countries. C&D waste is in relative terms more important in England than in Wales but in Wales its disposal is likely to have a more adverse environmental impact. These differences in the circumstances of C&D disposal and the structure of the construction industry are not well reflected in policy documents. Despite devolution Wales is hampered in being able to deliver on its policy commitments to a more sustainable waste management policy because many of the most important levers to stimulate action remain in the hands of the UK government.

Keywords: Wales, Waste policy, Construction and demolition waste.

Caroline Cohen is a Research Associate in the BRASS Centre at Cardiff University. She is currently working on a European Union funded project exploring innovate actions in the waste and construction sectors.

Andrew Flynn is a Senior Lecturer in the Department of City and Regional Planning and also a member of the BRASS Centre at Cardiff University. His research interests lie n the area of environmental regulation and the role of devolved administrations in delivering sustainable development.

John Ryder is based in the Small Firms Research Unit at Cardiff Business School. He is a chartered civil engineer currently undertaking research looking at the role of knowledge in the management of change in the construction industry, focusing particularly on SME's and their environmental performance.

INTRODUCTION

The challenges of dealing with the total amount of waste that is produced, and finding environmentally and socially acceptable disposal routes for waste is a pressing public policy issue. The creation of devolved administrations in Wales and Scotland and regional bodies in England has helped to create a patchwork of responsibilities and approaches to waste management that has potentially further complicated an already complex policy area. One of the key issues that arises is to what extent devolved administrations are capable of developing and implementing sustainable waste management strategies that are appropriate to their contexts. There are two questions to be considered here: do policy makers recognise the specific circumstances of their areas and do they have the necessary levers to deliver on their strategy? In order to tackle these issues in this paper we focus in particular on construction and demolition waste (C&D) in Wales and reflect on how waste composition and policy may differ from England.

Waste management in Wales has historically been characterised by an over-reliance on landfill as a disposal route. C&D waste[1] has often found its way into landfill sites. Whilst this may be in some cases the best practicable environmental option for the disposal of such waste it is nevertheless contributing to the rapid filling of a limited number of landfill sites in Wales. In 1998/99, C&D waste contributed about 15% of Wales' total waste. In this paper we examine whether the statistical importance of C&D waste in Wales is reflected in the Welsh Assembly Government's (WAG) approach to waste management and in other policy initiatives promoted in both England and Wales. We are concerned to explain why one of the most important waste streams does not receive an equivalent policy response.

Our paper is divided into three main sections sections. Below we point to the commitment of WAG to sustainable development and show how that is reflected in its waste policy. We also point out how WAGs vision is undermined because devolution has not provided it with key powers to fully pursue its agenda. The following section describes the structure of the construction industry and assesses how its domination by SME's may influence the production of C&D waste and also make it difficult to promote better environmental practices. The final section briefly reviews some of the key instruments to promote improved C&D waste management.

WASTE IN WALES: DEVELOPING A SUSTAINABLE WASTE STRATEGY?

To understand the development of waste policy in Wales it is essential to appreciate the policy context of devolution and the Welsh Assembly Government commitment to promoting sustainable development. Unlike the Scottish Parliament, the Welsh Assembly has no primary legislative powers, and no power to vary taxes. Nevertheless, to describe it as a 'glorified county council' as some have done is to seriously underestimate its significance. The Assembly's powers and policy remit is wide ranging. Moreover, under section 121 of the Government of Wales Act 1998, there is bestowed on the Assembly a responsibility for sustainable development that is unique for an elected body in Britain. The section states that "The Assembly shall make a scheme setting out how it proposes, in the exercise of its functions, to promote sustainable development."

---

[1] (mostly bricks, concrete, hardcore, subsoil and topsoil but it can also include timber, metal, plastics and occasionally special waste materials)

If the political and policy changes in Wales are matters of substance then they should be reflected in the development of waste management strategy. On two counts we should expect to see differences emerging between Wales and other parts of the UK:

1. Devolution should ensure that Welsh policy makers are sensitive to the circumstances surrounding waste management in Wales; and
2. The Assembly's commitment to promoting sustainable development should mean that its waste strategy is imbued with a philosophy of sustainability.

A key research question is, therefore, to what extent can we observe differences in waste policy and management between Wales and other parts of the UK? As Table 1 below shows there are subtle differences in the relative importance of different waste streams between the two countries. The volume of C&D waste is important in both countries but relatively more so in England.

Table 1   Waste production (1998-99) in Wales and England

WASTE STREAM	ENGLAND & WALES*	%	RANK	ENGLAND*	%	RANK	WALES*	%	RANK
Industrial	51, 920,000	12.5	4	46,931,000	11.9	4	4,989,000	21.5	3
Commercial	23,600,000	5.7	6	22,459,00	5.7	7	1,141,000	5	6
Municipal	28,320,000	6.9	5	26,990,000	6..9	6	1,330,000	5.7	5
C&D	70,800,000	17	3	67,515,000	17.2	3	3,285,000	14.1	4
Mines and Quarries	118,000,000	28.4	1	112,000,000	28.6	1	6,000,000	25.8	2
Sewage Sludge	28,320,000	6.9	5	27,915,00	7.2	5	405,000	1.7	7
Agricultural	94,400,000	22.6	2	88,325,000	22.5	2	6,075,000	26.2	1
Total	415, 360,000	100		392,135,000	100		23,225,000	100	

*Quantity produced (tonnes per annum)
Dredging waste has been excluded from the table because figures are available for England and Wales.
Source: http://www.environmentagency.gov.uk/ /105385/ [5]

Within this paper it is not possible to scrutinise all aspects of waste policy and so in this section we briefly review the key principles underlying policy and in the next section we concentrate on C&D waste. We have chosen to focus on C&D waste because in volume (and potentially environmental) terms it is significant and raises quite specific management challenges.

Vision And Ambition

The waste strategy for Wales, "Wise about Waste" (WAG 2002) [4] is the first waste policy specifically for Wales. It draws upon the earlier and joint England and Wales Waste Strategy 2000 (DEFRA 2000) [1]. From the outset the Welsh waste strategy recognises that "[w]aste is Wales' biggest environmental problem" (WAG 2002, pv) [4] and is "designed to move Wales from an over-reliance on landfill to a position where [by 2012] it will be a model for sustainable waste management" (WAG 2002, pvii) [4]. Such a commitment is all the more laudable as Wales is starting from such a poor base: it landfills around 4m tonnes of waste each year and only recycles 7% of municipal waste. The ambition, though, clearly fits with the Welsh commitment to sustainable development (see WAG 2002, para 1.4) [4].

The Waste Strategy 2000 [1] also recognises the link between sustainability and waste and notes "if we are to deliver sustainable development it is crucial that we begin to tackle our growing mountain of waste" (DEFRA 2000, Foreword and para 2.3) [1]. Policy development is termed a "step change" towards "more sustainable waste management" (DEFRA 2000,

Summary) [1].   Therefore in terms of the analysis of the waste problem and the policy approach there is much similarity between the overall direction being taken in England and Wales.

Principles And Levers

The starting point for both the English and Welsh and Welsh analysis of waste management is that the current levels of disposal to landfill are unsustainable. This is because landfilling represents a waste of resources (i.e. materials), there is a gathering shortage of landfill space for waste and landfill is a major source of methane (a greenhouse gas). Moreover, the EU Landfill directive sets ambitious targets for the reduction of biodegradable municipal waste sent to landfill and so will force a change in British waste management practices.

So how is greater value to be extracted from waste? There are two steps envisaged:

1. Decoupling, that is breaking the link between economic growth and waste production
2. Better materials management by putting waste to "good use – through substantial increases in re-use, recycling, composting, and recovery of energy" (DEFRA 2000, para 2.7) [1].

How is change to be achieved? The Waste Strategy 2000 [1] identifies a number of levers for change (see Chapter 3) and these include: the Waste and Resources Action Programme (WRAP); producer responsibility (e.g. in relation to packaging and the End of Life Vehicles Directive); the Landfill Tax; the Aggregates Levy; Best Practice Programmes; and limiting landfill.

A major challenge for WAG is that "it does not control some of the crucial levers required to make a fundamental change in the way waste is managed in Wales" (WAG 2002, para 1.7) [4]. For example, WAG cannot set tax levels so is unable to vary the level of landfill tax or the aggregates levy if it wished to do so and neither can it introduce or amend primary legislation. So WAG finds itself in a highly dependent situation as it recognises: "[f]or a number of the Assembly Government's aspirations to be delivered, action has to be taken in conjunction with UK Government" (WAG 2002, para 1.7) [4]. The one set of relationships where WAG can exercise some authority is with regard to local government and so not surprisingly local authorities emerge as key partners in the Welsh strategy. Many of the targets and actions are aimed at local government, which has also been given considerable additional funding to meet their responsibilities. For instance £79m is being provided by the Assembly for the period 2001/2-2004/5 principally to assist local government to meet the municipal waste minimisation, recycling, composting and reduction in landfill targets.

In this brief review of waste management policy in England and Wales we have been able to show that the Welsh commitment to producing a strategy that is informed by a commitment to sustainable development is one shared by England. Moreover, although both countries face similar waste management problems and issues, there are differences in the relative importance of different waste streams. The analysis of the waste problematic and policy prescription is, though, similar.   However, despite devolution Wales remains highly dependent on the UK government for the successful realisation of its own policy. This is because most of the key levers to influence better waste management remain in the hands of central government. So what effect, if any, will this have on the approaches to C&D waste? Before we can answer this question we briefly outline the structure of the Welsh construction industry because it will be the key source of C&D waste.

The structure of the Welsh construction industry and its policy implications

The construction industry in Wales, like that for other parts of the UK, is marked by its dominance by SME's (1-250 employees) (7916 firms in 2001) and many of these are micro firms (1-3 employees). The Welsh (i.e. those companies having a registered office and a primary trading office in Wales) construction industry as a whole is smaller than that of England in terms of maximum company size. It has no companies of the very large size (over 1200 employees) whereas England has 51 (0.03% of the total number of English companies). Wales has the same proportion of companies in the range 300-1199 employees as England (0.1% of the totals). Registered office and trading office data, show that there are some 245 construction companies from outside Wales (i.e. from England, Scotland, Northern Ireland, the Republic of Ireland or the British Crown dependencies) which have a trading address in Wales (equivalent to 3% of the total of 7916 home Welsh companies) and of these 31 have a larger turnover than the largest home Welsh construction company. Assuming that these larger construction companies from outside Wales do actually undertake work in Wales rather than just having a nominal trading address in Wales, it is reasonable to conclude from this latter point that there is a UK-wide construction market, at least as far as the larger companies are concerned, rather than just a Welsh construction market open only to Welsh companies.

Two points are important to note here. First, that it has proved notoriously difficult to promote improved environmental practices amongst SME's and so a sector dominated by such firms is likely to have a poor record in dealing with its wastes. Second, that market differentiation leads firms to engage in different types of construction projects and this may well affect the way in which they manage their waste. It is usual in the construction sector for the larger companies to seek to win work on the larger and more complex projects where there is less competition from the smaller companies and where their competitive advantage lies in their expertise and capitalisation. One of the effects of this is that Welsh companies operating in Wales are in effect constrained in the actual Welsh construction market (i.e. one that is open to competition from non-home Welsh companies) to either obtaining work as members of the larger companies' supply chains or bidding for smaller scale local projects.

The structural features of the construction industry allied to the relative underperformance of the Welsh economy help to explain much of why there are differences in the relative importance of C&D waste in Wales and England and also of its use and disposal. The production of C&D waste in Wales lags behind that of all of the English regions. The North East produces the lowest amount of C&D waste, some 4.75m tonnes compared to that in Wales of 3.28m tonnes. At the other extreme the South East (excluding London) produces about 4 times as much C&D waste as Wales about 13.12m tonnes (Environment Agency 2000, p27) [2]. Table 2 below shows the use and disposal of C&D waste in Wales.

Table 2   Use and disposal of C&D/inert waste in Wales

FATE	QUANTITY, 000's tonnes	PERCENTAGE
Recycled as aggregate and soil	747	23
C&D waste and soil re-used in landfill	421	13
Inert material recovered on exempt sites	1299	40
C&D waste and soil landfilled	818	24
Total estimated production	3285	100

Source: WAG, 2002, Part Two, P118 [4]

More detailed analysis of the figures on C&D waste provides a fuller picture of the Welsh context. When the fate of C&D waste is compared with that of the nine English planning regions it is apparent that only Yorkshire and Humberside shares with Wales the unenviable position of having a higher proportion of C&D waste going for landfilling rather than recycling. Indeed, for every English region with the exception of the North West (where inert material covered on exempt sites is the most popular form of end use) recycling of aggregates and soil is the most common option, whereas in Wales it ranks third (DEFRA 2000, para 8.50)[1]. This is probably because of the relatively low cost of primary aggregates in Wales and limited facilities for the processing and storage of construction and demolition waste.

What we have been able to show so far is the significance of C&D waste within the wastes produced in England and Wales. We have then gone on to show how the use and disposal of C&D waste is also spatially structured. What is rather surprising therefore is that strategy documents pay so little attention to C&D waste and of the spatial variation in its uses. The Waste Strategy 2000 (DEFRA 2000, para 8.43)[1] is typical in that it notes that C&D waste represents a significant proportion of total waste generation but largely limits its comments on C&D waste to a sanguine review of the potential of the construction industry to achieve greater efficiency in its use of materials, so minimising waste production (for example, para 8.47)[1]. It is notable that there is considerably more guidance on the reuse of C&D waste in relation to the planning system than the waste management system.

Although the Welsh follow-up document Wise about Waste [4] largely follows the London lead in encouraging industry to reduce its waste (see WAG 2002, paras 5.149-151)[4] and relies heavily on a mix of voluntary measures (such as eco-design), economic incentives (the Aggregates Levy) and control measures (Minerals Planning Policy Wales), it does set its own targets. These are:

- By 2005 , to re-use or recycle at least 75% of C&D waste produced;
- By 2010, to re-use or recycle at least 85% of C&D waste produced (WAG 2202, para 5.153)[4].

So what might be done to promote better management of C&D waste? And are the available levers likely to help WAG meet its ambitious targets? In the next section we review the mix of regulatory 'sticks' and fiscal and voluntary 'carrots' that apply to C&D waste.

PROMOTING BETTER WASTE MANAGEMENT: CARROTS AND STICKS

Regulatory measures

At the outset it is important to consider the nature of the EU waste legal framework, given the importance of its impact at the Member State level.   The European Commission has identified a number of priority waste streams because they pose a potential threat to the environment.   Inert and C&D waste is one of these. What is rather surprising therefore is that C&D waste is dealt with as part of the EUs general approach to waste management, perhaps most notably the Landfill Directive (99/31/EC) rather than as a specific waste stream like Packaging (which has its own directive 2000/53/EC)

The lack of specific regulation of C&D waste at the European level is mirrored at the UK level. Construction waste related legislation concentrates essentially on the diversion of waste from landfill, largely instigated by the EUs Landfill Directive, as well as on the use of secondary aggregates, with the aim of encouraging construction waste segregation, re-use and recycling.

Command and control has been the traditional method of environmental protection. The lack of measures targeted at C&D waste may be a significant omission if other measures to reduce and better manage C&D waste are also found wanting. The WAG does not have the power of being able to pass primary legislation and so it cannot close this regulatory gap on its own.

TAXES AND MARKET DEVELOPMENT

There are two main taxes to promote better waste management and resource productivity in the construction industry in the UK. Both the Landfill Tax, which was introduced in 1996, and the Aggregates Levy, which came into force in 2002, seek to ensure that material recovery becomes the most competitive solution for waste management, and in turn enhance market opportunities for recyclates. The landfill tax has encouraged producers to look for alternative uses for construction waste and has contributed to the diversion of C&D waste from landfill. According to a recent UK Government study (Survey of Arisings and Use of Construction and Demolition Waste in England and Wales in 2001, www.planning.odpm.gov.uk/consdemo)[3], almost half of the C&D wastes produced in England and Wales are re-used or recycled. Out of the total waste stream, more than 38 million tonnes was recycled as aggregate in 2001; compared to 22.7 million tonnes in 1999. Whilst these figures are laudable it is difficult to know how much they should be attributed to the landfill tax. The level of the tax for inert waste is far cheaper (£2) than for active waste (£13). At its current level, the tax would not seem to provide a big incentive to minimise C&D waste.

With regard to the effectiveness of the Aggregates levy[2] and the new Aggregates Levy Sustainability Fund (ALSF) to address C&D waste management, again the results are mixed. There are isolated examples of C&D waste minimisation and recycling in projects such as Trent Valley GeoArchaeology study and the Archaeology South East study on Lydd quarry (www.english-heritage.org.uk) [6]. The levy has also opened some market opportunities for companies that can use recycling aggregates in their activities. For instance, RMC has reduced its operational cost by using recycled green glass as road building materials (http://news.bbc.co.uk/1/hi/sci/tech/1907207.stm) [7].

However, a study for the Office of the Deputy Prime Minister (ODPM) has pointed out that there are limits to market development for C&D waste. One of the study's findings was that the "scope for further recycling of C&D waste for use outside landfills and registered exempt sites appear to be limited by the fact that much of the C&D waste that was not being recycled as aggregate was not physically capable of forming aggregate, because it was wholly or largely made up of soil" (www.planning.odpm.gov.uk/consdemo, chapter 7)[3].

Novel instruments for environmental protection appear to be meeting significant challenges in promoting better management of C&D waste. Once again the WAG finds itself at a disadvantage in seeking to overcome any shortcomings as it has no powers over taxation.

---

[2] The Aggregate levy charges £1.60 for every ton of newly quarried materials such as sand and gravel. The chief aims of the levy are to internalise the external environmental costs of aggregate production; to encourage the use of alternative materials and development of new recycling processes; and to promote more efficient use of virgin aggregate.

Voluntary initiatives

At the industry level, there is a wide range of best practice cases which provides advice on aspects of the management of C&D waste including the reduction of construction waste, waste segregation and re-use as well as the use of secondary aggregates. For instance, the Building Research Establishment (BRE) has launched a construction waste material exchange scheme on the internet; the Construction Industry Research and Information Association (CIRIA) provides a database of construction-related recycling facilities in Great Britain that accept or sell materials and promotes localised recycling of materials. These kinds of initiatives provide incentives for construction companies to engage in recycling activities as there are clear cost savings to gain from reduced costs for waste disposal, reduced landfill taxes and transportation costs.

The Department of Trade and Industry launched in April 2000 a strategy for sustainable construction "Building a Better Quality of Life" that sets out some clear targets to improve the construction industry building practices. Resource efficiency is recognised as a fundamental element and the strategy underlines the importance of "designing for minimum waste" and "re-using existing built assets" (see www.property.gov.uk/services/construction/gccp/100700.pdf)[8]. The Office of Government Commerce has adopted an action plan, "Achieving Sustainability in Construction Procurement" so that the public sector can become a leader and exemplar in green procurement in construction by promoting the use of materials with low environmental impact or made from renewable resources. All UK government departments have agreed to implement the action plan by March 2003. As government procurement accounts for 40% by value of the UK construction market, the public sector plays a key role in the promotion of resource productivity and waste minimisation. The specifying of recycled and reclaimed materials in tender documents will encourage the segregation and recovery of materials. The strategy was developed in partnership with industry bodies, and may have been unduly sympathetic to business interests as it places no direct requirements on clients or suppliers of building materials to use environmentally friendly materials or to improve their C&D waste management.

As Table 3 below shows there are a plethora of initiatives in place in England and Wales to tackle C&D wastes, from waste minimisation programmes to re-use and recycling guidance. Yet, current support programmes tend to operate independently of each other and there seems to be little co-ordination among stakeholders.

There is a crucial lack of integration and coherence between organisations, which undermines C&D waste management, recycling and re-use. For instance, the Waste as Resources Action Programme (WRAP) is working on the promotion of the use of recycled aggregates but does not have targeted programmes for C&D wastes and does not work in collaboration with other key organisation to achieve its aims. It is doubtful whether policy guidance and voluntary agreements represent strong enough incentives to promote better resource and waste management and will succeed in motivating businesses to improve their practices.

Table 3   C&D incentives and initiatives

TYPE OF INCENTIVE	ENGLISH AND WELSH INITIATIVES	WELSH INITIATIVES
Waste minimisation guidance	• CIRIA, Construction Industry Research and Information Association • The Construction Confederation • DTI, Department of Trade and Industry Construction Best Practice Programme	• Edexcel training courses (partnership between Environment Agency, Fforwm, Groundwork Wales, Penarth Management Consultants) • Business and Environment Challenge Scheme
Waste recovery and recycling support	• CIRIA reclaimed and recycled construction materials handbook • BRE internet Materials Information Exchange • Government's Aggregates Information Service (AIS) • WRAP Waste as Resources Action Programme	• Objective 1 and 2 Fund projects: the Centre for Research Into the Built Environment project • Arena Network waste exchange
Transport efficiency	• CIRIA recycling sites map (materials recovery and recycling)	
Planning Guidance	• PPG45 for England only	• Minerals Planning Policy Wales • Technical Advice Note TAN 21 on Waste

Wales: lack of strategic guidance but innovative business support

There are few initiatives and programmes that take place at the Welsh level to address C&D waste management. This is not surprising as the Welsh waste strategy has given priority to household waste management which represents 6% of Wales total waste to the detriment of other waste streams. Although some C&D waste recycling targets have been set in the strategy, senior officials have recognised during an interview that specific arrangement to tackle C&D - which accounts for 15% of Wales total waste - have yet to be made.

So far, there is a basic lack of institutional infrastructure to address C&D waste as a specific issue. For instance, no sub-group has been established in the Waste Wales Forum to tackle that issue despite the recognition of the need for one in Wise About Waste (p68)[4]. Moreover, the Welsh Regional Waste Planning Groups tend to follow the WAG agenda and so they too have paid little attention to developing an infrastructure for C&D waste.

Although the Welsh strategy is weak on policy guidance for C&D waste, it should be noted that the Assembly Government have dedicated some Objective 1 and 2 Funds specifically to support the implementation of waste recycling and re-use activities in the construction sector. For instance, the Centre for Research in the Built Environment (CRIBE) seeks to raise awareness of best waste management practices in construction SME's. In its quest to divert

C&D waste from landfill and to promote markets for recyclates, the Assembly Government is looking at the way in which this kind of initiative could be mainstreamed throughout Wales. The Assembly will publish a project report, which will provide guidance for building professionals, in May 2003.

CONCLUSIONS

In our review of C&D waste in Wales we have taken regular opportunities to compare the situation with England. Throughout we have sought to make five points. First, that Wales and England have similar approaches to their waste management strategies. The commitment to sustainable development that underlies the work of WAG is also echoed in the English approach to waste management. Beyond the rhetoric, though, there are subtle but significant differences between the two countries that could have significant implications for C&D waste management. Our second point is therefore that there are differences in waste composition between the two countries. Third, C&D waste is in relative terms more important in England than in Wales but in Wales its disposal is likely to have a more adverse impact on the environment. Fourth, these differences in the circumstances of C&D disposal are not well reflected in policy documents and neither do the documents pay attention to the structure of the construction industry which, dominated by SME's, is not likely to be sympathetic to receiving environmental messages. Fifth, Wales is hampered in being able to deliver on its policy commitments to a more sustainable waste management policy because many of the most important levers to stimulate action remain in the hands of the UK government. Where WAG has invested heavily with its local government partners in raising the profile of waste management it has concentrated on household waste and marginalised C&D waste. Consequently the targets that the Assembly has set itself on the reuse and recycling of C&D waste may prove difficult to meet.

REFERENCES

1. Department of the Environment Transport and the Regions (2000) Waste Strategy 2000 England and Wales, The Stationery Office.
2. Environment Agency (2000) Strategic Waste Management Assessment 2000: Wales.
3. Office of the Deputy Prime Minister www.planning.odpm.gov.uk/consdemo/
4. Welsh Assembly Government (2002) Wise About Waste: The National Waste Strategy for Wales, Parts One and Two, June.
5. http://www.environmentagency.gov.uk/commondata/105385/
6. www.english-heritage.org.uk
7. http://news.bbc.co.uk/1/hi/sci/tech/1907207.stm
8. www.property.gov.uk/services/construction/gccp/100700.pdf

WASTE MINIMISATION IN AUSTRIA: AN OVERVIEW ON THE PRESENT SITUATION AND POTENTIALS FOR IMPROVEMENT

S Salhofer

BOKU University

Austria

ABSTRACT. Waste minimisation includes measures for waste prevention (reduction of the generation of waste and/or the hazardous character of waste) as well as recycling (mainly feedstock recycling, the reuse of materials as secondary raw material, in some case also thermal recycling, the use of the calorific value of waste materials). On the other hand, waste streams can be defined by the waste producers, which are here categorised as households (typical processes are food preparation, the consumption of clothing, furniture and other goods) and companies (all types of industrial and commercial enterprises, public institutions etc.) The paper compares the activities and effects of waste prevention and recycling in Austria both for the household sector and companies. The prevention of household waste shows only small potentials, but the recycling of household waste was introduced rather effective in Austria. For industrial or commercial waste a reliable database in not available, thus the effects of waste prevention and recycling can be shown only on a case-to-case basis.

Keywords: Waste minimisation, Waste prevention, Recycling, Austria, Household waste, Industrial waste

S Salhofer is an Associate Professor at the BOKU University of Natural Resources and Applied Life Sciences, Vienna. He is a member of the Institute for Water Provision, Water Ecology and Waste Management, in the Department of Waste Management.

INTRODUCTION

In general, waste minimisation has been given a high priority in integrated waste management. European waste laws (e.g. in Austria or Germany) clearly define waste prevention as the first option for solving waste problems. This hierarchy is also found in the European Community Strategy, where prevention has been given the highest priority, followed by recycling and lastly by waste disposal. Waste minimisation includes measures for waste prevention as well as recycling. Waste prevention targets a reduction in the amount of waste generated (quantitative prevention) as well as in the hazardous character of waste (qualitative prevention). Recycling includes feedstock recycling (the reuse of materials like paper or glass as secondary raw materials) and in some cases also thermal recycling (the use of the calorific value of the waste materials).

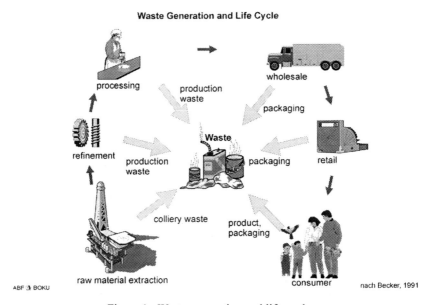

Figure 1   Waste generation and life cycle.

Waste streams can be classified according to the waste producers, defined here as households and commercial operations. The latter includes all kinds of industrial and commercial enterprises as well as public institutions and the like. From an LCA point of view ("cradle to grave"), waste is produced throughout the life cycle of any product from the stages of extraction, refinement and processing of raw materials (ores, crude oil etc.) to production (mostly in several stages) to distribution (wholesale, retail) and the usage phase. Waste generation throughout the life cycle of products is shown schematically in Figure 1. Products and product packaging end up with the consumer and are disposed of as household waste after the usage e-phase. The reason for making a differentiation in the waste producers is that the processes of waste production are different for households and commercial operations. In households, typical processes are consumption processes, which include food preparation as well as the usage of clothing, furniture and other goods. In commercial operations processes of production and distribution are relevant for waste generation. In the following section, waste prevention and recycling are considered for both the commercial and the household sectors.

FACTS FOR WASTE MINIMISATION IN AUSTRIA

Prevention Of Household Waste

Regarding only the quantitative aspect, the prevention of household waste can be measured in specific quantities (in kg/cap.yr) of the waste generated. In most cases the quantities originating from households are not known because municipal waste, which is documented in waste statistics, also includes a certain amount of commercial waste, which is collected and treated together with household waste. In Austria, the quantities of municipal waste are different from region to region, but in general the quantities show an increasing tendency (see Table 1).Thus the prevention of municipal waste has not been very successful.

In a case study on behalf of the City of Vienna [1], we analysed the potential for the reduction of waste quantities by waste prevention. First of all we conducted a broad review of the literature including data obtained through a survey on the activities for waste prevention from 76 European cities. In a next step, the top ten measures were selected, ie. activities where quantifiable data were obtained. The only measures we took into consideration were measures, which that can be implemented on a municipal or national level. Not considered were Oother measures such as resource taxes, strict requirements for the design of products or product bans, which can be implemented on a supranational level only. Qualitative aspects for reducing the content of pollutants in waste (heavy metals, organic pollutants etc.) were excluded from this analysis.

To estimate the waste prevention potential we used results from comparable cities and regions as well as calculations (e.g. effects from obligatory quotas for reusable beverage packaging). It should be mentioned that the results are valid only for this specific situation with respect to waste quantities and the state of implementation of individual measures. For example, in one Austrian region, 40% of the families with babies used reusable diapers, as a replacement for disposable diapers, whereas in other regions, the rate of participation was near zero. We found that the highest potential for waste prevention could be achieved by the enforcement of reusable beverage packaging (national packaging tax or a license model) and the prevention of unwanted advertising material. Optimistically calculated, 30 kg/inh.yr (6% of municipal waste in Vienna) could potentially be prevented if all ten measures would be implemented.

The results of this study show that due to the existing boundary conditions there are only a few possibilities for preventing waste at a communal level. Many of the measures offer only a small potential for waste prevention, or their implementation could lead to legal problems. Furthermore relatively high financial costs, resistance from affected economies and substitution effects should be expected. Further impediments to implementing waste prevention are [2]:

- changing production and consumption patterns; for example, decreasing costs of products increase consumption;

- an excessive emphasis placed on recycling as compared to prevention measures; from an administrative point of view departments have been set up to monitor waste collection, recovery and treatment whereas practically no human resources have been dedicated to prevention activities; and

- a lack of methods for measuring waste prevention effects.

Table 1 Generation of household waste (in kg/inh.yr) in Austria 1989-1999 [3]

YEAR	1989	1990	1991	1992	1993	1994
Residual & bulky waste	274	267	239	225	209	184
Hazardous household waste	1	1	2	2	2	2
Total of recyclables	41	52	64	73	80	98
Biogenic waste	3	4	5	12	23	35
TOTAL	320	324	309	313	314	320

YEAR	1995	1996	1997	1998	1999
Residual & bulky waste	181	188	187	182	190
Hazardous household waste	3	3	3	3	3
Total of recyclables	102	109	117	122	131
Biogenic waste	43	45	53	54	59
TOTAL	329	344	360	361	383

From the point of view of the municipalities, the main obstacle to waste prevention is the fact that municipal waste organisations can only implement waste minimising strategies once waste has been generated. Decisions made by producers and consumers on the design, purchase and use of products can scarcely be influenced by municipalities. Dette and Jülich [4] evaluated activities for waste prevention and minimisation in the member states of the European Union. They drew the conclusion that a strategy based on product orientated environmental policies is the correct answer to the waste prevention challenge.

Recycling Of Household Waste

Recycling of household waste was introduced quite successfully in Austria. In the 1980s and particularly in the 1990s, collection schemes for recyclables, biogenic waste (compostables) and hazardous household waste were put in place. In 2000, ~48% of municipal waste was collected for recycling. An analysis [5] has compared the recycling rates for regions in Germany, Austria and the UK. Recycling rates similar to Austria were found in Germany (39% collection for recycling), but rather lower ones were found in the UK (12%). Details are shown in Table 2. All the regions considered show the same tendency: rural areas achieve higher recycling rates than urban ones.

It is supposed that the framework of objective conditions (situational, social) is basically different in these two types of areas. According to sociological findings, there are several tight interdependencies between space and individual behaviour. This means that a specific behaviour, which may be appropriate in a certain environment, might not be the optimal behaviour in a different setting. Thus the conclusion could be drawn that recycling appears to be an appropriate practice in rural areas, but is less appropriate in high-density areas and multi-family dwellings. The specific problems in multi-family dwellings (lower recycling rates and awareness) are described in more detail in [6].

Table 2 Generation and separate collection of municipal waste (kg/inh.yr) in Germany, Austria and the UK [5]

	GERMANY			
	NATIONAL	NRW	NRW-CITIES	NEW-RURAL
Residual waste	210	278	353	192
Bulky waste	39	42	50	35
Glass	34	31	29	32
Paper	70	66	64	67
LWPM*	17	17	16	18
Biowaste	41	73	31	122
Hazardous waste				
Other recycling		8	8	8
TOTAL	411	515	551	474
Recycling rate	39%	38%	27%	52%

	AUSTRIA			UNITED KINGDOM		
	NATIONAL	AUSTRIA VIENNA	RURAL	ENGLAND	METRO-POLITAN	NON-METRO-POLITAN
Residual waste	160	280	121	408	456	380
Bulky waste	27	18	51	86	86	87
Glass	22	15	25	8	6	10
Paper	58	69	52	19	16	24
LWPM*	15	11	20	2	1	2
Biowaste	53	46	56	16	6	22
Hazardous waste	3	2	3			
Other recycling	22	59	0	20	15	24
TOTAL	360	500	328	559	586	551
Recycling rate	48%	40%	48%	12%	7%	15%

NRW = Northrine Westphalia
*LWPM includes tins, cans and packaging made of plastics and compound materials

Prevention And Recycling Of Commercial Waste

If we compare waste quantities by the generator, commercial operations clearly dominate the regional waste generation in Austria. Households produce only 6% of the overall waste, while 94% is commercial waste. Table 3 shows the main categories of waste produced in Austria in 1999. Excluding sewage sludge, construction debris, excavation materials and other mineral waste (which represent wastes of a different quality compared to household waste), the share of household waste is approximately 20 %, while nearly 80% of waste is produced in commercial operations. This leads to the conclusion that the prevention and recycling potentials in the commercial sector are higher than in the household (municipal) sector.

Table 3   Waste generation in Austria in 1999 [3]

CATEGORY	1000 TONNES	KG/INH.YR	%
Hazardous waste	1000	124	2
Household waste	3100	383	6
Recyclables in businesses	2200	272	5
Waste wood	3800	470	8
Sewage sludge	2300	284	5
Other non-hazardous wastes	4600	568	9
Construction debris	7500	927	15
Excavation material	20000	2471	41
Other mineral wastes (excl construction debris	4100	507	8
TOTAL	48600	6006	100

In Austria, there are many waste minimisation activities in industrial, commercial and other sectors. They are supported by:

- the legal obligation for commercial businesses to set up waste management plans (WMP);
- sectoral waste management plans, which have been developed for agriculture, food production, chemical industry, electroplating, hospitals, and others [7],
- consulting programs, frequently supported by public subsidies, e.g. the "Businessplan Vienna" (environmentally friendly procurement, subsidies for the implementation of EMS and WMP),
- research and development programs. Examples are the program "Fabrik der Zukunft" (factory of the future), fostering environmentally friendly production and products, the program "greenlogistics", linking logistics and environmental targets and other cleaner production programs.

The effects of waste prevention in the commercial sector cannot be measured on a general basis. Only case studies and sectoral studies can provide some estimation. The main reason is that reliable databases for the quantities and the changes (especially before and after specific programs) are not available. For the same reason, recycling effects cannot be quantified at a comprehensive level. In order to provide a database, we developed a waste generation model for commercial waste in a case study for the City of Vienna [8].

Important elements of this model are to define sectors with similar processes and waste (e.g. waste from the catering sector is different from the metal processing sector in terms of quantity and quality) and to find a link to existing business statistics. In each sector, a random sample of commercial operations was selected; data on waste quantities generated by these were investigated and mean values were calculated. Additional existing WMP from large size enterprises (> 100 employees) were evaluated. As a result we found (partially statistically significant) data for the production of recyclables, non-recyclable waste (residual waste) and sector-specific waste in most of the sectors. The results were used to identify sectors where higher waste minimisation potentials were supposed.

Results on the effects of industrial/commercial waste prevention and recycling activities are partially known from a number of case studies of individual companies as well as sectoral studies. In the case of the VIW (Philips Videowerk Wien), an electronics assembling company in Vienna with approximately 2,500 employees, a WMP reduced the amount of non-recyclable waste from 540 to 130 t/yr [9]. The implementation of the WMP was strongly supported by the management of this company, as (in addition to positive environmental effects) considerable financial savings were also realised.

In a sectoral study, a sample of 195 operations for the education sector and 110 businesses for the catering sector in Vienna [10] was visited to gather data on waste generation and disposal. Each sector was divided into sub--sectors, which were evaluated separately. For each sub-sector, optimal situations (a smaller sample of well functioning operations) were compared to the average, and thus potentials for waste minimisation were calculated. The results show a recycling potential of 20% in the education sector and 33% in the catering sector.

A further study of the automobile service sector in Austria [11] found that there are high potentials for waste prevention and recycling here as well. In this a case study of a sample of 31 car repair shops in Austria, the impacts of waste minimisation were evaluated by comparing waste quantities and disposal costs before and after the implementation of Environmental Management Systems (EMS) and WMP. The operations established waste prevention measures (e.g. reuse of cleansing liquidsagents), recycling measures (separate collection of recyclables) as well as organisational measures (e.g. contracting with waste disposal companies and price negotiations). The total amount of non-hazardous waste was reduced by 3% and the amount of recyclables increased by 6%; here recycling seems to be more effective. As for hazardous waste, the overall quantity was reduced by 37%, and recycling remained at the same level (-1%); here waste prevention was more effective.

In the next step, the effect of WMP and the introduction of EMS were compared. In Austria, the (legally required) WMP for commercial operations includes an analysis of processes and waste generation within the operation as well as the development of a waste strategy and a disposal plan. All of the investigated operations have established at least two WMP to analyse their current situations and to derive and organise environmental improvements. Half of these operations went one step further by implementing EMS according to the international ISO 14000 or the European EMAS Standard. These systems are more complex than WMP and include not only all the environmental aspects (water, air pollution etc.) of an organisation, but also certified procedures for continuous improvements in environmental performance.

The difference between these two instruments is that EMS provide a more detailed documentation and organisation for continuous improvement, and also emphasises the training of employees. A variance analysis reveals the differences in effectiveness of EMS and regularly implemented WMP. The results of this comparison are shown in Table 4. For waste reduction, EMS indeed appears to be more effective, emphasising the importance of education, training and motivation to achieve the goal of better environmental performance, particularly in service industries. The difference between WMP and EMS in terms of cost savings is insignificant, which means that any measures involving cost savings are independent of the system implemented and are more influenced by price negotiations.

Table 4   Reduction of waste and disposal costs in car repair shops, comparison of the effect of waste management plans and environmental management systems [11]

	QUANTITY, kg/car	COSTS, EUR/car
WMP	0.42	6.43
EMS	1.35	7.5
Significant?	Yes, (p=0.0201)	No (p=0.7164)

Overall, this case study of car repair shops shows a small reduction in the overall waste generation, a one-third reduction in hazardous waste as well as a remarkable improvement in recycling and a reduction in waste disposal costs.

CONCLUSIONS

In the household sector, waste prevention seems to have only a small potential. The main reason is that municipalities do not have many tools to influence consumption patterns. Municipal waste management mainly sets measures on waste after it is generated. To influence waste generation and waste prevention, product orientated policies are needed. As the Austrian results show, recycling of municipal waste can be quite effective. Source separation reduces the amount of waste for disposal and the environmental impact of waste management. It is a challenge to set up waste collection schemes, which are convenient for users and are considerate of the specific social situation and practices within a single area.

For commercial waste, higher potentials for waste prevention and recycling are assumed. Case studies and sectoral studies show both a reduction in waste as well as (in many cases) positive financial effects for the commercial operations. An obstacle to implementing measures in commercial operations is that there is only a weak legal basis for fostering environmental measures in commercial operations. Voluntary programs (like EMS) seem to be successful, but they are implemented by only a small number of operations. Furthermore, there is no reliable database on commercial waste production; thus measures and effects cannot be evaluated properly. It seems to be worthto the effort to intensify research in this field.

REFERENCES

1. SALHOFER S, GRAGGABER M, GRASSINGER D & LEBERSORGER S: Prevention of Municipal Waste, An Analysis of Measures and Effects. In: Christensen T.H., Cossu R., Stegmann R. (Ed.) Proceedings Sardinia 2001, Proceedings of the Eighths International Waste Management and Landfill Symposium; Vol. 5, pp. 99 – 108, Cagliari, Italy. 2001, 2001

2. ONIDA M Innovation in prevention of quantitative and qualitative waste production. In: Centro di Ingegneria per la Protezione dell' Ambiente (C.I.P.A.) [Hrsg.]: Innovation in Waste Management. Proceedings of the iV European Waste Forum, Milano 30. Nov. – 1. DezDec. 2000, 2000, pp. 91 — 105.

3. DOMENIG M: Nicht gefährliche Abfälle in Österreich (Non-hazardous waste in Austria). Materialien zum Bundeabfallwirtschaftsplan 2001. In: Federal Environmental Agency – Austria (Ed): Monographien, Band 140, Klagenfurt, Austria, 2001.

4. DETTE B & JÜLICH R R: Waste prevention and minimisation. Study commissioned by the EC, DG XI. Final report. Darmstadt, Germany, 1999

5. SOTHEN FV, SALHOFER S & PARFITT J: A Comparison of Municipal Waste Recycling Strategies in Germany, Austria and the United Kingdom. In: Proceedings of the WASTE 2002 Conference, 24 – 26 September 2002, Stratford-upon-Avon, UK, pp. 620 – 628, Coventry, 2002

6. LEBERSORGER S, GRASSINGER D, BAUER B & SALHOFER S: Encouragement of the Source Separation of Waste in Multiple FamilyDwellings. In: Christensen T.H., Cossu R., Stegmann R. (Ed.), Proceedings Sardinia 2001, of the EEighths International Waste Management and Landfill Symposium; Vol. 5, pp 119 – 128 Cagliari, Italy, 2001

7. AUSTRIA FEDERAL MINISTRY OF AGRICULTURE, FORESTRY, ENVIRONMENT AND WATER MANAGEMENT (ED.) Federal Waste Management Plan 2001., Vienna, Austria, 2001.

8. SALHOFER S: Modelling Commercial / Industrial Waste Generation: A Vienna, Austria Case Study. Waste Management and Research, Volume Vol. 18, Number 3, pp. 269 – 282, Copenhagen, 2000.

9. KANZIAN R & SALHOFER S Abfallwirtschàftskonzept Philips Videowerk Wien (Waste Management Plan VIW). In: Bundesministerium für Umwelt, Jugend und Familie [Hrsg.] Abfallwirtschaftskonzepte, Tagungsband zur Enquete am 23. April 1992, Schriftenreihe der Sektion V, Band 5, Wien, 1992

10. GRAGGABER M, LÄNGERT-MÜHLEGGER H & SALHOFER S Potentiale und Maßnahmen zur Abfallverringerung in Bildungswesen und der Gastronomie. In: MA 22 [Hrsg.]: Beiträge zum Umweltschutz, Heft 62/2000, Wien, 2000

11. SALHOFER S & STARKE R: The Ecological and Economical Impacts of Environmental Management Systems and Waste Management Plans in Commercial Operations. In: Proceedings of the VIIIth International Symposium Waste Management, Zagreb 2002, 13 – 15 November 2002, pp. 247 – 265, Zagreb 2002 Croatia, 2002

WASTE GENERATION ISSUES IN PETROLEUM PRODUCING COMMUNITIES IN NIGERIA

O S Dalton
Edo State University
Nigeria

ABSTRACT. The increasing global concern on the environment demands that waste should be properly managed in order to minimize and possibly eliminate their potential harm to public health and the environment. This paper looks into the various classes of waste and their sources in petroleum producing communities in Nigeria and discusses possible treatment and methods of disposing of them. From research, it was discovered that the main sources of wastes in petroleum producing communities in Nigeria are from the Petroleum Industry (Upstream Operation). These include drilling mud/fluids and the drill cuttings which come as a result of well drilling processes. Others are the tank bottom sludges, which come as a result of crude oil sedimentation in the storage tank, the oily water phase, accidental oil spillage, well blow-out, leakage of the pipelines and storage tanks. Also, gaseous wastes are produced as a result of gas flaring, stock flare and the exhaust from heavy vehicles during the exploration process. This paper intends to summarise the consequences of the waste generated in the oil producing communities of Nigeria, and the methods adopted by oil companies in managing these wastes.

Keywords: Waste, Upstream, Exploration, Environment, Petroleum or oil, Communities.

Oseghale Sunday Dalton is currently studying Mechanical Engineering at Ambrose Alli University in Nigeria.

INTRODUCTION

The discovery of crude oil in 1957 in a town called Oloibiri in the Niger-Delta (South-South geo-political zone) of Nigeria brought relief to many Nigerians, especially those living in the region. Though many saw it as an economic instrument, little did they know that it also had its own dangerous implications. It was not until the early 1970's that people started experiencing the by-product of the post war abundance and technology: uncontrolled pollution of the air, land and water. The atmosphere became polluted by intense industrial activities. Land and water were also not spared.

In this paper, it is the wish of the author to focus on the main source of waste in this community, having proven reserves of crude oil and natural gas of over 27 billion barrels and 120 trillion standard cubic feet. From research and general observations, it was discovered that the main source of waste in oil producing communities in Nigeria is from the petroleum Industry (Upstream Operation). This consists mainly of solid (or semi-solid) materials, liquid and gaseous substances resulting from activities and processes necessary for the successful operation of the petroleum industry.

In recent years, due to the consequences of the wastes generated in the host communities, the Federal Environmental Protection Agency (FEPA) through the Department of Petroleum Resources (DPR) called for better management of these wastes or more scrupulous methods that did not endanger the environment. It was stated that proper management of wastes must be matched or accompanied by a proper analytical process to ensure that what is being disposed of as waste does not turn out to be hazardous. This is to say that all solid, liquid and gaseous wastes must be rendered harmless before being disposed of.

The oil industry is divided into three stages of operation, namely:

- Exploration
- Production
- Transportation

Exploration usually consists of surveys, such as seismic, gravimetric and magnetic measurements to determine the subsurface structure and to estimate the potential for petroleum and gas accumulation. The problems encountered with seismic activities are not necessarily environmental pollution, but rather safety problems associated with the use of explosives. Furthermore, exploration and development activities involve drilling and well completion, as well as production and transportation. Each phase differs to some extent, in terms of the type and quality of waste discharge. Most exploration activities may be found on shore, near shore, (including swamps, coastal waters, estuaries, rivers etc) and offshore. These physical locations influence the manner in which the operations are conducted, and in which the effluents are treated and discharged. Then, the ultimate destination is either the ground, water or the atmosphere which comprise the habitat of the people living in petroleum producing communities.

DEFINITION OF SOME TERMS

Before going into detail, let me define some terms as used in this paper.

Waste

Waste is any substance, solid, liquid or gas that remains as a residue or an incidental by-product of the processing of a substance for which no use can be found by the organisms or systems that produce it.

Upstream Operation

This covers all the activities related to exploration, discovery and extraction of petroleum and gas and their treatment, transportation and delivery to designated export terminals or otherwise to the processing plants eg refineries. Also included are the auxiliary services related to petroleum and gas production, transportation and export.

Solid Waste

The main sources of solid wastes in Upstream Operation are drill cuttings, tank bottom sludges and contaminated soils which are the combination of both sludges and other impurities which may tend to stay in the tank bottom where they are deposited.

Drill cuttings

These consist of various rocks, particles and liquid released from geological formations in the drill hole. The cuttings are coated with drilling fluids. These drill cuttings are discharged to a rig shale shaker where the cuttings are separated from the drilling fluid. This separation step does not completely remove drilling fluid from the cuttings. Consequently, the composition of the cuttings will have similarities to the drilling fluid Also present are polynuclear aromatic hydrocarbons, organic acid extractables and heavy metals from spent drilling fluids and cuttings which may be toxic and hazardous. Even sanitary waste is not left out. Domestic sanitary waste originates from toilets, sinks, showers, laundries and galleys. The volume and concentration of this type of waste varies widely with time, facility occupancy and operational situation. The pollutants are oxygen consuming organic matter and floating solids etc.

Liquid Waste

The main source of liquid waste is the expressed liquid when oil is pumped under pressure by machines (busters), drilling fluids (muds), deck drainage, well treatment and oil spillage. During the production process, oil is pumped up by buster machines and collected into a storage tank (flow station). This petroleum is contaminated with sand and other impurities which are then allowed to settle out. Here tank bottom sludge is formed and only water is separated.

- Drilling Fluids (muds) are suspension of solids and dissolved materials in base water or oil that are used to maintain hydrostatic pressure-control in the well, remove drill cuttings and stabilize the wall of the well during drilling or work over operation. Water-based and oil-based mud are used in Nigeria. Water-based mud consists of natural clays and additives (organic and inorganic) like lignosulphate (containing chromium) and lead compounds. Oil based mud contains oxidized asphalt, organic acids, alkalis and stabilizing agents.
- Deck drainage results from precipitation run-off, miscellaneous leakages, spills and wash-down of platforms or drill ship decks and floors. It often contains petroleum-based oils, from miscellaneous spills and leakages of petroleum-based oils from wash-down operations, spilled drilling fluids and other production chemicals used by the facility.
- Well treatment wastes are spent fluid that results from acidification and hydraulic fractioning operations to improve oil recovery. Work over fluids and completion fluids are also considered to be treatment wastes.
- Oil spillage from exploration and development may result from well blow-out and the characteristics of the crude will depend on the crude oil composition.

Gaseous Waste

These are atmospheric emissions from the rigs which consist mainly of exhausts from diesel engines supplying power to meet rig requirements. These emissions may contain a small amount of sulfur dioxide. (dependent upon fuel sulfur content) and exhaust smoke (heavy hydrocarbons). An unexpected over-pressure formation encountered during drilling (ie formation pressure above normal hydrostatic pressure gradient of 0.465psi/ft) may result in a blow-out or gas discharge. If the mud used does not provide adequate hydrostatic head balance over the reservoir pressure, transient emission of light hydrocarbons and possible hydrogen sulphide may occur. Apart from major wastes which are deliberately discharged and planned, several factors during production operation and processing sometimes result in hazardous unplanned oil discharges. These are what are referred to as 'accidental discharges' which are principally due to system leakage or mechanical failures. The "system" here refers to the network of gas pipelines, compressors, vessels, the control loops etc, which are being operated under high pressure for petroleum transportation and distribution between the well and flow stations. Nigeria has a pipeline network of over 5,000km. When such leaks occur, it results in uncontrolled release of oil and gas at high pressure into the land and atmosphere. Leaks may result from outside forces, such as equipment (eg road diggers) operated by another party, earth tremors or movement, floods, weathering and disputes over encroachment on rights-of-way by villagers and outsiders. Another factor may be service failure, including material failures (due to ageing, wrong choice of materials or cracks), construction defects and corrosion (due to chemical attacks).

Table 1   Rates of generation of major wastes from oil and gas drilling operations

DISCHARGE CATEGORY	EXPLORATION/APPRAISAL	DEVELOPMENT
Drilling Fluid	Well depths less than 10,000ft (3050m), 300-600 tons/well 240-489m^3	
	Well depth greater than 10,000ft (3050m) 500-850tons/well 400-680m^3/well	250-500tons/well 200-400m^3/well
Drill Cuttings	100-1600tons/well	900-1400tons/well

Table 1 gives the rate of major discharges from off-shore oil and gas operations. For gas flaring it is estimated at 2 billion standard cubic feet per day out of a total daily production of 2.5 billion standard cubic feet.

CLASSIFICATION OF WASTES

Broadly speaking, wastes can be classified as follows:

- Corrosive types are waste material which contain high amounts of acidic or alkaline substances.
- Toxic wastes are those with a tendency to harm plant or animal life through either external or internal action which produces deleterious environmental or physiological effects.
- Flammables are wastes which contain combustible materials.

CONSEQUENCES OF WASTE GENERATION

Hazardous waste generation from the petroleum industry causes considerable environmental pollution. In addition to hydrocarbons, industrial wastes may also contain phenols, ammonia, sulphides, cyanide and metals, all of which may constitute health hazards. As a result of these, it is worthwhile to review the effects of wastes generated from upstream operation on agricultural land, aquatic life and the ecosystem in general.

Consequences Of Waste On Land

Effect of wastes on land can be studied from the agricultural stand-point and be looked at from two angles.

- Effects On The Land - Many fertile or arable lands have been rendered useless, recreational areas have been abandoned and certain neighbourhood refuse (wastes) were dumped and have become the source of various epidemics.
- Effects On The Crop Plants - Oil by itself in soil is not toxic to plants directly, but it exerts adverse effects on plants indirectly by creating conditions which make nutrients (eg nitrogen) unavailable to plants, while the adverse condition in the soil make elements toxic to plants more available.

Consequences Of Waste On Sea And/Or Rivers

The effect of petroleum on aquatic organisms, mangroves and communities which rely on them are very diverse and complex. However, a careful consideration of the biological and ecological effects of petroleum hydrocarbons is important for prevention and control of damage.

Pollution of water means interfering with various uses of water, recreation, drinking, navigation, irrigation, fishing, farming etc. On the surface of water, oil may limit oxygen exchange, entrap and kill surface organisms and damage the gills of fishes. However, there are only a few reports of oil pollution causing severe biological damage.

Consequences Of Waste In Air

Climatic conditions, combined with human and mechanical activities often worsen the effect of air pollution on human health and wellbeing. We are all aware that polluted air makes the eyes water, and stifles the lungs.

Fumes of sulfur and nitrogen oxide, sulfuric acid and photochemical oxidants irritate the respiratory system causing coughing, chest discomfort and impaired breathing. Carbon monoxide interferes with the ability of the red blood cells to carry oxygen. Heart and nerve tissue are particularly susceptible to oxygen deficiency. Carbon monoxide pollution can seriously impair coronary and central nervous system function.

ECONOMIC CONSEQUENCES OF WASTE GENERATION

Apart from the consequences of waste generation already highlighted. Waste generation also has economic consequences such as wastage of large acres of land and energy.

Wastage Of Land

Solid wastes occupy space on land. Many acres are wasted by dumping of solid wastes and landfill processes. Other activities which occupy a large space are petroleum exploration and exploitation. These involve the following:

- Implementation of Oil Concession - A survey of the petroleum concession map of Nigeria (from the Department of petroleum resources) shows that virtually all land in petroleu producing areas has been conceded to various petroleum prospecting companies. Thus the duration of oil mining leases (OML) appears long when abandoned and prematurely dried up wells are considered. As a result, the land may be in "limbo" and not put to other uses including agriculture as long as the original leases of prospecting companies last.
- Shooting of Seismic Lines - The shooting of seismic lines off shore and onshore need not disrupt agricultural activities, but could lead to a suspension of the use of the land by some rural farmers .The extent to which this could happen depends on the consistency and intensity of shooting.   Seismic activities could prevent efficient use of the land. Wildlife habitats are disturbed by shooting with a resultant decrease in the number of catches/kill per hunting hour during game hunting. Vibration due to the shooting of seismic lines tends to shake bodies of water (onshore or offshore). Therefore, aquatic animals (eg. fish) in such water bodies are disturbed. Some may even die as a result of such vibration.
- Digging of Oil Wells - Many wells are dug during the exploration and exploitation (appraisal/development) activities of petroleum and gas production. The objective of exploratory and exploitation activities are to locate petroleum and gas if they exist. However, if the petroleum and gas do not exist and are therefore not located, the land ought to revert as soon as possible to its pre-exploration or other highest value and best use. The problem with oil wells, whether abandoned or active, is not the drilling space, which is usually negligible, but the amount of land consumed in the digging and operation of wells which includes the following:
  - Land area for equipment.
  - Land area for working space.
  - Land area for gas flaring (where applicable).
  - Pipe line space (where applicable).
  - Access road

WASTE MANAGEMENT/TREATMENT APPLIED BY VARIOUS PETROLEUM COMPANIES IN NIGERIA

Waste minimization goals are both necessary and desirable but most oil companies in Nigeria still create waste products that will ultimately need to be treated to destroy the waste or render it harmless to the environment. This is due to improper or inefficient waste management/treatment adopted by some petroleum companies. Some companies adopt proper techniques used to manage hazardous wastes which have broad acceptance by the government, industry and public alike. These methods include:

- Chemical waste treatment.
- Biological waste treatment.
- Physical waste treatment.

Chemical Treatment

Chemical treatment involves the use of reactions to transform hazardous waste streams into less hazardous substances. Chemical treatment can be useful in promoting resource recovery from waste,in that it can be employed to produce useful by-products and residual effluents that are environmentally acceptable. Chemical treatment is a far better method of waste management than the traditional method of disposal to landfill. The majority of petroleum companies in Nigeria do not adopt this method fully, due to high costs and restrictive regulations.

There are many different forms of chemical treatment used in this method of hazardous waste management. These are:

- Solubility - The fact that water will dissolve many organic and inorganic substances containing various chemical elements, and with various structural configurations forms, the basis of this method.
- Neutralization - The neutralization of acidic and alkaline waste streams is an example of the use of chemical treatment to treat waste that has been characterized as corrosive and, therefore, hazardous under the Federal   Environmental Protection Agency through Department of Petroleum Resources guidelines. Neutralization of an acid or base is easily measured by pH; an acid-base reaction is one of the most common chemical processes used in waste water treatment, according to the equation

$$\text{Acid} + \text{Base} \longrightarrow \text{Salt} + \text{water}.$$

- Precipitation - Often undesirable heavy metals will be present in liquid waste streams. If the concentration of heavy metals is sufficiently high, the waste stream is designated as hazardous according to the Federal Environmental Protection Agency via Department of Petroleum Resources toxicity characterization which states that the metal must be removed. The usual method adopted by these petroleum industries for the removal of such an organic heavy metal is chemical precipitation. The metals will be precipitated at varying pH levels, depending upon the metal ion, resulting in the formation of an insoluble salt. The neutralization of an acidic waste stream can cause precipitation of heavy metals and allow them to be removed as a sludge residue by clarification, sedimentation, or filtration.
- Disinfection - The purpose of disinfecting waste is to destroy associated hazards that cause diseases. Most pathogenic and other micro-organisms are removed by conventional chemical treatment. However, chlorination is often used in disinfecting waste water in Nigeria.

Biological Waste Treatment

Organic wastes can be stabilized by biological treatment processes through the metabolic activities of heterotropic micro-organisms. These micro-organisms convert organic matter to end products such as carbon dioxide, water and methane gas.

The biological treatment process adopted by some petroleum industries in Nigeria are:-

- Activated Sludge Process - The activated sludge process is the most popular of the biological treatment processes because it is the most efficient and occupies the least amount of space. Complicated petroleum wastes have been successfully handled by this and other biological treatments.

- Lagoons - In this process chemical wastes have been successfully deposited and biologically stabilized in lagoons. The chief disadvantage of this technique is the large area of land that is needed. Lagoons in some chemical plants cover as much as 300 acres.

Physical Waste Treatment

Wastes are also treated by physical methods. These methods are sedimentation, filtration, flotation, reverse osmosis and submerged combustion. Others are landfill disposal and incineration.

Landfill has been used for the disposal of many solid and semi-solid organic wastes including oily sludges. However, this approach is becoming increasingly costly and only postpones the problem. The "environmental guidelines and standards for the petroleum industry in Nigeria" from the Department of Petroleum Resources (DPR) provides adequate specification for landfill. This includes the requirement that the bottom of the landfill must be at least 1.5m above the seasonal high water table and the top (ie after covering up the waste) must be at least 1.5m below ground level. The disadvantages of this method are that wastes may blown out and there may be leakage into underground and surface water supplies.

DISCUSSION

Based on our past experience, it has been established that the environmental impact of the petroleum industry results from activities and processes necessary for the successful operation of the industry. Familiar factors in the petroleum producing areas which have had some effect on the environment and which have generated hostile reactions from individuals and the general public include:

- Destruction of vegetable and farmlands during exploration, siting of locations, as well as laying of pipelines.
- The continuous presence of light, heat, noise and emissions from gas flares.
- Oil pipeline leaks, failure of storage tanks and effluents from general upstream operation.

Pollution control cannot be compartmentalized. For simplicity and ease of comprehension, I have decided to restrict discussion to oil spillage and specifically to crude oil, which has caused hardship to unfortunate Nigerians living in petroleum producing regions of the country. Pollution from refineries and refined products and hundreds of other chemical products from petroleum is a "different ball game".

There is limited documentation of the operational experience or the impact of oil pollution in Nigeria, but the basic problems are the emotional and physical effects on the health of the inhabitants. These are later followed by more chronic problems. Firstly, any emergency situation creates panic, fear, insecurity and threatens survival. Hence, the rural society is, in most cases, disorganized following a huge spillage. People could be displaced without shelter, food etc, families may be separated and an artificial situation substituted. Frequent respiratory tract infection due to gas flares arise. These have attendant psychosocial problems which must be treated as an emergency.

RECOMMENDATION/CONCLUSION

In order to protect and preserve our environment from pollution caused by petroleum-related operations, each petroleum producing company should enhance their existing oil spill contingency plans and continuously update and extend the baseline data using any further information that the company may collect from their areas of operation. Co-operation amongst companies, national authorities and local authorities would be a further step towards making significant progress. A computerized oil spill record system for its wide applicability is an advantageous tool that would give a practical answer to waste problems.

As for waste management strategy, waste should be divided into three broad categories in order to achieve better results - domestic, industrial and hazardous wastes. The first two categories of wastes are non-hazardous. They are normally encountered in municipal waste management. Hazardous wastes (sometimes called special or chemical wastes) require specific treatment and disposal methods. For hazardous waste it will be required that generators of the waste should provide facilities for treatment/disposal or alternatively enter into contracts with specialized firms to provide such services. The contract for disposal of hazardous waste will not, however, relieve the generator of his "cradle to grave" responsibility and he must ensure that the wastes are disposed of safely with adequate records and proofs. Subsequently, the modern waste management practice should be adopted which follow the principles of Reduce, Reuse, Recycle and Recover.

Environmentalists in the various petroleum companies should be encouraged by their respective managements to actively participate in the activities of the Nigerian Environmental Society; (a society that aims to get all the industries in the country to practice good environmental management) and also they should endeavour to attend a symposium of this kind. More so, the various petroleum companies should ensure that they invest in research and development to improve on the already existing environmental control devices and processes. Furthermore, effective measures should always be taken in the event of obvious problems, while awaiting full proof of the cause and effect.

Even with minimization of waste, treatment of the waste to minimize harmful effects and correct methods of disposal, still these cannot eliminate hazards to the environment. This is because these wastes are continuously being produced. The most important thing that can help to protect the environment from pollution and then encourage public health, is by educating the people to understand the basic consequences of wastes found in their environments, whether they are the ones producing them or whether they are being produced by neighbouring industries. This can obviate the individual from being exposed to pollutants and, thus, the public can handle their own produced wastes appropriately.

REFERENCES

1. DEPARTMENT OF PETROLEUM RESOURCES (DPR), NNPC Environmental Guidelines, 1990.

2. NNPC AND FMW AND HOUSING., The Petroleum Industry and Nigerian Environment Proceedings of 1997 seminar.

3. ALLABY M, Macmillan Dictionary of the Environment Third Edition

4. HARDAM SINGH AZARD, Industrial waste water Management Hand Book.

5. CHEREMISINOFF AND YOUNG, Pollution Engineering Practice Handbook

6. ISMAN W E AND. CALSON G P, Hazardous materials.

7. THE PETROLEUM INDUSTRY AND THE NIGERIAN ENVIRONMENT, Proceeding International seminar1999.

8. O.Y. ABABIO, New Certificate Chemistry Science Series.

9. NIGERIAN SOCIETY OF ENGINEERS, Proceeding of Seminar on Municipal Wastes Management organized by the Warri Branch 2001.

10. LAMID ALIU YAMAH , Paper Presentation on Waste Management (Chemical Treatment) And Remediation of Tank Bottom Sludges, Oily and Contaminated Soils.

11. FEDERAL REPUBLIC OF NIGERIA, Obasanjo's Economic Direction, 1999-2003.

WASTE AND RESOURCE STRATEGIES IN SWITZERLAND

M Chardonnens

Swiss Agency for Environment, Forests and Landscape
Switzerland

ABSTRACT. Ensuring an environmentally sound disposal of waste is one of the requisites for the long-term sustainability of the standard of living in industrialised nations. A country with a high population density and a limited space, Switzerland has tried since the middle of the 1980s to bring innovative solutions to the management of its waste. The implementation of this policy by the cantonnes, the municipalities and by the economic sector, resulted in an increase in the percentage of recycled municipal waste, from 25% in 1988 to over 45% in 2000. To achieve the objective of non-polluting final storage, the Confederation decreed that all combustible non-recyclable waste must be incinerated as of the 1st of January 2000. At the end of 2002, this objective was fulfilled to 97%. Wastes are not only materials without value: they also constitute a resource. Switzerland for example intends to systematically recover the aluminium, copper and zinc from incinerator residues. Through efficient dismantling schemes, the recyclable portions of waste electrical and electronic equipment are diverted to appropriate recycling channels. Because of the permanent ban on the agricultural recovery of sewage sludge that will come into force at the end of 2006 and of the temporary ban on meat and bone meal (due to BSE), phosphor recovery has become a pressing concern. Other important resources still remain to be mined from biowaste and construction waste. In the longer term, an efficient resource management policy must possess instruments, including economic incentives, to act at the source, at the design stage of production, so as to orient the flow of materials.

Keywords : Waste management, Switzerland, Recycling, Resources, New waste technologies

Marc Chardonnens graduated in engineering from the Swiss Federal Institute of Technology in Zurich and obtained his Master in Public Administration at the University of Lausanne. He entered the waste division of the Swiss Agency for Environment, Forests and Landscape (SAEFL) in 1987. Since 1995 he has been head of the Waste Treatment Section, where his activities centre on municipal and construction waste, as well as on the development of the treatment technologies for these wastes.

INTRODUCTION

All products and consumer goods end up as waste sooner or later. But the path leading from the extraction of raw materials to the manufactured products is not only littered with pollution and wasted resources. More prosaically, human activity does not consist merely of transferring matter from a primary hole, the mine, to a final one, the landfill. The transformations that the materials undergo in the process are multiple and allow for many possible points of intervention.

Obviously, it is impossible to immediately control all the stages in the life of a product. Priorities must be defined and those stages in the chain likely to cause the most damage must be attacked first. After having made sure their means of existence, industrialised societies have gradually been confronted by the end of lives of their products, in other words, by waste. In effect, the environment and public health were becoming steadily more and more threatened and damaged. Treatment techniques adapted to every kind of waste were then developed and they are being improved year after year. But for the last twenty years, waste has been perceived not simply as valueless and disposable, but also as a resource. Therefore recycling and recovery techniques now fortunately complement "end of pipe" treatment strategies. Today, we try to have an influence even closer to the source by extending producer responsibility to the entire life-cycle of the products marketed. This acts as a restraint on the traditional aims our industrial societies, these being growth, higher incomes and continually increasing consumption.

As a small country devoid of raw materials and with a high population density, Switzerland was rapidly confronted with the effects of the proliferating waste. Even if the political awareness of the challenge posed by waste was expressed only in the middle of the 1980s, the path travelled since then is a good illustration of how such a policy can be implemented, with its successes, but also its difficulties and setbacks.

SWISS WASTE MANAGEMENT POLICY

The "Guidelines of Waste Management in Switzerland" [1] were published in 1986 and constitute the founding document of waste management policy in Switzerland. The guidelines stated, among other principles, that waste disposal systems should generate only two groups of materials, namely those which can be recycled and those which are safe for final storage by landfill. According to this concept, waste disposal in landfills is clearly restricted to residues whose emissions will not harm the natural environment and do not require further treatment. In 1991, the Swiss federal government enacted the Technical Ordinance on Waste [2]. It contains provisions on waste recycling and on operating incineration facilities, landfill sites and composting plants. The implementation of these requirements has significantly reduced environmental pollution caused by waste management in Switzerland over the past fifteen years.

Waste Recycling

Switzerland has considerably improved waste recycling in the last decade: the proportion of separately collected municipal solid waste (MSW) from households and businesses increased from 25% in 1988 to 46% in 2001, amounting to more than 300 kilograms per capita per year. An important factor boosting recycling rates has been the application of the polluter-pays principle, e.g. in the form of pay-per-bag charges which are now paid by about two thirds of the population. The main areas of recycling are paper and cardboard with 160 kg per capita per year, biowaste (90 kg) and glass (40 kg). The remaining 15 kg consist of PET bottles, tin and aluminium cans, batteries and textiles.

Table 1   Recovery of different types of waste in Switzerland in 2001

	COLLECTED QUANTITIES, t/yr	COLLECTED QUANTITIES PER CAPITA, kg/inh.	RECOVERY RATE, %
Paper and cardboard	1,168,420	161	69
Biowastes	650,000	90	-
Glass (bottles)	293,700	40	92
Textiles	40,000	5	-
PET (bottles)	26,000	4	82
Tin cans	12,000	2	70
Batteries	3,460	0.5	67
Aluminium cans	2,580	0.3	92
TOTAL	2,196,160	303	

The basic principle for waste recovery is stipulated in the federal law relating to the protection of the environment [3]: a waste must be recycled only if recycling is less polluting than disposal and the consequent production of a new product.

For example, various life-cycle assessments have shown that producing paper from recycled fibres is 30 to 50% less polluting than using new fibres. Formerly, newspaper produced in Switzerland contained some 75% wood pulp and 25% cellulose. Today, newspaper contains on average 85% old papers and practically no cellulose. A recent study on the composition of the waste in our rubbish bags showed that there still is some potential for paper recycling, around 10 kg per capita. The recycling rate compared to the annual paper consumption could thus be increased from 69% to nearly 75%.

Waste Incineration

Landfilling of combustible waste has been banned since 1 January 2000 because the processes occurring in landfill sites generate polluting gases and leachates over decades. Switzerland could only impose this measure because two essential prerequisites were fulfilled :

- the wide-ranging development of waste recovery.
- the constant upgrading of the waste incineration plants to state of the art facilities.

Concerning this last point, the requirements of the 1986 Federal Ordinance relating to air pollution control [4] made it necessary to equip waste incinerators with efficient flue gas cleaning systems to trap sulphur oxides (SO2), hydrochloric (HCl) and hydrofluoric (HF) acids and heavy metals. New requirements were set in 1992 to limit nitrogen oxides (NO_x) to 80 mg/m^3 and reduce dust emissions to less than 10 mg/m^3. These measures also allowed to reduce the emissions of dioxins and furans to less than 0.1 ng/m^3 TE, the limit recently set by the European legislation. Having regularly had to adapt to state of the art, Swiss waste incineration plants now contribute less than 1% of the total atmospheric pollution for most pollutants. In the 1980s they still were among the main sources of emissions for HCl or dioxins for example (Figure 1).

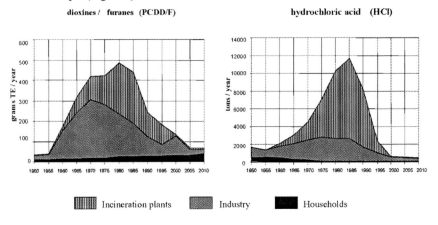

Figure 1   Air emissions in Switzerland

In Switzerland, waste incinerators are also required to recover the energy generated as electricity, or heat if the conditions allow it. The average energy yield of Swiss waste incinerations plants is 40% of the calorific value of the waste. The energy produced amounts to 1.7% of the net energy consumption in Switzerland, and would cover the annual needs in energy of 275'000 households.

In 2002, the 28 incinerators in Switzerland treated slightly more than 3 million tonnes of combustible waste. This amounts to 97% of the total combustible non-recyclable waste; only 100'000 tonnes were still directly landfilled. The objective of 100% incineration should be reached in 2004, when a new plant will be brought into service in the canton of Berne.

Landfills

In Switzerland, landfilling must obey the principle of preventive limitation of emissions to the environment. The 1991 Technical Ordinance on Waste instituted a system of three complementary barriers. The site must first satisfy strict geological requirements, to prevent infiltration of pollutants into the subsoil. Very impermeable (1 x 10^{-7} m/s) clay layers several meters thick are required. The second barrier constits of the technical construction measures: lining of the bottom and sides, recovery and treatment of leachate and gas. Finally, the third barrier relates to the nature of the waste itself, which must first be treated so as to reduce its impact on the environment as much as possible. Depending on the type of waste, it must therefore be sorted, incinerated or solidified prior to its final storage. The aim is twofold: eliminate organic carbon, which is very reactive in landfills and to extract as much as possible of the organic and mineral pollutants. This thereby both reduces their pollution potential and preserve non-renewable resources as heavy metals.

Switzerland has defined three types of landfills, depending on the type of waste that can be disposed :

- Landfills for inert waste for the disposal of mineral waste with a low content of pollutants, such as concrete, cement, tiles, and some excavation materials.

- Landfills for stabilised residues which are used for the disposal of highly polluted waste such as filter ash or non-recoverable galvanic sludge. Before their final storage, such wastes must satisfy strict elution criteria, generally requiring cement solidification.

- Bioactive landfills for the disposal of waste with a moderate amount of pollutants, such as incinerator bottom ash, or mineral materials from soil remediation sites that do not require further treatment.

Costs Of Waste Management

In Switzerland, waste management is generally financed according to the polluter pays principle. For MSW, the municipalities apply a tax proportional to the amount of waste produced, which is commonly called "pay-per-bag tax". A quick calculation shows that waste treatment costs every person an average of about 70 euros per year. 45 euros serve to cover the cost of collecting and incinerating the non-recyclable combustible waste, while the other 25 euros finance the separate collections (paper, glass, biowaste, metals, etc.). Disposal of our MSW thus costs some 20 euro cents per day and inhabitant.

Prepaid disposal fees for separate collections complete the pay-per-bag system. The advantage of having a tax included in the retail price of a product is clear: it ensures that once a product has become waste it will be returned to a proper recycling channel. Prepaid disposal fees have thus been introduced on a voluntary basis on electrical and electronic appliances, on PET beverage containers and on aluminium cans. The Confederation also introduced mandatory prepaid taxes on batteries and glass beverage bottles, when the actors in the branch were not able to reach a voluntary agreement. A similar tax is being discussed for paper, which will transfer the costs of collection from public funds to the consumers.

At the macroeconomic level, the costs of disposal of all the waste produced in Switzerland – municipal, hazardous and construction waste, as well as sewage sludge – amount to about 1.7 billion euros, or 0.6% of the GNP. This is similar to the other European countries which dispose of their combustible waste by incineration, such as the Netherlands or Germany.

TOWARDS A RESOURCE STRATEGY FOR SWITZERLAND

Although the recycling and treatment of solid waste are relatively efficient in Switzerland, important material and energy resources still remain in the wastes. In the face of our dwindling natural resources, this point becomes of increasing importance for the future. To create incentives for waste avoidance and for the sustainable use of raw materials, one should take into consideration the entire life-cycle of products, starting from their design, through their manufacture and utilization, to their recycling and disposal. Switzerland aims to improve the use of available resources in products and wastes wherever it is ecologically appropriate and economically justified.

Organic Wastes – New Quality Requirements For Compost

Since the entrance in force of the Ordinance on Substances [5] in 1986, composting of unsorted MSW is not possible in Switzerland anymore. At that time consciousness emerged that such highly polluted composts rapidly endangered soil fertility. Today compost can only be produced from separately collected biowaste.

However, agriculture operates more and more according to the "zero risk" principle, in answer to the expectations of consumers. It has therefore become necessary to clarify the situation regarding compost quality, in particular where organic pollutants are concerned. Detailed investigations will be carried out over the next two years to take a clear stock of the situation. The results will possibly be translated into recommendations restricting the composting of certain types of contaminated biowaste. For example it is expected that the levels of polycyclic aromatic hydrocarbons (PAHs) in road-side cuttings could be problematic.

Compost, however, is not only a produce susceptible to containing pollutants. It undeniably also presents interesting agronomical qualities, both for soil structure and plant nutrition. These qualities are only very incompletely documented in the literature. Opinions differ for example on the quality and value that must be given to the organic matter of compost. Another project will therefore aim to bring as objective answers as possible concerning the positive effects of composts in their various agronomic applications. Special attention will be devoted to the so-called digestates produced by anaerobic digestion, since very little is as yet known about their properties, in particular their phytotoxicity.

Sewage Sludge And Meat And Bone Meal – The Phosphorus Cycle In Danger

On 1 May 2003, new legal prescriptions came into force, prohibiting the recovery of sewage sludge in agriculture as of 1 October 2006. In 2002, 25% of sewage sludge still went to agriculture, amounting to some 50,000 tonnes of dry weight. Distrust towards sewage sludge has considerably increased recently, particularly due to the risks of mad cow disease (BSE) propagation through agricultural channels and the presence of increasing amounts of organic contaminants in the sludge. Both the consumer and the agricultural lobbies were clearly in favour of abandoning this fertiliser.

Fear of BSE also lead to the banning of meat and bone meal as animal fodder in 2001. This measure was taken to halt the BSE epidemic and should only be temporary. However it already appears that the ban will not be lifted before the year 2010.

Abandoning the recovery of these products also entails the loss of essential resources. Aside from the organic matter of sewage sludge and the proteins of bone meal, important and concentrated pools of phosphorus are not returned to the natural cycles (Table 2).

At the same time, the available phosphorus reserves of our planet are estimated at less than 100 years. It therefore falls to waste management to develop appropriate phosphorus recovery techniques.

Table 2   Phosphorus in waste in Switzerland

TYPE OF WASTE	QUANTITY, t/yr	CONCENTRATION, kg/t	LOAD, t/yr
MSW	2,500,000	0.6	1,500
Sewage sludge (DM)	200,000	28.0	5,600
Bone meal	20,000	80.0	1,600
Meat meal	45,000	30.0	1,350
Compost (DM)	210,000	3.5	735

Electrical And Electronic Scrap – A Financial Solution

For years, end-of-life electrical and electronic appliances were simply disposed of by bulk incineration. With the development of information technology, the waste quantities have been steadily increasing. The Confederation therefore enacted an ordinance in 1998 that set the framework and environmental requirements for the treatment of electrical and electronic wastes [6]. The central elements of this ordinance are the obligation on consumers to return their appliances to the retailers and for retailers the reciprocal obligation to take them back, on the sole condition that they sell appliances of the same type. The appliances are then brought to treatment plants authorised by the cantonnes and dismantled. Harmful components containing mercury, cadmium and certain toxic organic substances can then be removed and the copper, aluminium, iron and even small amounts of gold and other precious metals recovered. Today, some 7 kilos of electrical and electronic waste per capita are disposed of in appropriate channels in Switzerland, a cut above the European Union directive.

The financing of the disposal of this waste was quite problematic at first. A tax was habitually imposed on the consumer when he returned the appliance for disposal, which did nothing to encourage their return. A financing solution was progressively implemented in the form of voluntarily established prepaid disposal fees. These fees were first levied on office equipment, as of 1994. In 2002, a prepaid fee was finally introduced on hobby electronics and in 2003 on domestic appliances. The prepaid fee amounts on average to 1 to 1.5 euros per kilo. Since the financing has been settled, collections of waste appliances have significantly increased.

Automotive Shredder Residues (ASR) – New High Temperature Process

After having been emptied of the various liquids they contain, automobile wrecks are taken to scrap metal dealers, where they are shredded and the metal fraction is recovered. Shredder residues represent some 60,000 tonnes annually in Switzerland. They are composed of a mixture of plastics, synthetic foams, varnishes, contaminated with various heavy metals, such as copper, lead, chrome, nickel or cadmium. The present mode of disposal by thermal treatment in incineration plants or export to German blast furnaces is unsatisfactory. As a result, the foundation « Autorecycling Schweiz » which groups automobile importers proceeded to evaluate various systems for the disposal of ASR in Switzerland. The Reshment process, presented by the consortium formed by Métraux Services and CT Umwelttechnik was chosen. This high temperature process recovers 90% of the zinc and 60% of the copper content of the ASR and vitrifies the process residues. The plant which is to be built will have a capacity of 100,000 tonnes per year and will also be capable of treating other hazardous waste containing heavy metals, such as filter ash or galvanic sludge.

Between 1998 and 2000, the SAEFL carried out an in-depth study of sixteen vitrified residues of high-temperature incineration processes. Specific elution tests (distilled water, 90°C, residues pulverised to 100 μm, duration 72 hours) showed that the retention of pollutants inside the vitrified matrix resulting from these processes was very good. The long-term durability of samples, estimated from thermodynamic calculations, is comprised between that for medieval glasses (<1,000 years) and that for high-level nuclear waste glasses (>10,000 years). SiO_2-Al_2O_3-rich/CaO-MgO-poor samples exhibit higher thermodynamic stability (Figure 2). In the future, it should become possible to store such residues in landfills for inert wastes, which is economically advantageous.

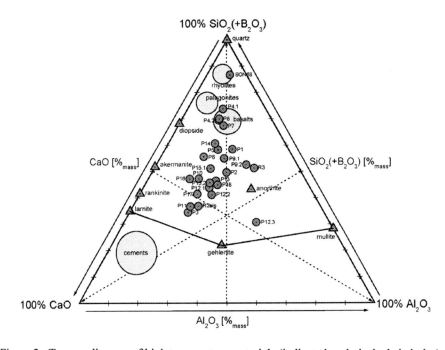

Figure 2   Ternary diagram of high temperature materials (indicated as dark shaded circles)

Bottom Ash From MSW Incineration – A Heavy-Metal Mine

Although non-ferrous metals such as aluminium are collected separately from household wastes in Switzerland, some metal pieces still find their way into the garbage. After incineration of the MSW, these pieces of aluminium, copper and brass end up in the bottom ash. An inspection of these metal pieces confirmed that most of them are rather small with typical sizes of 4mm to 20mm. Moreover, many of these metal pieces appear to have originally been parts of composite products, for example embedded in plastic. Efforts to separately collect non-ferrous metals obviously reach their limits here.

Metal pieces can be recovered quite easily from the bottom ash by means of dry mechanical processing. The larger the pieces, the cheaper it is to recover them. As metal pieces smaller than 4mm cannot yet be economically recovered, those above 4mm have been somewhat randomly defined as "particulate" and those smaller than 2mm as "disperse".

Surprisingly, it was shown that the "particulate" non-ferrous metal content in Swiss MWI bottom ash is around 3%. This translates into approximately 20,000 tonnes of non-ferrous metals that are dumped annually with the bottom ash - much more than is recovered through the separate collection of aluminium cans, batteries and the like.

In order to evaluate the possibility of tapping this resource, a pilot plant for the dry processing of bottom ash was constructed. The plant operated according to the principle of "selective comminution". Essentially, the bottom ash was fed into a ball mill and ground. The mineral matrix of the bottom ash (including glass and ceramics) was consequently pulverized. However, metals are ductile and they are therefore not significantly comminuted in a ball mill. After milling, the material was screened at 4mm. The oversize consisted almost entirely of metal pieces, while the undersize was free of metal pieces larger than 4mm. In preliminary tests, the undersize of the screen has been shown to be a rather potent hydraulic binder and is expected to be significantly less hazardous when landfilled, than conventional bottom ash. Further investigations into this matter are planned.

Construction And Demolition Wastes – A Considerable Material Flow

In Switzerland, the total amount of construction materials incorporated into the built environment correspond to about 300 tonnes per capita. Renovation works will increase in the coming years, and this will result in large amounts of construction wastes, which should rise from some 12 million tonnes annually to nearly 15 million tonnes in 2010. This represents a material flow five to six times higher than the municipal waste flow, without taking excavation material into account (30 to 40 million m^3 per year).

In 1997, the SAEFL published a directive to encourage the recycling of inert construction materials [7]. The two main points of this directive are: on one hand the minimal quality requirements for the various recycled fractions, on the other the conditions set for the utilization of these materials (such as their use with or without road cover). These requirements resulted in the development of deconstruction techniques far more efficient than before. However, 17% of construction wastes, amounting to 2 million tonnes, still go directly to landfill, so there still is a large potential for recovery. One of the obstacles is the relatively high prices of recycled products such as recycled concrete, which make it difficult for it to compete against natural gravel and sand. One must also consider the strong prejudices of certain actors in the construction sector, who consider that the quality of recycled materials is still below par. A lot of communication work remains to be done in this field.

One point remains outstanding : that of new materials which incorporate substances that are not free from contaminants (such as blast furnace slag). Depending on their degree of contamination, these products could in the medium term compromise the environmentally sustainable recycling of construction materials. The SAEFL intends soon to lay out rules in this field, to avoid future damages. The strategy consists in designing new constructions keeping in mind right from the start their potential longevity and the consequences of their demolition.

PERSPECTIVES

A global assessment of Swiss waste management policy will be undertaken during the next two years. It will indicate, among other things, in which way resource-oriented strategies may complement the traditional waste management policy measures. The strategies adopted to date clearly go in the direction of making better use of the resources contained in our wastes. However, the instruments that conventional environmental legislation has placed at our disposal will soon have exhausted all their possibilities. In the future one will have to consider whether new instruments for resource recovery should not also be developed. Economic ones in particular, in the form of tax incentives, would place the responsibility for the entire life-cycle of the products on the producers and distributors of goods.

Nevertheless, it is true that a small country like Switzerland cannot singly influence an entire production chain. Furthermore, the stages preceding the marketing of a product often generate more pollution than its use and disposal. Concerted international action is therefore necessary. Contradictory objectives may however appear when one reasons in terms of global ecological budget. Let us consider cars for example: it is recognised that during the life cycle of a motor vehicle, it is the petrol consumption which generates the highest pollution load. To reduce this one should develop lighter vehicles. However, European legislation requires that the recyclable fraction of cars must reach 85% of the total car weight by the year 2015. This places lighter vehicles, and in particular the dreamed "two-litre" car, at a serious disadvantage. A product policy really designed in terms of our resources still remains to be invented. This is a true challenge for every environmental policy of our modern societies!

REFERENCES

1. SAEFL, Guidelines of Waste Management in Switzerland, Bern, 1986, 37 p.

2. SWISS FEDERAL COUNCIL, Technical Ordinance on Waste, Bern, 1990, 32 p.

3. SWISS PARLIAMENT, Law relating to the Protection of the Environment, Bern, 1983, 32 p.

4. SWISS FEDERAL COUNCIL, Ordinance relating to Air Pollution Control, Bern, 1985, 108 p.

5. SWISS FEDERAL COUNCIL, Ordinance on Substances, 1986, 97 p.

6. SWISS FEDERAL COUNCIL, Ordinance on the Return, the Taking back and the Disposal of Electrical and Electronic Appliances, Bern, 1998, 7 p.

7. SAEFL, Directive relating to the recycling of mineral demolition waste, Bern, 1997, 20 p.

AN INTEGRATED APPROACH TO MARKET DEVELOPMENT FOR GLASS CULLET

A Dawe

E Ribbans

Waste and Resources Action Programme

United Kingdom

ABSTRACT. This paper describes the work being undertaken by WRAP (the Waste & Resources Action Programme) to develop stable and efficient markets for glass cullet. WRAP – which was established by Government to promote sustainable waste management, in particular by building markets for recycled glass, plastic, wood, paper and aggregates – takes an integrated approach to market development activity, aiming to break down barriers that disrupt market flow and to encourage conditions that will draw materials along the supply chain. WRAP's glass programme is working along the entire supply chain. Through its R&D programme, it is developing diverse higher value applications for glass, e.g. as a fluxing agent in brick manufacture, as a partial substitute for cement, as aggregate in concrete products, and for use as a water filtration medium. These applications will require equally diverse glass cullet specifications – and WRAP is therefore helping to establish processing facilities to meet the anticipated demand. It is also helping local authorities acquire the knowledge and resources needed to supply these facilities with sufficient cullet. At the other end of the supply chain, WRAP is working to inform and convince manufacturers, specifiers and consumers of the qualities and benefits of purchasing such products.

Keywords: Glass recycling, Cullet, Market Development, Recycling policy, Recycling legislation

A J Dawe is the Materials Sector Manager for Glass at WRAP (The Waste and Resources Action Programme). He has responsibility for developing the strategy for the glass programme and ensuring the resulting activities are successfully implemented.

E Ribbans is now Deputy Editor of the Guardian Weekly but was formerly WRAP's Communications Manager.

INTRODUCTION

The UK faces a major challenge in improving its record on recycling. As one of the worst performers in the European Union – recycling just 11 percent of its municipal waste and burying 80 percent – it has a particularly long climb out of its waste management hole. The Government's *Waste Strategy 2000*, which led to the setting of statutory recycling targets for English local authorities, has been followed by similar national strategies for Scotland, Wales and Northern Ireland. These targets are supported by a raft of other measures aimed at boosting recycling, including the Landfill Tax - scheduled to rise (from the present £14 per tonne) by £3 per tonne per year from 2005/6 – and the EU Packaging Directive that was transposed into UK law in 1997 and requires increasing levels of packaging waste recovery and recycling.

Increasing the volume of recycling in the UK also requires the carrot of opportunity – economic as well as environmental. Collected materials must have sufficient value to attract demand in preference to traditional virgin alternatives. (In this respect, fiscal measures such as the Climate Change Levy – from which high energy consumption companies have negotiated an 80 percent rebate in exchange for year-on-year energy reductions – and the Aggregates Levy that acts to reduce natural aggregates in favour of recycled aggregates such as glass, have a useful incentivising role to play). With demand in place, supply must be of reliable quantity and quality. The creation of stable and efficient markets – in time to meet legislative and policy deadlines – requires scientific and technological research, cultural change, investment in infrastructure, sound financial mechanisms, and the rapid dissemination of experience and good practice from a largely immature sector. In short, market development in recycling cannot be left entirely to the market.

This is why WRAP (the Waste & Resources Action Programme) – another key element in the package of measures that sprang from *Waste Strategy 2000* – was established. A not-for-profit company in the private sector, and funded by DEFRA, the DTI and the devolved administrations of Scotland, Wales and Northern Ireland, WRAP's brief is to play a strategic role in building new and stronger markets for recycled resources.

Of WRAP's eight programmes, three focus on generic areas: procurement, financial mechanisms and standards and specifications. The other five are focused on materials, namely: glass, paper, plastics, wood and (in England and Scotland) aggregates.

This paper will discuss the merits of market development as a means of improving recycling levels with specific reference to the activities of WRAP's glass programme, which aims, among other things, to raise the overall recycling rate of municipal glass to 35 percent by 2003/4.

It is estimated that around 4 million tonnes of various glass products enter the waste stream each year, of which the UK recycles less than 1 million tonnes – most of which is accounted for by container glass that is re-melted to produce new bottles and jars. For container glass, the biggest pressure to increase recycling rates comes interestingly from the industry itself. Already saving £8.5m a year in raw materials and energy by using recycled glass, the container manufacturers estimate that with ready supply they could double their current yearly intake of 587,000 tonnes of recycled glass. WRAP's initiatives to secure the collection of an extra 350,000 tonnes of glass each year are designed to help tackle this market-obstructing shortfall.

Container glass, of course, accounts for only part of the UK's total waste glass arisings. Lamp glass as well as television and computer monitor glass feature among the items of waste electrical and electronic equipment whose safe disposal will be targeted under the Waste Electrical and Electronic Equipment (WEEE) directive. Slated for transposition into UK law by August 2004, it stipulates that up to 80 percent of these waste materials (depending on product category) should be recycled by 2006. Add to this the car windscreens, headlamps and bulbs being banned from landfill under the EU's End-of-Life Vehicle (ELV) Directive, which calls for at least 85 percent of vehicle waste (by weight) to be re-used, recovered or recycled from 2005, and factor in proposals for a revised Packaging Directive that set a 60 percent recycling target for glass by 2006, and the urgency of finding sustainable market solutions for waste glass is clear.

PRINCIPLES OF INTEGRATED MARKET DEVELOPMENT

The development of markets for recyclable materials has become a recognised method for improving recycling performance. It relies on market forces turning a demand for a material into supply, thereby taking useful secondary resources out of the waste stream. However, this simplistic model is seldom achievable without attending to barriers that are present along the supply chain.

Potential users of recycled materials must firstly be convinced of the benefits of using them in preference to virgin raw materials. Benefits essentially fall into two main categories: economic and environmental. The economic benefits can arise either directly from reduced material costs or from improved life cycle costs. This may be illustrated by glass containers being recycled back into glass containers (closed loop recycling). The benefit to the glass manufacturer is significantly greater than any savings that may be made over raw material (batch) costs, extending to reduced energy consumption, due to the reduction in thermal energy required to melt the glass, and longer furnace life.

The economic balance must also take into account any financial incentives arising from Government policy. This might include packaging recovery notes (PRNs), which are used in the UK to demonstrate compliance with the requirements of the Packaging Directive, or the aggregates levy, which applies a tax on primary aggregates but is not levied on secondary aggregates.

Environmental benefits may also translate themselves into economic ones. This may arise directly through energy savings, particularly where targets may be met in order to avail of negotiated rebates to the Climate Change Levy or through carbon trading. Alternatively, the benefits may be more long term, such as maintaining raw material reserves, capital savings from reduced emissions or through improved stakeholder relations.

Examples of how recycled glass may contribute to raw material preservation include the replacement of special graded sands for water filtration and sports turf applications, the extension of special clay reserves used in brick manufacture through incorporating glass as a fluxing agent and of nephylene syenite and feldspar reserves for sanitaryware production. Recycled glass may even be used to produce decorative tiles with the appearance and properties of terrazzo marble, reducing the need to despoil the local environment in quarrying the material.

The arguments to the potential end user are frequently strong, the challenge then becomes ensuring the supply of material for such applications. This starts with collection of the material from the consumers which, in the case of glass containers, includes collection from householders and licensed premises.

The barriers to collection of glass from householders through local authorities are summarised in Table 1. Similarly, barriers for licensed premises include: lack of incentive (except those premises obligated under the Packaging Regulations), lack of space, cost, staff training and the availability of a collection service.

The quality of the collected material will frequently determine its potential uses and its value. For example, colour separated glass will command a higher price than mixed colour material and clear glass will usually have a higher value than green or even brown in view of the high demand from the glass manufacturing sector. In conjunction with BSI, WRAP is developing a Publicly Available Specification (PAS) for the quality of unprocessed glass cullet that should help to bring greater uniformity and clarity to this issue.

Once the material has been collected it is usually transported to a processing facility. Transport not only has significant environmental impact but also adds expense that puts the viability of transporting glass long distances in doubt. This may be partially overcome by using sea freight or making efficient use of the rail network, but the other alternative is to supply to more local markets. The UK glass manufacturing sector is concentrated in the north of England and the Scottish central belt with only one container plant close to London.

To supply local markets, the processing infrastructure must be available locally. The capital and operating costs of plant for this must be appropriate to the scale of the operation (size of local market) and the value of the end market for the material.

Overall processing capacity in the UK is sufficient to meet the anticipated increase in container glass recycling over the next 4-5 years. However, this capacity is primarily linked to the container glass manufacturers, making it difficult to ensure sufficient supply for the development of alternative applications.

Furthermore, the level of processing which can be provided is inadequate for some end uses, for example, the generation of fine glass powder as a partial cement substitute or fluxing agent, or graded granular material for water filtration and sports turf applications. An integrated market development strategy must take this into account and ensure sufficient supply of material in order to give industry confidence to change its processes to accommodate the use of recycled glass.

Once manufacturers have become convinced of the benefits of incorporating recycled materials into their products, their customers and specifiers must also be convinced. This is seldom a simple operation, except in instances when it is not necessary to advise them of such additions, and attention to standards, specifications and product certification is frequently required before market acceptance can be gained.

Table 1   Possible barriers to increased glass collection

BARRIER	COMMENTS	POSSIBLE SOLUTIONS
Perception of lack of markets	Although markets do exist for glass of all colours, the legacy of the green glass mountain is still strong in some people's minds	Increased awareness. Generation of local markets
Lack of infrastructure	The UK's glass bank density is very low compared with other European countries. Also the quality of the amenity often does not inspire the public to use it. Lack of bulking facilities and vehicles or boxes for kerbside collection may also be barriers	Increased investment in facilities. Kerbside collection with other dry recyclables
Inappropriate contractual arrangements	Contracts frequently do not require an incentive to recycle and are too inflexible. They are also normally long term and provide an excuse to do nothing.	Contract renegotiation. Increased training of LA officers. Dissemination of model contracts
Lack of funding	Inadequate provision of funds for infrastructure and promotion	Funds available through DEFRA and devolved assemblies, NOF etc.
Unattractive economics	Low market prices (often perceived), and low landfill tax rates. Also licensed premises being unwilling to pay for glass collection. Colour separation adds cost especially for commercial and kerbside collection. LA's not commercially aware	Increased awareness. Increased landfill tax. Establishment of higher value markets able to use mixed colour glass
Lack of public participation	This would be caused by poor public awareness, apathy, lack of incentive and confidence that the glass will be recycled. Poor servicing and maintenance of glass banks.	Public awareness campaign. Kerbside to make it easy. Reverse vending. Improved organisation and financing.
Lack of information	Local Authorities do not always appreciate where to go for advice on, for example, glass collected kerbside. Also collection data is published too late to be helpful.	Information provision and training
Quality requirements	An understanding of quality requirements, notably acceptable contamination levels is sometimes lacking	Improved communication. Clearer specifications
Health & safety issues	Noise during collection from licensed premises and personal protection when handling glass are areas of concern	Reduce noise (and transport costs) by using glass compactors. Provision of personal protection equipment

INTEGRATING RESULTS FROM THE GLASS R&D PROGRAMME

Brick Fluxing Agent

Several of WRAP's R&D projects are exploring the commercial and technical viability of incorporating recycled glass in non-container applications. Development of these markets is considered to have potential to (a) absorb the coloured glass that is becoming disproportionately prominent as overall waste glass collection rates increase, (b) decrease consumption of ultimately finite raw materials and (c) reduce energy costs and emissions.

One such project, led by CERAM Building Technology, is investigating the manufacture of bricks using up to 10 percent of fine glass powder (below 90 microns) as a fluxing agent. Laboratory tests on bricks and pavers made by Blockleys, Hanson and York Handmade, have shown promising results. When fired at normal temperatures (between 1000°C and 1100°C), the bricks containing glass have greater strength and frost resistance – in some cases vastly so - than those in which none of the clay has been substituted for glass. This suggests it should be possible to manufacture bricks of similar physical properties at significantly lower temperatures – for Blockley pavers in the region of 1020°C rather than 1100°C – helping brick makers to secure hefty rebates on the Climate Change Levy as well as reduced energy costs and emissions of hydrogen fluoride. Testing at lower temperatures is now under way.

Figure 1   Stacks of bricks entering a typical brick kiln

With a total 7 million tonnes of clay used every year in brick production, the savings on raw materials could be substantial, too. If only half of UK manufacturers were to substitute 5 percent glass in their bricks, it would reduce the extraction of clay by 175,000 tonnes a year. However, a change in production processes will be achieved only if industry is confident of a reliable supply of ground glass. In Britain there are currently no dedicated facilities for processing recycled glass into powder to fulfil the anticipated demand . Without an assured market - and with increasing pressure on many glass processors to service the container sector, and with some collection companies promoting low value, aggregates markets - investment is unlikely to be forthcoming within the necessary timeframe. WRAP is therefore working on increasing capacity for recycled ground glass - for brick-making and other new applications - via a number of initiatives, including a competition for grant assistance for at least two regional plants capable of producing a combined total of 160,000 tonnes of recycled glass powder per year. Through its recycling manager's training programme, WRAP is also helping local authorities to establish more effective and efficient collection systems that will deliver increased supplies to processors.

Market pricing for ground glass will become more firmly established as the R&D projects continue, but it will certainly attract a significantly higher value than in conventional aggregate application (£0 to £6 per tonne) and, with further processing, is likely to command upward of £35 per tonne. Furthermore, energy, emissions and quality benefits will be quantified in order to determine a comprehensive cost-benefit analysis.

Water Filtration

Another promising new market for recycled glass is in water filtration systems. Scottish-based company Dryden Aqua has developed an "advanced filtration medium" (AFM) that utilises 0.5mm to 2mm reprocessed glass particles as a direct replacement for sand in all pressure or gravity-flow filters. Not only does recycled glass substitute for a primary raw material, it has also demonstrated a number of superior chemical and physical qualities (see Table 2) in all the applications examined. For example, because it carries a permanent surface negative charge, AFM can remove particles much smaller than sand and is able to reduce organic levels by surface adsorption. Surface catalytic properties also make AFM self-sterilising. Furthermore, AFM does not contain free silica, making it safer to use than sand.

Figure 2   Typical waste water pressure filter installation

Table 2   Summary of AFM properties against sand

PROPERTY	SAND	AFM
Particle size removal, smallest particle	10 to 15 microns	5 to 10 micron
Organic adsorption	None	Yes, but needs to be quantified
Combined chlorine (THMs) content of processed water for chlorinated applications such as drinking water and swimming pools	THMs will be present	Lower due to dissolved organic removal
Bacterial levels of process water	Will become biofouled and contribute to bacterial levels	Self-sterilising, lower bacterial levels
Replacement of filter media	Need to be replaced due to attrition and contamination of media	Likely to last 2 to 3 times longer than sand
Backflush water requirements	Excessive	Approx. half the volume of water required
Catalytic activity/cracking of organics, removal of toxins	None	Yes, but needs to be quantified
Health and safety of dry product regarding dust	Contains free silica, class 2 carcinogen US EPA	100% fused silica, not as dangerous as free silica
Boiler feed water contamination	Silica contamination	No contamination

The current UK market for filter sand is estimated to be at least 70,000 tonnes per year – with large quantities of the sand imported from the Netherlands, Germany and Spain. The British Geological Survey reports that 1.1 million tonnes of silica sand is mined in the UK each year, of which a high proportion is used as filler in ceramic and brick manufacture as well as water filtration. With AFM competing directly with sand in existing and new applications for pressure filtration, the estimated potential UK demand for AFM is 217,500 tonnes per year (Table 3). This represents a significant diversion of waste glass from landfill. And the added bonus – given the relative predominance of green glass in the waste stream – is that coloured glass particles seem to be even more effective than clear ones. This is because the chromium used in green glass and the iron added to brown can split oxygen into single, highly reactive "radicals" that prove deadly to microbes.

Dryden Aqua presently exports more than half of all AFM produced, and expects this market to increase over the next five years.

The key industrial sectors in which a potential market for AFM has been identified, include:

- Potable water treatment
- Sewage effluent, tertiary treatment
- Industrial effluent discharge compliance
- Industrial process water
- Swimming pool industry

Table 3 Potential UK AFM requirements

INDUSTRIAL SECTOR	ESTIMATED POTENTIAL AFM REQUIREMENTS, tonnes per annum	ESTIMATED PRESENT SAND USAGE, tonnes per annum
Potable water treatment	30,000 to 50,000	30,000 to 50,000
Sewage effluent, tertiary treatment	40,000 to 60,000	10,000
Industrial effluent, discharge compliance	50,000	Nominal
Industrial process water, cooling towers, boiler feed water, & grey water recirculation	50,000	5000
Hydroponics & agriculture	Not known	Not known
Swimming pool industry	5000 to 7000	5000 to 7000
Aquaculture industry	500 to 2000	200
Total	175,100 to 217,500	45,050 to 67,050

Through its R&D programme, WRAP is supporting Dryden's investigation into the development of AFM for potentially widespread use in industrial-scale filtration of drinking water. Not only would this application offer a market for up to 50,000 tonnes of recycled glass per year, but AFM would also offer performance advantages that should reduce the need for chemical additives.

With regard to the treatment of sewage effluent and industrial effluent, sand (which suffers from biofouling issues that leads to bed blockage and filter failure) is used on only a relatively small scale. Given water company budget constraints and the lack of alternative low-cost technology – along with the fact that the majority of systems are operating at peak capacity and are struggling to comply with effluent regulations - there would seem to be a strong case for the use of AFM.

UK industry uses a huge amount of process water, which comes in many cases from rivers and lakes. Although this means it is out-with the public water supply systems, it can end up in the sewage treatment works. As industry is charged on the basis of quantity and quality of water discharged, there is a strong incentive to improve or recycle it. In either case, there is a potentially massive market for AFM.

In the UK, there are over 100,000 private pools, 2,518 public swimming pools, 13,000 health clubs with spas and small pools, 3,700 school swimming pools, 800 holiday parks and 200 hospital and army swimming pools. AFM again offers proven benefits to the swimming pool and spa industries where there are many serious water quality issues. Figure 3 illustrates the problem of bio-fouling through long term bacterial accumulation where sand is used relative to the clean surface presented by glass grains after 5 years in operation.

While AFM costs between £250 and £350 per tonne – against roughly £80 to £120 for filter sand – a cost benefit analysis conducted by Entec shows that, due to operational savings, AFM still compares favourably. For example, the improved filtration and water quality benefits gained by using AFM should enable water companies to comply more easily with performance criteria with the need to upgrade treatment plant infrastructure. The extent of these benefits will be quantified by the trials on both drinking and waste water systems. Cost reductions should also be possible as increased demand allows economies of scale.

 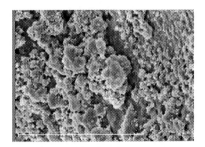

Figure 3   AFM (left) and sand (right) photographs after five years in a swimming
pool filter, 50μm scale

WRAP is supporting the development of this market not only through the works trials in
drinking water and sewage treatment facilities, but also through obtaining approval from the
Drinking Water Inspectorate, facilitating the construction of a 20,000 tonne per annum
processing facility, establishment of the market data presented above ensuring a quality
supply of recycled glass for processing and working with Water Authorities to ensure that
acceptance and change takes place.

GLASS REMELT APPLICATIONS

We should not forget that probably the most sustainable method of using recycled glass
products is the closed loop recycling process e.g. using glass containers to make new glass
containers.

Consumers and policy makers alike frequently do not appreciate the high recycled content of
glass containers – many green wine, beer and whiskey bottles produced in this country
contain over 90% recycled glass. And this is glass which has itself already been recycled
many times before. When the environmental and economic benefits of reduced energy costs
and reduced emissions are added to the reductions in raw material consumption, it becomes
evident that this 'closed loop' recycling is the most sustainable option.

However, container glass recycling rates in the UK are still significantly lower than most
other European countries and this reflects the glass industry's historic ambivalence toward
glass recycling. It is only with the introduction of environmental legislation and the creation
of alternative markets that the industry is now adopting a responsible and collaborative
approach to improving the situation. As collection rates increase, the amount of green glass
collected will quickly exceed the amount that can be used in green container manufacture.
This glass will provide the ideal material for the water filtration and brick flux applications
discussed above. Similarly, it is anticipated that a higher proportion of bottles will be
collected mixed rather than colour separated for both practical and economic reasons. This
will arise from the increasing proportions of glass collected from both households using
kerbside collection schemes and from commercial premises. Although technology is
becoming available for sorting mixed glass into separate colour fractions, it is unlikely that
all such glass will be able to be treated in this way for many years to come. Suitable
applications for this cullet are also required. The proximity principle, in simple terms - using
recovered materials locally to avoid the environmental burden associated with transport, is
also relevant with nearly all glass container plants located either in central Scotland and
Yorkshire. It is in some cases better to collect, process and use recycled glass in local
applications rather than transport it hundreds of miles to the processing plants which supply
the glass container industry.

REFERENCES

1. DRYDEN, H., Glass as an active ingredient. Water and Waste Treatment, March 2003, pp 16-19

2. http://www.wrap.org.uk/funded_project_search.asp

3. http://www.cwc.org/glass.htm

4. http://www.drydenaqua.com/afm/index.htm

5. SAMPSON G. et al, Advanced Filtration Media (AFM): Clean water market analysis. To be published May 2003

6. SMITH A.S., Brick making: the ultimate waste repository, Industrial Minerals and Extractive Industry Geology. Geological Society of London Special Publication. 2002 pp 323-326

ASSISTING THE MALTESE WORKFORCE DEAL WITH THE CHALLENGES OF SUSTAINABLE DEVELOPMENT

A Mifsud   C Ciantar   V Cassar

Ministry for Resources and Infrastructure

Malta

M Plant

Nottingham Trent University

United Kingdom

ABSTRACT. For Malta's waste management strategy to succeed in attaining the proposed changes and targets, there is the need to achieve the cooperation and collaboration with all sectors of society. The Maltese workforce is a sectoral group that has the potential to play a major role in waste management. From research being carried out amongst hotel employees, it is evident that workers are willing to be actively involved in the design and planning of training on waste management. Most of the employees interviewed stated their interest in having training that ought to be relevant and useful to them. Almost all the employees interviewed believe that when training is designed through an open process between management and staff, it is highly probable that all the employees would attend thus achieving optimum value of the training. For some of the employees, the waste management issues that were of great concern to them centred primarily on the waste collection service whilst others felt largely concerned with the need for a total change in Maltese culture by means of education. All the employees interviewed said that should the hotel implement a waste management plan, each worker would feel the importance of contributing towards its success only when such a plan is preceded by adequate training on waste management. However, many were apprehensive about the lack of space for separate collection bins in some areas of the hotel and believe that the management ought to address this situation during the initial stages of the waste plan. The Direct Attention Thinking Tools (DATT) by de Bono were used in the design of the interview questions as this helped the researcher direct the employees to the specific issues needed for the research project. The DATT tools also assisted the employees in prioritising their ideas, in finding the strengths and weaknesses of their suggestions as well as identifying other members of staff who would be involved in the waste plan and how these would be affected.

Keywords: Creative thinking, Employee training, Social transformation, Change, sustainability, Waste management, Environmental education.

A Mifsud, C Ciantar and **V Cassar** work in the Works Division of the Ministry of Resources and Infrastructure for the Government of Malta.

M Plant, works in the Department of Secondary and Tertiary Education at Nottingham Trent University.

INTRODUCTION

The Role of Education

The society that we now live in is primarily concerned with progress and development yet we are living with the knowledge that the world is facing an environmental crisis. Some are aware of the tremendous interconnectedness between economic development, technological and scientific progress, nature and resource conservation, human health, political stability, population growth and cultural diversity. It is only when there is the realization of the interrelationships among these issues that society feels the need to address the causes leading to the environmental problems that we are facing today. The developed world eventually encapsulated these ideas into one term: *sustainable development*.

The IUCN (1980, 1991) [1] has set out the following guiding principles to sustainability:
- respect and care for the community;
- improve the quality of human life;
- conserve the Earth's vitality and diversity;
- minimise the depletion of non-renewable resources;
- keep within the earth's carrying capacity;
- change personal attitudes and practice;
- enable communities to care for their own environments.

The above principles do in fact reflect and aim to address the set of interconnected issues mentioned earlier. Korten (1990) [2] summarises these principles by stating that a 'sustainable society' is one in which:

"each generation recognises its obligation for stewardship of the earth's natural resources and ecosystems on behalf of future generations. The transformed society must use the Earth's resources in ways that will ensure sustainable benefits for our children."

It is clear that, in order to achieve sustainable development, there is the need for change in today's world resulting in a transformation of society. Environmental education and the more recent term education for sustainability are essential tools in assisting society take up the challenges of change, social transformation and sustainability. If real sustainability is to take centre stage in our society then education in all forms and sectors has a vital role to play. Every individual needs to take part in this change and ought to be given the opportunity to act as a change agent since "You cannot learn without changing, or change without learning." Kosko (1994) [3]. Business organizations are not spared from the growing global concern of sustainable development. They too ought to shoulder their responsibility by addressing these concerns. To cope with changes implied by sustainability

"it will be necessary for organizations, their managers and workforce to accept the importance of a 'learning' mindset in which staying attentive to the shifting needs of society and the dynamics of environmental change will be as important as staying close to the needs of customers." Roome, Oates (1996) [4].

In view of this the pedagogic approach or strategy for environmental education should itself be socially sustainable with meaningful rather than token empowerment, participation and ownership. The strategy needs to be supported by the methodologies used in environmental education that include the clarification of values, critical reflection, creative thinking and learning as a continuous process for all.

Applying Creative Thinking to Sustainable Development

The challenges inherent to sustainable development have been outlined above. However, it is now appropriate to shift focus onto the *process* and *tools* needed by our society, particularly business organizations, to address these challenges. Creative thinking applied to environmental education has the potential to be one of the desired cornerstones for the transformation of society.

Through his work in the field of creative thinking, Dr. Edward de Bono has shown that the study and practice of thinking is a positive factor that develops personality and self-esteem, increases efficiency and contributes to greater success in all forms of endeavour. He claims that "one of the main purposes for the use of creative thinking is to find better ways of doing things" de Bono (1992) [5] and the practical need for creativity falls into two areas:

"1. Where there is a real need for a new idea and we cannot proceed without a new idea. It may be a problem, a crisis, or a conflict. Other approaches have failed. Creativity is the only hope.
2. Where there is no pressing need for a new idea but a new idea offers opportunity, advantage, and benefit." de Bono (1992) [5]

Drawing directly from his background in medicine and psychology, de Bono gained insights into how the mind works as a self-organising system. The latter inhibits the mind's ability to generate creative thinking. To overcome these limitations he designed deliberate thinking frameworks that are practical and easy to use. Six Thinking Hats® and Direct Attention Thinking Tools (DATT)™ are just two of the thinking programmes developed by de Bono. A brief description of each is given in the next section.

Direct Attention Thinking Tools (DATT)™

The ten thinking tools in the DATT™ programme have been developed to assist people to make good decisions and solve problems. They can be seen as being similar to software for the mind and can be applied to any situation. DATT™ can help organisations to plan, design, re-engineer, create or restructure products/systems and continually improve.

Six Thinking Hats®

The Six Thinking Hats® programme is purposely designed to maximise team performance. It is based on the concept of Parallel thinking and encourages co-operation, exploration and innovation. The Six Thinking Hats promote co-operative and co-ordinated thinking since all parties are thinking in parallel, using the same style of thinking simultaneously.

Consequences and Sequels

How to look ahead and see the consequences of an action, plan, decision or rule. Ways to see future repercussions and possibilities.

Alternatives, Possibilities, Choices

Give yourself the benefit of alternatives. Find out if there are any possibilities and then make a choice.

Plus, Minus, Interesting

Look at the positive and negative aspects of decisions and actions, before it's too late.

Other People's Views

Gain the power of broad vision by exploring other people's viewpoints. Who are the people that are going to be affected by your action?

Recognise, Analyse, Divide

Break a larger concept or situation into smaller, more manageable parts. Getting down to the core of a situation makes it easier to deal with.

Key Values Involved

Your values have a big impact on your thinking. Make sure your actions match or fit your values.

Consider All Factors

Explore all the factors related to an action decision, plan, judgement or conclusion. Learn to focus your attention on a particular factor.

First Important Priorities

Guide actions, choices and decisions. Get the right things done. Target, deliberately, what is most important and what must be done first.

Aims, Goals, Objectives ➡ O

Focus directly and deliberately on the intention behind actions. Prevent your thinking from drifting and keep your mind on the objectives.

Decision/Design, Outcome, Channel

Action

Direct your attention to the outcome of your thinking and the action needed to follow. Learn how to be confident in your decisions.

Figure 1   The Ten Direct Attention Thinking Tools (DATT)™ by de Bono

Yellow Hat

What are the benefits? What are the positives and values? It reinforces creative ideas and new directions.

Black Hat

It points out the difficulties. What are the points of caution?   What are the risks? What could be the possible problems?

Blue Hat

It helps in managing the thinking process. What is the agenda? What should be done next? Which hats are needed?

Green Hat

It makes time and space for creativity. What are the alternatives?   How can difficulties be overcome?

White Hat

It assists in obtaining information. What information is already available? What is missing and how can this be obtained?

Red Hat

It gives permission to express feelings and intuition. What are the feelings right now?

Figure 2   The Six Thinking Hats® by de Bono

AIMS AND OBJECTIVES

Aims of the Research

Firstly, it is hoped that the research will provide information on the thinking skills that are required by employees if they are to become more responsible about waste management issues.   Creative thinking helps them to be proactive in waste related issues that concern them. The *knowledge-behaviour gap* is usually dealt with in behavioural science whereby, in simple terms it means that knowledge and information alone do not necessarily translate into changed behaviour. Linking the information about waste management to personal benefit is more important. Martin Holdgate (1996) [6] claims that: *"People need to feel 'ownership' of problems and solutions, and to derive benefits from the new policies they adopt as a result of the programme of education and communication."*

Secondly, the research being conducted amongst the hotel employees (see next Sextion) aims to formulate a suitable course design methodology to ensure that the employees own the training programme. One of the main methodology tools being used is the application of de Bono's Direct Attention Thinking Tools programme. The thinking tools have provided the framework for the interview questions the researcher has carried out amongst a representative sample of the hotel employees. By means of the thinking tools, employees were assisted in placing their thinking focus on (i) the design process and content for training that is most suited to their needs as well as the requirements for operation of the hotel; (b) the waste management issues in Malta that are of concern to them and (c) practical measures leading to improved waste management within their respective departments.

The Research Context

There has been much research in the field of environmental education within the formal education sector and the debate on how the young can be given the appropriate skills and attitudes leading to environmental care is still unfolding. Undoubtedly educating the young in order to help them be better prepared to deal with sustainability is imperative. Yet, there is a vacuum for those members of society who are no longer of compulsory school age. A particular sector that has been targeted by means of this research is Maltese workers at a leading five star hotel. The hotel, which is situated in one of Malta's prime tourist resorts, employs over two hundred people. It has an international franchise and is part of another three local hotels. By means of ten line managers heading its various departments, communication between hotel management and staff is efficient and effective. Besides, it is not rare for those in top management positions to be seen in the operational sections of the hotel such as the stores, kitchens etc. Such practice has led to a generally positive relationship between management and staff. This emerged very clearly during the research interviews with the members of staff and will be explored in depth in the Results section.

Research Methodology and Research Methods

Action research

The research methodology used is action research enquiry since the researcher is directly engaged in the practice under study with the main objective being to induce beneficial change. The action research enquiry is based on a study of singularity because the research will be carried out at a particular place and at a particular time. This will be done by:

- Analysing the existing situation in the research context chosen through interviews with employees at the hotel.
- Evaluating the information collected.
- Analysing key concepts of education for sustainability, Huckle & Sterling, eds. (1996) [7] and de Bono thinking programmes.
- Devising a process – through action research employees will be involved in (i) the identification of environmental concerns particular to their place of work, and (ii) the design of a training programme that could help them address these concerns.
- Proposing practical guidelines on how an educational programme from, by and for the employees can be designed.

The action research approach offers an excellent medium to assess and redesign the pedagogic strategy and it is for this reason that such a research methodology has been adopted for this project.

Research methods

Waste Audit: This has been carried out by *Progett Skart*, a Project Team within the Ministry for Resources and Infrastructure. The report is already available and gives a good insight into the existing waste management practices at the hotel. It will assist the employees in identifying achievable targets besides evaluating, monitoring and reporting progress. This report will prove beneficial, as it will serve as a baseline on which improvements and changes will be measured against.

Interviews: The employees were selected with the assistance of the Human Resources Manager at the Hotel. The interviewees are a representative sample of the total staff complement and have been specifically chosen to reflect the opinions and ideas of each department and grade (horizontal and vertical representation).

1. Group Interviews – two group interviews with four employees in each have been conducted.
2. Workers' In-Depth Interviews – five workers were selected for individual in-depth interviews.
3. Management's In-Depth Interviews – five people holding managerial positions have been identified for in-depth interviewing.

Research Diary: It enables the researcher to log in observations as the research process develops. The researcher takes note of the personal experiences during the time spent working within the hotel setting that would be of relevance to the objectives of the research study. Some of these observations could be of significance to the researcher's own professional development particularly to the role as a change agent. Furthermore, other data can be recorded in the research diary. This is data emerging from the human dynamics within an organisation that is generally difficult to capture by other more conventional means of data collection such as questionnaires.

RESULTS

Since data is currently still being collected and analysed, it would be presumptuous to draw up definite conclusions at this stage. However, sufficient information gathered from the initial set of interviews is already giving indicative preliminary results. Furthermore, observations noted in the research diary are providing valuable support data to the questionnaire responses.

The Interviews

The researcher conducted the interviews at the hotel during staff working hours. Some last minute employee substitutions had to be made due to the fact that a couple of the interviewees were on sick leave. The group interviews were of about ninety minutes long. Surprisingly, it is worth noting that all the members of staff spoke freely without any inhibitions. They eagerly participated in the discussion and most of them showed great passion and enthusiasm for the issue of waste management. The Interviewer had to ensure that equal time was given to each employee by preventing an interviewee from taking over the discussion. They seemed very confident when answering the first set of questions dealing with course design methodology for employee training. In fact, they were able to generate several new ideas whilst placing them in a list of priorities. The second set of questions concerning local waste management issues was also received very positively by the employees. Here, they discussed at great length the problems facing Malta's waste management situation. The interviewees encountered some problems during the last part of the interview. Here they were asked to give suggestions on how the hotel could achieve improved waste management through their own departments. It is to be noted that the employees' perception of waste management seems to centre on waste collection. Therefore, most of the employees interviewed stated that an improved waste management plan for the hotel should comprise, almost exclusively of, separation of waste at source. This indicates the need for education on waste management whereby the basic concepts of waste reduction and reuse feature more prominently in the course content. When prompted by the interviewer most employees mentioned the use of paper on both sides as the classic example for waste reduction. Only two employees from the Food and Beverage department suggested other ideas such as the reuse of glass jars in the kitchen as well as the making of breadcrumbs from dry bread.

Course Design Methodology

The first five questions during the group interviews addressed the process through which employee education can be managed. All employees felt satisfied with the way they were given the freedom and space to give suggestions or be involved in the process of employee training. There seems to be an open communication structure between all the staff and the hotel management. On being asked to list factors that should be taken into consideration when courses are being planned and designed, responses varied from being held at low-season and during working hours, to the importance that content should be relevant to staff and the employees' right to refuse the course. However, they placed: during working hours; at low season; and interesting and relevant content at the top of their priority list. This means that though they are not too pleased to attend courses after working hours, they still feel a commitment to the smooth-running of the hotel since they are less inclined in attending courses during the peak seasons. The staff commitment towards the hotel's efficiency and performance will be one of the key values required when introducing change within the hotel organization. Implementing a waste management plan is about change and employees ought to believe in this change because they are able to link it to personal and organisational benefit. Therefore, their commitment to the hotel organization will determine the success of the waste management plan. It was worth noting that the staff showed great motivation to be actively involved in the course content. All those in the group interviews felt strongly inclined to select courses where the content was interesting and beneficial to them both at a professional and personal level. Similarly, they all stated that the main benefit of having courses *for* employees designed *by* the employees is that since the courses are well-suited to staff needs no resources are wasted on uninterested staff. However, they were apprehensive that there could be the possibility of poor attendance should courses be offered solely on a voluntary level.

Local Waste Management Issues

It was interesting to note the responses regarding the waste management issues in Malta that are of concern to them were very different in the two group interviews held. One group mentioned:

- The need for a change in culture whereby Maltese society becomes more proactive in waste management issues.
- Law enforcement – the public's general perception on law enforcement is a very negative one, therefore both local and national authorities must work extensively on this point in order to prove that they intend to keep law breakers at bay.
- The provision of waste facilities such as Bring-In sites and separate waste collection service to the public across the island. Furthermore, these facilities need to be made available in both public and private organizations too.

- On-going education for all members of society by means of training courses, public lectures, publications, media and communication campaigns.

Whilst the second group listed:

- Misuse of bring-in sites by the public, since some individuals dispose of their bulky refuse next to the bell bins at the existing Bring-in sites. This is done despite the fact that Malta has a free door-to-door bulky refuse collection service by central government in collaboration with Local Councils.
- Stray dogs and cats tearing up garbage bags awaiting collection because some householders take out their waste at the wrong time of day.
- Waste collection contractors' lack the sufficient knowledge on waste resulting in providing very primitive service.
- Widespread waste management illiteracy partially due to inadequate educational programmes as well as lack of interest from the public.
- The public's lack of compliance to door-to-door collection times.

Hence, it is evident that the first group focussed on broader issues of national concern that could have long-term and far-reaching consequences. The second group appear to be primarily concerned with the issue of waste collection methods. However, both groups made reference to the need for a more informed society in order to attain the desired changes for Malta to be able to enjoy a waste management system that is less damaging to human health and to the environment.

Practical Waste Management Measures

The last set of questions during the group interviews required a higher level of creative thinking since the interviewees were asked to give some practical tips for improved waste management in their own department. Tips given by those on the job are valuable information for an educator who is planning a waste management course at the hotel. It would be futile for the course tutor to suggest a list of practical waste management tips to an audience that would immediately find flaw in several of them since they are more knowledgeable about the day-to-day running of their own working environment.

Through this set of questions, employees were also assisted in listing those people who would be involved in the waste management changes and how they would be affected. Most of the staff interviewed believes that if all the employees took on the responsibility to implement practical waste management measures, then the required changes would be easy to adopt. Besides, whilst the changes will involve everyone, they would not be regarded as a burden or additional work since all members of staff would be doing their bit. However, they feel that the management will need to (i) plan for an increase in space requirements since more bins for the separate collection of waste would have to be put in place; (ii) allocate necessary funding for initial expenditure such as bins and employee training; (iii) revise existing purchasing contract conditions and waste collection contract.

FUTURE WORK

It is important to point out that this research would not have been possible without the enthusiasm, commitment and support of the management at SAS Radisson Hotel, Malta. On completion of the research project, it is hoped that a basic manual will be produced. This will include practical guidelines on how an educational programme on waste management *from*, *by* and *for* hotel employees can be designed. The main objective for the compilation of the manual is to serve as a template that various other hotels may use as a reference. A key principle that should be emphasised is that there cannot be one standard methodology that fits all business organizations. Each hotel together with its employees ought to be treated as a different case since any waste management programme including employee training ought to reflect the specific needs of those involved by highlighting the relevance of its changes to their own lives.

REFERENCES

1a. IUCN (1980) *World Conservation Strategy: living resource conservation for sustainable development*. Gland, Switzerland: IUCN.

1b. IUC/UNEP/WWF-UK (1991) *Caring for the Earth: a strategy for sustainable development*. London: Earthscan Publications.

2. KORTEN, D. (1990) *Getting to the 21st Century: voluntary action and the global agenda*. Connecticut: Kumarian Press.

3. KOSKO, B (1994) *Fuzzy Thinking*. London: Flamingo.

4. ROOME, N, OATES, A (1996) "Corporate Greening" in J. Huckle, S. Sterling (eds.) *Education for Sustainability*. London: Earthscan Publications.

5. DE BONO, E (1992) *Serious Creativity*. Glasgow, Great Britain: Harper Collins Publishers.

6. HOLDGATE, M (1996) *From Care to Action: Making a Sustainable World*. IUCN. London: Earthscan Publications.

7. HUCKLE, J. & STERLING, S., Eds. (1996) *Education for Sustainability*, London: Earthscan Publications.

COAL AND CLEAN COAL BYPRODUCTS RESEARCH AND UTILIZATION IN CHINA

P Wang

S Zhong-Yi

Shanghai Research Institute of Building Sciences

China

ABSTRACT. Arisings of coal byproducts are increasing dramatically with economic development. In China the amount of fly ash discharged is more than 100 million tonnes per year, but in recent years about 60 million tonnes per year has been utilized. This situation and some technologies for the utilization of fly ash in China, particularly in Shanghai, will be introduced in this paper. The requirements for environment protection has meant that clean coal byproducts are becoming increasingly common in China. The development of technologies for utilizing the clean coal ash is our new task. This paper explores some utilization methods, not only for FGD gypsum by wet technology, but also the clean coal byproducts rich in sulfur dioxide (SO_2), which will be also introduced in the paper.

Keywords: Fly ash, Clean coal ash, FGD Gypsum, Concrete.

Wang Pu is a Professor in the Shanghai Research Institute of Building Sciences. He is Director of the China Fly Ash Utilization Center, National Director of fly ash utilization R&D Project of UNDP/UNIDO (1983-1992). He has been involved in fly ash utilization R&D and management for about 30 years.

Zhong-yi Shi is a senior engineer in Shanghai Research Institute of Building Sciences. His Research interests include utilization of mineral admixtures for concrete.

INTRODUCTION

China has some of the largest coal resources in the world. Utilizing coal as fuel for electricity generation is a keystone of Chinese energy policy. In China the coal by-products increase dramatically with Chinese economic growth (see Table 1). For example, the amount of fly ash discharged by China has been more than 110 million tonnes since 1996. Such huge amounts of fly ash arisings year after year could potentially be a disaster for the environment. Fortunately fly ash application research and utilization programmes were also developed in parallel. Shanghai is the first Chinese city to have produced fly ash since the 1930's, and also the first city of fly ash utilization in China. Since its rapid economic growth, Shanghai's fly ash discharge has increased at a greater rate than other Chinese cities and provinces. On the other hand, due to large scale construction activity in Shanghai, shortage of construction materials and other resources is always very serious.

Therefore, recycling fly ash in construction materials is very important and necessary in Shanghai. Over the past 6 years Shanghai's fly ash utilization rate has been more than 100% - the highest rate in China. According to 2002 statistics, the total annual coal byproducts arisings in Shanghai was 4,550,000 tonnes, whilst the utilization amount was 4,930,000 tonnes, a new record in Shanghai. According to our National environment bulletin, the two main air pollutants are coal ash dust and SO_2, which produces acid rain. This is a major concern for the Chinese Government which, at the beginning of the 1990's, started to set up desulfurization (FGD) equipment in power stations. Up to the end of 1999, 1.68GW of such equipment have been put into operation and 5.00GW of equipment is under construction. Since there are a variety of FGD technologies, the qualities of the byproducts are different. How to use these by products is a major task facing researchers and users.

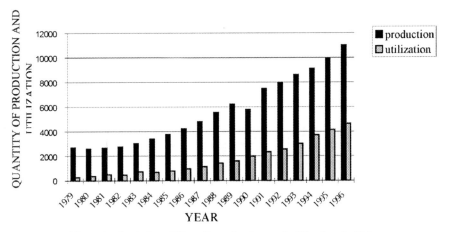

Figure1   Statistics of fly ash production and utilization in China

Fly Ash Utilization Statistics for Shanghai and China

Figures 1 to 3 show to us that the rate of production and utilization of fly ash in China increased in parallel.

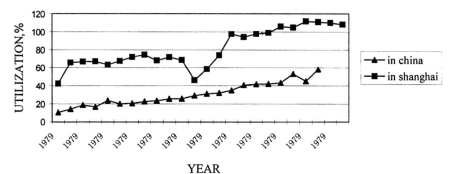

Figure 2   The statistics of fly ash utilization rate in Shanghai and in China, %

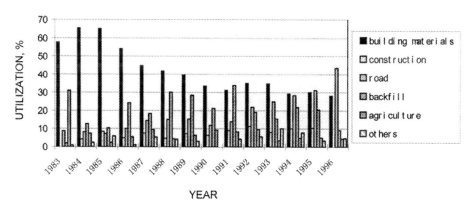

Figure 3   Statistics of fly ash utilization

Recently, since the start of the construction of the Three-Gorge Yangtze river hydro-power station and other projects adding Class I fly ash to concrete to improve the quality has become more and more important. For example, the total volume of concrete in the principal dams of the three-gorge is more than 27,000,000 M^3. Class I fly ash has been used in all the dams, exploiting the corrosion protection properties and hydration heat. Therefore, up to November 2002 the dams of Three-Gorge have used 1,200,000 tonnes of Class I fly ash.

USING FLY ASH IN CONCRETE

Pumped Concrete

We have been doing a number of research projects on pumpable concrete with fly ash. Pumpability can be improved significantly with fly ash. Many of the highest pumping heights in the world are required in Shanghai. Table 2 lists some of the pumped concrete projects in Shanghai. Their pumping heights are also listed.

Table1   Statistics of Shanghai Coal By-products, 1000T

ITEM	YEAR																	
	1984	1985	1986	1987	1988	1989	1990	1991	1992	1993	1994	1995	1996	1997	1998	1999	2000	2001
Pro-amount	1240	1330	1570	1750	1880	1990	2520	2760	2840	2910	3270	3860	4350	3710	3280	3280	3730	4100
Utilization	841	958	1170	1190	1350	1370	1160	1610	2100	2850	3080	3780	4310	3930	3440	3670	4130	4510
Wall materials	229	214	226	230	194	184	153	163	177	217	100	183	414	131	143	141	133	67
Cement add.	174	219	219	237	237	260	255	273	288	353	352	475	383	599	675	585	871	907
mortar	68	91	114	141	199	219	176	194	245	334	404	457	673	699	591	751	1010	1440
Road const.	223	343	366	339	373	394	325	403	841	1340	1370	1800	1880	1790	1890	1980	1740	2020
Back fill	/	/	/	/	/	/	/	288	248	423	672	537	482	611	2.1	67	41.4	37
Others	147	91	244	243	346	312	184	290	305	181	182	324	467	102	141	209	338	47
Util rate,%	676	721	745	681	719	687	461	584	740	97.6	94.2	97.8	99	106	105	112	111	110

Table 2   Pumping concrete with fly ash projects in Shanghai

NAME OF PROJECTS	STRENGTH OF CONCRETE, MPa	PUMPING HEIGHT (ONE STAGE),m
Yangpu Bridge	60	108
Nanpu Bridge	50	186
Shanghai TV tower	60	< 150
	30-40	350
Shanghai Jingmao Bldg	60-30	384

Mass Concrete With Fly Ash

It is well known that fly ash can reduce the heat of hydration of fresh concrete. In China fly ash was already being used in 1959 in concrete for dam construction. Since the 1980's a number of tall buildings have been constructed in China, particularly in Shanghai. A thick slab foundation is needed for such structures. For example, in Shanghai the thickness of a tall building's foundation slab is often $\geq 2 \sim 3$ metres, some projects even ≥ 4 metres. Fly ash mass concrete is widely used in such slabs. One slab of 28,000m^3 fly ash mass concrete has been cast in one site and another one is 10,000m^3 without construction joints. Its thickness is 4 metres.

Fly ash concrete is also used in radiation shielding of nuclear power plants. For example, a nuclear power plant with a shielding wall thickness of one metre met its severe quality requirements via the use of fly ash.

As a result of using fly ash, the hydration temperature of mass concrete is reduced. Table 3 shows the results of temperature measurements.

Table 3   Temperature raising comparison between fly ash concrete and normal concrete

TYPE OF CONCRETE	PROPERTY					
	Initial Temp, °C	Maximum Temp, °C	Arriving Time, h	Duration, h	Increasing Temp, °C	Temp at 14 days, °C
Normal Concrete	15.0	34.2	37	5	19.2	27.9
Fly Ash Concrete	16.7	31.6	48	3.5	14.9	23.7

Impermeable Concrete

It has been shown that the impermeability of concrete can be raised by the addition of fly ash. By this principle, our research results show that fly ash concrete has good impermeability and resistance to seawater attack. Table 4 compares the impermeability for a Shanghai subway tunnel segment made with fly ash concrete and normal concrete.

Table 4   Comparison of C50 tunnel concrete segment impermeability

TYPE	CEMENT kg/m³	FLY ASH kg/m³	CONCRETE STRENGTH, MPa		PERMEABLE PRESSURE, 2.8Mpa		PERMEABILITY COEF, m/s
			28d	90d	Max permeable height, mm	Average permeable height, mm	
Normal concrete	500	/	51.0	62.8	54.0	39.4	4.8×10^{-13}
Fly ash concrete	462	72	53.5	64.2	40.2	1.0	6.4×10^{-14}

Roller Compacted Concrete with Fly Ash

In China roller compacted concrete with fly ash began to be used as a pavement material in the 1980's, but since the 1990's it has been used for high grade highway pavements with severe abrasion resistance requirements. For example, in 1992 the National Highway No.329 had used fly ash RCC with a special admixture. The level of replacement of cement by fly ash was 33%. Its compressive strength was 46MPa and its flexural strength was 6.4MPa. Two years later it was demonstrated that its abrasion value was only $0.58g/cm^2$ whilst normal concrete was $0.73g/cm^2$. The result was obtained by a core sample analysis.

Fly ash high volume utilization

Fly ash embankment

In 1985 Shanghai carried out a trial on a section of the first 1Km of expressway in China fly ash embankments. The research results showed that fly ash of particle size between 0.005-0.5 mm possess characteristics like fine sand and silt, such as non-plasticity, large inner frictional angle and small adhesion. Up to now this trial section of expressway is in good condition. Based on these results, one of Shanghai's Local Standard "Fly ash embankment design and construction"" has been drafted It is the basis to disseminate Shanghai's experience to the whole of China. Right now in China fly ash embankments are being used in constructing many National expressway projects. It is estimated that the Shanghai-Nanjing and Beijing-Shenzhen Expressways have used about 17 million tonnes of fly ash in embankments.

Utilization of fly ash in structural backfill

Since 1992, China's conditioned fly ash has been used for foundation fill and stabilization techniques using fly ash have made significant progress. For instance, moist fly ash was placed in woven bags for dam construction and an ash disposal site was divided into small parts and then filled with wet ash forming separate zones. The foundations for a 260,000m^2 coal stockyard in Shanghai Harbour were sited on a former ash pool containing deposits of wet ash between 4-5m deep. The foundation had only a very low bearing capacity. Therefore, the foundation had to be stabilized to meet its bearing capacity requirement. Foundation stabilizing technologies included 5 methods. Experts held comprehensive assessments of research and considered that for a such great area it would be better to use a dynamically consolidating plastic drainage blanket. The project foundation stabilization has been accomplished with a total utilization of 2 million m^3 of fly ash, shortening the construction period by half and saving over 20 million RMB. The project quality was assessed by the state as excellent.

Use of fly ash in production of building materials

Since the 1950's, a large amount of research into making building materials with fly ash have been conducted in China. The scope of applications has been steadily increasing and this mode of use has comprised a significant proportion of the total utilization. There are more than 20 kinds of fly ash building products. All these have contributed to the formation of a relatively stable area of fly ash utilization. The most important fly ash building product is fly ash-lime block (LFB). Its main constituents are fly ash, lime and a small quantity of gypsum, which combine together as cementitious material, with some light aggregate such as cinder, non-quenched slag or expanded clay. It resembles a kind of lightweight concrete block but steam-cured at 100°C. Its natural volume weight is 1,600 kg/m^3 and a compressive strength ranging from 10 to 20 MPa. This product started being produced in Shanghai in the beginning of the 1960's. By 1990 about 15,000,000m^2 of residential housing had been built with this block.

In order to save energy and soil, and increase the amount of fly ash used in building materials, ministries and state commissions are taking measures to restrict manufacturing of clay brick,   and encourage production and development of new masonry materials such as sintered fly ash brick, lightweight fly ash aggregate and other building components. In China sintered brick with 30 to 50% fly ash has been produced since the 1960's. Fly ash with cement to produce parpaing block is also produced in some provinces, particularly in the coastal areas.

Research and application on high calcium fly ash (HAF)

In the 1980's China started to produce high calcium fly ash. The China Fly Ash Utilization Technology Center and Shanghai Research Institute of Building Science have carried out several research projects, some of which were fundamental research but dealing with practical applications. The main conclusions of this research are as follows:

Shanghai's HFA can be categorized as a medium lime HFA, and present as crystalline mineral components are cement clinker minerals such as C_2S and C_3A, much free lime, and a certain amount of anhydrite. In physical terms it has a high density and a low water requirement.

HFA suffers from poor soundness due to the presence of crystalline free lime. However, this can be controlled by restricting its content in cement and concrete, or by the addition of mineral admixtures Within a certain range of water cement ratios, HFA is effectively hydraulic in nature. Taking as an example Shanghai Shidongkou's HFA, the water cement ratio should be less than 0.4. The strength of cement mortar containing less than 35% HFA can be equivalent to that of normal cement mortar. The glass of HFA with cement can produce a reaction prior to an age of 28 days but it also contributes to middle stage and later stage strength development. Therefore it implies the glass of HFA has higher pozzolanic activity than that of normal fly ash.   Recently cement containing both HFA and normal fly ash as a pozzolanic addition has been produced.

APPLIED EXPERIMENT ON CLEANED COAL ASH IN CHINA

Summary of Application of Desulfurization Technology in China

In view of the fact that it is difficult to foresee a change in the reliance on coal as a primary energy source in the next 50 years, the effect of pollution of SO_2, NOx and CO_2 discharged by coal combustion to the environment has caused worldwide concern. 1998's statistic showed that 20.90 million tonnes of SO_2 were discharged in China annually, 0.5 million tonnes of which was discharged in Shanghai. Since about 90% of SO_2 emissions result from coal combusion, including 60% discharged by coal-fired power stations, clean power generation technologies must be enforced.

Long-term development of furnace desulfurization processes for medium and small industrial furnaces has made a number of advances. The move towards 1,000W grade circulated fluidized bed equipment has been a major breakthrough.   Circulated fluidized beds with an 80% desulfurization rate carry out multigrade-burn by spraying coal powder mixed with limestone or dolomite (acting as the desulfurization medium) into a fluidized-bed furnace where it is burnt. Limestone is heated to remove CO_2, forming porous CaO which reacts with SO_2 to produce calcium sulphate and thus desulfurizing the gases.

The China electric power departments have introduced new desulfurization technologies since the beginning of 1990. These include wet limestone-gypsum process, electron beam irradiation process, sea-water process, rotary spray-dry process and LIFAC process as shown in Table 7.

Table 7   Brief Introduction to 7 Flue Gas Desulfurization Processes

POWER STATION	DESULFURIZATION PROCESS	PRODUCTION DATE	MANUFACTURE	DESULFURIZED FLUE GAS AMOUNT/ID[4] ML.B	FLUE GAS AMOUNT RELATIVE TO UNIT CAPACITY/MW	DESULFURIZATION RATIO %	Ca/S
Luohuang Power Station	normal limestone-gypsum wet process	Sep.,1992 Apr.,1993	Mitsubishi Japan	2*108.7	720	95	1.02-1.05
Taiyuan No.1 Power Station	High-Speed levelling Limestone wet process	Oct.,1996	Hitachi Japan	60	150	80-85	1.1-1.2
Huangdao Power Station	Semi-Dry Rotational Spray-Dry process	Oct., 1994	Mitsubishi Japan	30	75	70-75	1.8
Chengdu Power Station	Dry Electron Beam Irradiation	July., 1997	Ebara Japan	30	75	>80	
Neijiang Gaoba Power Station	Dry Circulating Sulfurized bed	Sep., 1996	OSLFinland	38	100	90-95	(2.88)
Nanjing Power Station	Dry LIFAC Activator	Dec., 1998	IVD Finland	64	250	70-75	2.5
Sengzheng Western Power Station	Sea-Water Process	The First half year of 1993	ABB Norway	122	300	≥90	

Compared with the dry desulfurization process, the wet process is more complicated and the plant occupies a larger area. However, wet processes have the advantage of higher desulfurization efficiency, availability of sulfurized byproducts, and easier realization of comprehensive utilization. Since the desulfurized ash by dry processes contains sulfurized byproducts and residual desulfurizer etc, its chemical and physical characteristics are significantly different from that of common coal ash. Furthermore, there are considerable differences in ash performance with different desulfurization processes.

The Performance of Desulfurized Ash and Its Utilization

The constitution and physical properties of desulfurized ash produced by the LIFAC (Limestone Furnace Injection) dry desulfurization process of the Nanjing XiaGuan Power Station are shown in Tables 8 and 9. Besides normal fly ash components, it contains various calcium-based compounds such as $CaCO_3$, $CaSO_3$, $CaSO_4$, free CaO and $Ca(OH)_2$.

Table 8   The physical and mechanical performance of desulfurized ash

FINENESS(%) 45μm RESIDUE	WATER REQUIREMENT RATIO(%)	COMPRESSIVE STRENGTH RATIO % (28d)*	BENDING STRENGTH RATIO % (28d)*
22.6	105	85	94

Compared with class F ash cement mortar strength.

Table 9   Chemical Composition of Desulfurized Ash

ITEM NAME	LOI	SIO₂	F₂O₃	Al₂O₃	CaO	MgO	SO₂	Na₂O	H₂O	F-CaO	CaSO₄	CaSO₃	CaCO₃	Cl-
Content	4.28	48.52	8.00	7.61	14.07	1.12	1.88	0.32	1.10	9.63	3.19	0.08	2.36	0.41

The Shanghai Research Institute of Building Sciences, etc has investigated the utilization of desulfurized ash. Its conclusions are:

- Desulfurized ash used as an admixture of concrete should be used at lower levels than normal fly ash. When the replacement of cement with ash is kept at 20%, the compressive strength of concrete is equal or higher than that of normal fly concrete at 28 days.
- The optimum level of addition of desulfurized ash should be less than 10%. The cement has higher early strength compared with cement containing normal fly ash.
- Desulfurized ash used for building block gives a product with enhanced strength development properties.
- Ash produced from semi-dry desulfurization processes can be used as a raw material for high and early strength cement production.

The Performance of Application of Wet Process Flue Gas Desulfurized Gypsum (FGD-Gypsum)

The major components of FGD produced by the wet "Limestone-Gypsum Process" is gypsum ($CaSO_4.2H_2O$) whose purity is over 90%. The FGD gypsum has 10-12% free water fine orange powder. Its purity is higher than natural gypsum produced in China, and it will become an important gypsum resource in the near future. The chemical composition of FGD gypsum is shown in Table 10.

Table 10   The Chemical Analytical Results of FGD

RESOURCES	CaO	SiO₂	Al₂O₃	SO₃	Fe₂O₃	MgO	LOI	PURITY
Luohuang (Chongqing)	32.6	2.7	0.7	42.4	0.5	1.0	19.2	94

The properties of natural gypsum and FGD gypsum are shown in Table 11. Results indicate that FGD gypsum is good in quality, the performances of building products with FGD are superior to those containing natural gypsum. The hydraulic gypsum products prepared using FGD gypsum also have good prospects for development.

Table 11   Properties of Gypsum products with Different Source of Gypsum and FGD

NO.	RESOURCE	TYPES OF PRODUCTS	STANDARD CONSISTENCY %	INITIAL SETTING (Min)	FINAL SETTING (min)	BENDING STRENGTH MPa	COMPRESSIVE STRENGTH MPa
1	Natural Gypsum (Sanxi)	Plaster of paris	62	7	12	2.4	4.9
2	Chongqing Luohuang	Plaster of paris	60	7	12	3.6	8.4
3	Hangzhou bansan	Plaster of paris	56	6	12	3.3	9.1
4	Hangzhou bansan	Plaster of paris	55	6	8	6.1	15.0
5	Hangzhou bansan DSG Hydraulic	DSG Hydraulic Cementitious Material 1	50	120	140	3.9	15.3
6	Hangzhou bansan DSG Hydraulic	DSG Hydraulic Cementitious Material 2	50	140	150	2.95	29.5
7	Building Gypsum Hydraulic	DSG Hydraulic Cementitious Material I-7		337	357	7.5	43.8

*No.I-760 (60% cement +40% gypsum): sand =1=2.5 Result from cement mortar test

No.I-6 Result from building gypsum test.

It has been found that not only can FGD gypsum be used to replace natural gypsum for plaster of Paris and hydraulic gypsum products, but can also be used as cement retarder in conventional cement production.

Larfarge-Onoda Gypsum Ltd (Shanghai) substitutes FGD gypsum taken from different power plants for natural gypsum to produce gypsum board. Compared with natural gypsum, FGD gypsum has the advantages of high purity, finer particle size, low chloride content and it reduced energy demand, improving the quality of products as well as decreasing plant wear.

UK WASTE AND RESOURCES MANAGEMENT RESEARCH AND DEVELOPMENT: THE QUEST FOR CO-ORDINATION, INTEGRATION AND FOCUS

S Penney

SNIFFER

United Kingdom

ABSTRACT. The UK waste and resources research community is made up of various university departments, government departments, NGO's, consultants and research organisations. However until recently there has been very little joined up thinking and coordinated action within this community. The research which has been carried out has often suffered from duplication and a lack of integration. Strategic and integrated R&D activity has been the distant dream of a few visionaries. However the last few years has seen a growing realization by much of the waste sector that this approach must be challenged and that there must be a new dawn in the way that waste research is identified, carried out and disseminated. There has been the emergence of UK forums which have and are attempting to facilitate this change. Such forums include the National Resource and Waste Forum (NRWF) and the Forum for Waste and Resources Research and Development (FORWARRD). While the new dawn of integration, coordination and focus may not have arrived yet, the dawn chorus might be about to start singing.

Keywords: Policy, United Kingdom, Research and development, Coordination, Integration, Focus

S P Penney is a Chartered Waste Manager and has an MSc in Wastes Management from the University of Sunderland: he was also winner of the 1999 James Sumner award from the Chartered Institute of Wastes Management. Simon gained 5 years experience in waste regulation with the States of Guernsey before joining the Scotland and Northern Ireland Forum for Environmental Research (SNIFFER) as the research manager for the UK Waste Management and Regulation Programme and the SNIFFER resources theme.

INTRODUCTION

The UK waste research community is made up of various university departments, government departments, NGO's, consultants and research organisations. However until recently there has been very little joined up thinking and coordinated action within this community. The research which has been carried out has often suffered from duplication and a lack of integration. There is now however a growing realization that this approach must be challenged and that the way that waste and resources research is identified, carried out and disseminated must be radically changed. While there is no over riding government policy on coordinating such work two significant UK forums have emerged and are attempting to facilitate this change. The National Resource and Waste Forum (NRWF) has recently emerged into the research arena, as has the Forum for Waste and Resources Research and Development (FORWARRD): these forums are now acting as points of focus and coordination and are encouraging an integrated approach to delivering such research. The vision for coordinated waste R&D in the UK is however still not clear and requires to be bought into much sharper focus. This is a process that must happen both within and outside government and then this vision needs commitment and resources for its effective implementation. In addition to this there have been recent and significant changes to the way in which wastes management research will be funded in the UK.

A UK OVERVIEW

Public funding for waste research in the UK has undergone radical change in the last year although for many people this change has been a step in the wrong direction. Until April 2003 and in addition to other funding from central government, the devolved administrations or other publicly funded bodies a substantial fund was available for research through the landfill tax credit scheme, a scheme which operated throughout the UK.

The scheme was a fiscal instrument which made allowance for landfill operators to contribute private money to various specified activities instead of paying that money as tax. This money was distributed by distributive environmental bodies (DEBs), often arms length companies set up by the landfill operators. While the concept of this system was good, it did suffer a severe coordination problem, indeed this was the catalyst for the formation of FORWARRD.

The specified activities outlined by the scheme included research and development, educational activity or communication that encourages more sustainable waste management and the development of products and markets for recycled waste. During 2001 this scheme provided £33 million[1] towards research, pilot and demonstration studies. In some cases this allowed for the significant leverage of additional resources for sustainable wastes management research and development.

Following concerns about the running of the landfill tax credit scheme in relation to waste products and following a consultation the Government decided to alter this method of funding. The scheme was at the time of writing being transformed into a number of funds administered by central government and the devolved administrations with the aim of using LTCS money as a public expenditure scheme. Decisions about the exact amount and what this money will be used for in England are being taken in the light of the Strategy Unit report "Waste Not Want Not". There will also be new arrangements within the devolved administrations.

Such questions of funding have in recent years helped the UK waste research community to become acutely aware of the need for better coordination, integration and focus of work which is being carried out in the UK, throughout Europe and internationally. With the introduction of an increasing amount of European legislation and the launch of the National Waste Strategies never has there been a greater need for effective identification, coordination, development and delivery of sound research across the spectrum of issues that make up wastes and resources management.

Despite the fact that there are several waste strategies being implemented within the UK, it should be noted that it is the UK which constitutes the nation state in terms of EU law and it is as the UK that the state must deliver its legal obligations under EU law. This again underlines the need for coordinating research work across the UK.

While there remain numerous research centres, consultancies and government agencies involved in waste and resources research in the UK some key bodies which are specifically focused on improving coordination and integration of such research are now forming. These embryonic organisations include the National Resource and Waste Forum (NRWF) and the Forum for Resources and Waste Research and Development (FORWARRD). These two bodies are key forums that currently exist which allow stakeholders that are both internal to the wastes sector and external to it to come together, identify research needs and coordinate their endeavours at a strategic UK level. The government backed Waste and Resources Action Programme (WRAP) also facilitates research work at a UK level with dedicated officers now located in both Scotland and Wales. There is also the possibility that the role of this organisation may expand into some areas of research coordination.

While the NRWF (a government endorsed, cross-sectoral initiative) has its primary focus in promoting public engagement in the sustainable management of wastes, two key research needs have been identified and are being met through this forum. The first project is the production of a UK framework for household waste prevention, looking at both local and strategic national needs and the relationship between the two and is focused specifically on the needs of the individual home countries.

A second project being undertaken through this forum is the development of best practice for stakeholder engagement in waste strategy development; this work is looking at the experience of Scotland in its development of its National Waste Strategy as well as other case studies.

The Forum for Resources and Waste Research and Development was formed a little over 2 years ago in order to coordinate research being carried out under the landfill tax regime. The need for this body was identified at an Environmental Symposium on the 'Future Strategy for Waste Management Research', held in London during May 2000. Many of the bodies that were responsible for making awards to researchers identified that they had problems in finding out if the work they were being asked to financially support had in fact already been undertaken.

During its short history FORWARRD has developed three projects, wastenet, "A Framework for Prioritising UK Waste and Resources R&D" and a Peer Review iniative. The wastenet project has sought to develop a web based tool to aid access and coordination of waste related research[2]. Using a purpose built search engine in conjunction with more traditional IT search engine technology this system will be free at the point of use and available to everyone. The system will provide information on what research is being undertaken, where it is being undertaken, its objectives, how near to completion it is, key contacts and other such information.   Initially this project is intended to cover only UK waste and resources research but it has always been the wastenet vision that this system should extend both into the rest of Europe and internationally. An embryonic form of this system should be on the web within the next two or three months.

The report entitled "A Framework for Prioritorising UK Waste and Resources R&D" was published and launched at the launch of FORWARRD. This report bought together, perhaps for the first time the key research programmes being undertaken within the UK, both within government and externally to government. The report asked the questions: 'where are we now?'; 'where do we need to be?'; 'what is the knowledge gap and how do we get there?'. Ten key priority themes were identified in the light of current research programmes and future research drivers including policy appraisal tools, social aspects and integrated product design[3]. While this project was not directly sponsored by government, various government departments contributed to advisory panel discussions.

The Government sponsored Waste and Resources Action programme (WRAP) has a UK remit (although currently with a focus on market development for research) and sponsors numerous key projects to remove market barriers in order to increase recycling of paper, wood, glass and plastics. Current projects[4] are looking at the use of ground glass in the manufacture of clay bricks, removing barriers to recycling glass from licensed premises and methods to recycle preservative-treated wood into charcoal and recycled wood preservative. WRAP works closely with other market development focused organisations who are members of the UK Market Development Network, such as ReMaDe, which facilitates research and development activities.

Another key programme is FORESIGHT, this is backed by government, business and the science community and aims to provide challenging visions of the future to ensure effective strategies now[5].   During 2000 it published a report entitled "Stepping Stones to Sustainability", which outlined a number of short, medium and long term areas for research in waste and resource management [6]. These sorts of outputs are directed to business, research councils, government departments and charities, but they are not policy and FORESIGHT explicitly states that it does not produce national research strategies. Reference was made to this report in the FORWARRD report published late last year and referred to above.

There are also a number of government sponsored research councils that operate within the UK, the Engineering and Physical Science Council, Natural Environment Research Council, Economic and Social Research Council and the Biotechnology and Biological Research Council. In total these research councils distributed nearly £14 million during 2001 [7].

In addition to this, there is a great deal of other research that is being carried out in the UK by commercial companies, numerous and various university departments, consultancies, community and NGO level groups. In particular the University College Northampton, the SITA centre for Wastes management and the Shanks Caledonian centre for Wastes Management in Glasgow undertake significant amounts of work. Numerous MSc courses and PhDs are undertaken at universities such as Leeds and Southampton.

There is of course a wide spectrum of activities that constitute research, from PhD to near market consultancy. Perhaps the most common definition of research is the Frascati definition;

"Research and experimental development (R&D) comprise of creative work undertaken on a systematic basis in order to increase the stock of knowledge, including knowledge of man, culture and society, and the use of this stock of knowledge to devise new applications [8]."

Taking a wide definition of research and including pilot studies, demonstrations and other such activities it is possible to start to understand the massive challenge that faces any effort to coordinate, integrate and focus this work to commonly agreed objectives and within an agreed framework. In the course of the next few minutes I will look in more detail at some of the Government sponsored research programmes in this area from across the United Kingdom, make a few comments on them and offer a few personal thoughts on some possible next steps.

SCOTLAND AND NORTHERN IRELAND

It is appropriate that I start this tour of the UK with Scotland. The Scottish National Waste Plan (NWP) [9] launched in February this year identifies research and development as a central component. The Scottish National Waste Strategy, which will be implemented by this plan, was developed through an extensive stakeholder engagement process. The strategy divided the country up into 11 waste strategy areas, they were urban, rural and mixed in nature. It brought together relevant stakeholders from these areas and then undertook a best practicable environmental option (BPEO) process to identify the most appropriate waste management activities for the future waste management of these areas. The work has initially focused on the sustainable management of municipal solid wastes but is now expanding to look at other key waste streams. All of the area waste plans where then integrated into the National Waste Plan.

In support of the NWP the Scottish Environment Protection Agency, the Scottish Executive and others are carrying out a great deal of research. This research covers specific projects detailed in the area waste plans including the analysis of landfill capacity needs, social and behavioural research, priority waste streams (including clinical, hazardous etc), the development of plans for the non-municipal waste streams, special wastes management in Scotland and horizon scanning. The NWP makes specific mention of life cycle analysis, human health, household waste prevention, and composting, recycling, energy recovery and waste disposal as well as economic and fiscal elements of the equation.

The Scottish Executive is sponsoring a number of research projects, including the development of a financial support scheme for the recycling sector and incentives for changing household waste behaviour.

SEPA and the Scottish Executive undertake a great deal of work together and are integrated through the National Development Group which is a forum on which the coordinators for each waste area, the Scottish Executive and others sit. The main mechanism at the moment for integrating and coordinating their research with other work being undertaken within the UK is the National Resource and Waste Forum.

The Scotland and Northern Ireland Forum for Environmental Research (SNIFFER) acts as a coordinating mechanism for waste and resources research within Scotland and Northern Ireland. It identifies, manages and commissions research on behalf of its members – SEPA, the Scottish Executive, Environment and Heritage Service Northern Ireland, Forestry Commission and Scottish Natural Heritage – in the light of other research which is being carried out. Current SNIFFER 'waste' projects include the development of guidance for the management of Sanpro (human hygiene) waste, emerging waste technologies, consumer responsibility and the development of strategies for stakeholder engagement during the implementation of waste strategy. In addition to this work SNIFFER is also coordinating work on the management of radioactive wastes and several projects in air quality, contaminated ground and remediation.

The focus of government research concerning waste management in Northern Ireland has been specifically in support of the Northern Ireland Waste Management Strategy, which was launched during 2000 after extensive consultation with numerous stakeholders during its development by a team of consultants. The focus of this research has been on waste prevention and improving waste recycling and recovery to meet EU targets. Most of the waste research undertaken by the government in Northern Ireland is sourced through SNIFFER and is therefore produced in collaboration with the Scottish Executive and SEPA.

ENGLAND

The England and Wales Waste Strategy Policy was launched in 2000, although a Welsh Strategy has now been developed by the Principality and launched in its own right (Wise up to Waste: The National Waste Strategy for Wales). The key government departments based in England currently undertaking waste and resources research are the Department of Trade and Industry (DTI) and the Department for the Environment, Food and Rural Affairs (DEFRA), with publicly funded research programmes also being carried out by the Environment Agency, WRAP and the Research Councils.

The Strategy Unit, a high level government project group set up to look at progressing sustainable wastes management, published its report 'Waste Not, Want Not', at the end of 2002. Whilst it is not a policy document it is a key feature in the wastes management debate in England. This report comprehensively reviewed the major issues in wastes management, providing solutions and key recommendations [10]. The four strategic strands are: supporting local authorities; new technologies; research and development; and waste minimisation, recycling and education [11].

Environment Agency (EA)

The Environment Agency research programme has developed from the research programme which was undertaken during the 1980s by the then Department of the Environment. Historically the Environment Agency and its predecessor organisations have carried out research with a focus on technical and regulatory requirements. It has a history of widely respected research into landfill science. More recently the research programme has broadened to include composting, life cycle assessment, waste pre-treatment, household waste management and health effects from waste management.

The Environment Agency spends approximately £1.4 million p.a. on waste research and currently this money is divided into 5 key topic areas: landfill processes; industrial and hazardous waste; strategy and options; exposure assessment and risk; agriculture and biowastes. This research is carried out to meet the business needs of the organisation and is developed internally. The Environment Agency does however convene an annual waste research advisory panel, which brings together a number of key external research stakeholders to advise the Agency on its proposed research programme. In addition to research that is carried out into purely wastes management, work is also undertaken on contaminated land, sustainable development, lifecycle assessment, computer modelling and air quality all of which cross into the waste management subject area. The FORWARRD report[12] states that future priority areas for research include gaining a better understanding of economic instruments, general policy appraisal and understanding and managing health effects associated with waste management practice. During 2002 Environmental Resources Management undertook a major review of all EA, DEFRA and Department of the Environment research dating back to the early 1990s[13].

Department of Trade and Industry (DTI)

It is of course vitally important that industry in the UK follows sound environmental policy. The Department has a central role to play in this as it has in encouraging the development and growth of the whole environmental industry sector. In addition the DTI, through the Sustainable Development Directorate helps industry to account for the environmental, economic and social impacts of their work, the so called triple bottom line. It is also involved within the Office of Science and Technology which has a responsibility to review priorities for UK publicly funded research.

The DTI seeks to encourage the future uptake of clean technologies; supports waste minimisation through the Envirowise campaign and the application and improving of current waste technology[14].

Department for the Environment, Food and Rural Affairs (DEFRA)

Most of the work undertaken by DEFRA in the area of wastes management falls within the remit of the Environment Protection Group. Current or recent projects includes a study which is looking at international examples of best practice in waste prevention. In the short term, work on the management of waste oils may be carried out and there is the likelihood of future research into the four strategic strands outlined above. A number of these issues mirror the needs of government organisations and others in different parts of the UK. Some research in the more general areas of Sustainable Development is also carried out by the Office of the Deputy Prime Minister (ODPM) and the Department of Transport (DTR)

WALES

Wales released its own National Waste Strategy, 'Wise up to Waste', during 2002 and in support of this strategy has identified a number of key research and development needs [15]. Despite the many social, economic and environmental differences with the rest of the UK, many of the research requirements identified in Wales reflect the needs of other parts of the UK, again underlining the need for a coordinated and integrated approach.

In the FORWARRD report, the National Assembly for Wales[16] outline the need for better data and highlighted, inter alia, risks to human health, emerging technologies, computer based decision support tools, integrated product policy and resource economics.

Most current government-sponsored research into wastes management in Wales occurs in collaboration with the Environment Agency waste management and regulation research programme. For example, the Assembly has funded a pilot study of municipal waste composition in conjunction with the Environment Agency and local authorities. Based on an agreed protocol for analysing the various components of municipal waste, further research funded by the Assembly is in progress in 9 local authorities to provide waste composition data linked to socio-economic profiles and waste collection methods. Composition data on trade waste is also being collected. An additional initiative is a pilot waste minimisation scheme within the public sector. It was announced in March 2003 that a new centre of excellence to coordinate waste research in Wales was being established at Cardiff University[17].

CONCLUSIONS

This brief overview of government sponsored research and associated policy across the UK has highlighted the continuing need to put a robust framework in place in order to better coordinate and integrate waste and resources research.

The future of waste and resources research is likely to be yet even more complex and legally driven. The implementation of the National Waste Strategies for the UK and the research that is needed to support these is the key focus. It is vitally important that there is clear government policy to support this need. Such policy needs to be supported and understood at the highest levels of government both at strategic UK level and within and between the devolved administrations.

There is a debate to be had as to whether such policy should encourage the imposition of a coordinating framework or matrix of mechanisms or whether the approach should be one of bringing about a consensus solution. Once a solution has been identified there is the need to identify the best place for it to sit or be implemented whether by supporting existing mechanisms such as FORWARRD, FORESIGHT, WRAP and NRWF or by instigating a new initiative, further developing a project like wastenet or leaving it to enterprise or consultancy.

There is the need for real consensus to be found in the devolved administrations as to how they wish to be involved in this process. Do they wish to have a specifically Scottish, Northern Irish or Welsh mechanism or are the devolved administrations content to meet this need through their involvement with UK level initiatives? How do they wish to relate to each other? How do they wish to relate to the UK as a whole and to the various European and International agencies? These are all questions that remain as yet unanswered.

Is government the right stakeholder to coordinate such research, or is the job too big for government on its own and where are the boundaries to be drawn? It should also be noted that for every barrier identified at the UK level, there is a mirror issue at the European and international level.

A great deal of discussion has already taken place about better research coordination and the advantages of actually doing this. Better coordination, integration and focus of research will bring manifold benefits to all stakeholders. Such benefits include: added value; synthesis of ideas and concepts; increased creativity; the formation of both objective lead and enabling networks (allowing efficient and effective flows of information); more customer orientated research; the format of research outputs being improved to suit specific audiences and bridges being built between academic/blue sky work and application.

There is the real need for a clear vision to be set out as to what an effective and efficient coordinating/integrating framework would look like. A clear understanding is needed of the roles and responsibilities of the various stakeholders involved and it must be clarified which stakeholder or group of stakeholders has the responsibility to lead in the meeting of this need. Such a vision needs to be properly resourced, financed and accepted by all of the key stakeholders from across the UK. The strap line from the Scottish National Waste Strategy provides an apt conclusion to this brief overview of National Government Policy in the UK with reference to waste and resources research and development: "There is indeed No Option but Change".

ACKNOWLEDGEMENTS

The author would like to thank the following individuals for reviewing and commenting on the draft of this paper.

Stephan Jeffries (Surrey University),Gillian Neville (DEFRA), Ray Alderton (DEFRA), Ruth Wolstenholme (SNIFFER), Jill Ogle-Skan (SNIFFER), Roy Ramsay (EHS NI), Pamela Patterson (EHS NI), Harvard Prosser (NaW), Chris Coggins, Gordon Jackson (SEPA), Peter Brown (Scottish Executive), Nick Blakey (ESART), Peter Tucker (University of Paisley), Des Langford (DTI), Wendy Rayner (Enviros).

REFERENCES

1. POTTER A., GENTIL E, A Framework for Prioritising UK Waste and Resources Research and Development, Golders Associates, for the Forum for Resources and Waste Research and Development (FORWARRD) 2002, p.23.

2. AEA TECHNOLOGY, WasteNet - Phase 2, Developing the Concept of Implementation, 2002,p. i.

3. POTTER A., GENTIL E, A Framework for Prioritising UK Waste and Resources Research and Development, Golders Associates, for the Forum for Resources and Waste Research and Development (FORWARRD) 2002, Executive Summary.

4. WASTE & RESOURCES ACTION PROGRAMME (WRAP), Achievements Report 2001/2002, p.10.

5. FORESIGHT website 2003, http://www.foresight.gov.uk

6. POTTER A., GENTIL E, A Framework for Prioritising UK Waste and Resources Research and Development, Golders Associates, for the Forum for Resources and Waste Research and Development (FORWARRD) 2002, p.4.

7. POTTER A., GENTIL E, A Framework for Prioritising UK Waste and Resources Research and Development, Golders Associates, for the Forum for Resources and Waste Research and Development (FORWARRD) 2002, p.26.

8. OECD Paris, "The Measurement of Scientific and Technological Activities Proposed Standard Practice for Surveys of Research and Experimental Development" (The "Frascati Manual") 1994.

9. SCOTTISH ENVIRONMENT PROTECTION AGENCY (SEPA) & SCOTTISH EXECTUTIVE, The National Waste Plan 2003, Scotland, pp77-80.

10. WASTES MANAGEMENT JOURNAL, Chartered Institute of Wastes Management December 2002, p.41.

11. Personal Communication, April 2003.

12. POTTER A., GENTIL E, A Framework for Prioritising UK Waste and Resources Research and Development, Golders Associates for the Forum for Resources and Waste Research and Development (FORWARRD), 2002, p.44.

13. Personal Communication, April 2003.

14. POTTER A., GENTIL E, A Framework for Prioritising UK Waste and Resources Research and Development, Golders Associates for the Forum for Resources and Waste Research and Development (FORWARRD), 2002, p.44.

15. NATIONAL ASSEMBLY FOR WALES, Managing Waste Sustainably, Consultation Paper 2001.

16. POTTER A., GENTIL E, A Framework for Prioritising UK Waste and Resources Research and Development, Golders Associates for the Forum for Resources and Waste Research and Development (FORWARRD), 2002, p46.

17. Personal Communication, April 2003.

THEME FOUR:

LOCAL GOVERNMENT POLICY

HOUSEHOLDERS AS A MEANS OF INSTIGATING CHANGE IN WASTE MANAGEMENT PRACTICES

M Fenech

V Cassar   C Ciantar

Ministry for Resources and Infrastructure

Malta

ABSTRACT. This paper examines the motivations behind individual environmental action and how these can pressure Local Councils to introduce waste separation at source in their localities. It also seeks to understand the present position of Local Councils and the reasons why they are not showing a lot of interest. This paper points out to knowledge and 'frustration' with the present system as a means of instigating such action.

Keywords: Local Councils, Environmental action, Waste separation.

M Fenech recently joined the team responsible for the implementation of the waste management strategy in Malta, where she forms part of the team responsible for the introduction of waste separation in Malta.

V Cassar is a member of the Ministry for Resources and Infrastructure, Malta

C Ciantar currently heads the Strategy, Communications and Development Unit with WasteServ Malta Ltd. within the Ministry for Resources and Infrastructure.

INTRODUCTION

In Malta approximately 1.6 million tonnes of waste are produced annually. Of this 7.3% consists of Municipal Solid Waste (MSW), while industrial waste makes up 8.3% and construction and demolition waste make up 84.4% [1]. Waste management in the Maltese Islands consists mainly of uncontrolled landfilling. The existing landfills (Maghtab in Malta and il-Qortin in Gozo) are land-rise operations and represent major intrusions on the surrounding landscape. These landfills receive 92.3% of MSW, while the rest (7.7%) is composted [1]. These practices are obviously unsustainable because no conservation of material and land resources takes place and waste related problems created by the current generation will burden future generations (The development of an integrated solid waste management strategy was commissioned and funded by the European Commission on behalf of the Government of Malta under the Malta-EC fourth financial protocol. Construction and demolition waste constitutes a higher percentage than that normally found in other countries due to the amount of excavation that takes place during the construction of houses and other establishment in order to maximise the space available.). These waste practices recently received a greater awareness due to the national and international obligations.

As a result of this the "Solid Waste Management Strategy for the Maltese Islands, 2001' [2] was designed and published. The aim of this document is to provide direction and context to guide both Government and public sector in waste management issues over the planned period of time. It establishes a policy framework within which proposed physical development relating to waste management will be regulated by the Malta Environment and Planning Authority (MEPA) [3]. A fundamental objective of the strategy is the need to curb the growth in waste. In addition to this, it sets out standards for the treatment of waste prior to disposal and ban various forms of waste from landfilling unless it is previously treated. In accordance to EU Directive 99/31/EC on the Landfill of Waste, the strategy also aims to reduce the amount of biodegradable waste going to landfill.

The responsibility for implementation falls primarily on the Ministry for Resources and Infrastructure and its subordinate entities. Local Councils, as specified by Section 33 (1b) of the Local Council Act, are obliged to 'provide for the collection and removal of all refuse from any public or private place' [4]. In present practices, Local Councils, issue contacts based on separate collection. However, they still do not enforce separate collection because it is claimed that the Government does not have the necessary facilities to treat separated waste. Since the costs of providing for the collection and disposal services of MSW are entirely met through transfers from the Government budget to the Local Councils, there is no direct link between what waste a community generates and the services it pays for. Such financing prevents both local councils and householders to have any perception with regards to the true costs of providing these services and therefore provides no incentive to reduce or avoid the generation of waste [2] or to optimise the system for collecting waste. The government has now amended the contract conditions, issued by the Local Councils, in order to reflect the changes necessary to ensure future more environmentally acceptable management of waste. Local Councils will be obliged to collect separated waste to comply with the requirements currently planned for in the Strategy.

Amongst other aspects, the waste management strategy aims to introduce source segregation of MSW and separate collection of recyclables (including biodegradable) materials from MSW and to optimise the collection frequencies to reflect local conditions and collection methods by 2004. The system for source segregation follows a combination of bring-in sites or drop-off centres and colour-coded plastic sacks varying according to the different types of waste for kerbside collection. This collection system has already started in some parts of the island. In this system, households have bags delivered to their houses. These bags are of a different colour depending on the type of waste that is supposed to be disposed of in them. Once these bags are full they are collected on assigned days depending on the type of waste they carry. For example, biodegradable waste is collected on Mondays, Wednesdays and Saturdays, while on Thursdays recyclables are collected and on Tuesdays residual waste is collected. Waste from households that have not started segregating waste is collected as normal, that is, daily in mixed form.

Obviously a segregation system requires much more effort to organise than one where all waste is mixed and collected daily in any type of bag or bin. These changes would require the providers of MSW collection services to have the necessary technical capabilities and financial resources to adapt. Therefore Local Councils will need to introduce a standard procedure for pre-qualifying potential bidders for MSW collection contracts against stringent financial and technical criteria. In addition to this, the supervision and control of MSW collection services providers will have to be strengthened and local councils play a key role in this.

However, apart from the places where waste separation is gradually being introduced, Local Councils have demonstrated very little interest in the changes. Possible reasons for this could be issues that are of greater concern to the local population in that area - for example the Mayor of one council said that the priority of the Council was the size of the school in the locality. Other reasons range from lack of interest in environmental issues, convenience of the present system both for households and Local Councils themselves, the lack of knowledge with regards to the real cost of the system and the lack of need felt for change to a more environmentally friendly and efficient system.

INDIVIDUAL PRESSURE ON AUTHORITIES AS A MEANS OF INSTIGATING CHANGE – A LITERATURE REVIEW

Sometimes, institutions like government organisations and municipalities engage in activities that are directed at maintaining or extending the possibilities of environmentally unsound behaviour. Such activities serve to frustrate the intentions of individuals who want to engage in pro-environment behaviour. An example of this arises from the manner in which some products are offered in packages that many consumers consider to be oversized and unnecessary [5]. Such behaviour is frustrating for environmentally conscious individuals because of the powerlessness felt when people are unable to amend their own behaviour and instigate amelioration in the authority's behaviour towards the environment.

The physical limitations imposed by these authorities produce an absence of behavioural alternatives even when pro-environmental attitudes are present. In many cases, although individuals can amend their own behaviour, for a significant behavioural change to be achieved it has to take place within the context of the institutional programme and collective decisions. This occurs because the creation of a substitute for current environmentally unsound behaviour requires the introduction of schemes and procedures that give an opportunity for alternative behaviour. Often, public awareness, at the level of values and general environmental attitudes, is much more ahead of policy than authorities think. These attitudes and (when it takes place) the resulting actions, often have a potential power of influencing collective and political decision-making. These actions can either be directed at environmental improvements i.e. conducted for the recovery of environmental damage, the prevention of it or the interest in refraining from a behaviour that is a burden to the environment [5].

Therefore even when the right attitudes exist, an individual's behaviour may be constrained or hindered by the lack of opportunities imposed, which in cases like this form part of the external conditions. These constraints are of an objective nature because they influence the performance of an act. Objective constraints, sometimes, emerge from natural and socio-cultural surroundings, structural factors like the limitations of time, income and money, available technology, social information and the state of infrastructure [6]. In the case of waste management, where no system exists to accommodate waste separation in households, the objective constraints emerge from the state of infrastructure. Obviously these are not the only type of constraints. For example, an individual can constrained from action because the particular alternative did not occur to him or her in the situation. Therefore, the behavioural option must be salient for the individual in the situation. Finally, constraints can also be of a subjective form, that is, people must consider the behaviour to be relevant for themselves. Since, these constraints are based upon the beliefs of what is possible, permissible and pleasurable, they influence the willingness to act [6].

When constraints are of an objective form, the required changes are not in attitudes but in the general behaviour of the responsible institutions. Citizens demonstrate that they are aware that the solution for the environmental problem lies in a change in behaviour but because of the lack in infrastructure they have no possibility to introduce these changes. Institutions and authorities might not engage in such activities, not because they are not convinced of the benefits, but because they are not required to do so or because the implementation of these changes requires a lot of effort and they lack both the motivation and the resources for this.

Motivations that cause individual environmental action

Environmentally conscious individuals can either aim to change their own behaviour or that of others. When an activity aims to change personal action it is referred to as environmental behaviour, while when it aims to change the behaviour of other people, it is referred to as environmental action. The concept of environmental action refers to human activities on a small and large scale and the term can be used to indicate activities that range from a single person who tries to prevent the felling of a tree in front of his house during a short period of time, to the struggle against nuclear energy by many people in many countries over many years. When the latter takes place it is referred to as environmental movement [7].

These constraints and the purpose of particular activities, make environmental action problem driven [6]. Those preparing to carry out the action should decide upon the purpose and the procedure of the particular act. Actions of this type are targeted at a change - a change in one's lifestyle, in schools, in the local community etc. This implies that, an environmental action can include both indirect and direct actions, for example demonstrating against traffic conditions is considered as action as cleaning up litter and encouraging others to do so [7].

However, individuals act rationally in order to maximise utility and usually distinct intrinsic satisfactions are derived from participating in environmental actions. Therefore selective incentives are needed for individuals to engage in such activities. This is due to the nature of environmental quality as a collective good and therefore it is possible to enjoy the benefits without engaging in the co-operative behaviour necessary to bring them about [7]. Incentives, although they can be of an economic nature, they are often related to personal satisfaction resulting from the knowledge that something beneficial is coming from one's actions, like for example, better environmental conditions especially when those participating suffer from the immediate consequences of pollution etc. This satisfaction makes the potential costs of not participating greater than the time and, in some cases, the money invested in the activity.

There are different factors that lead to environmental action. In the following paragraphs the aspects of knowledge, perceptions, health hazards and the stage reached by the particular environmental issue are examined. These aspects were chosen because environmental driven action is often related to issues that are of particular concern at that moment and because these actions are influenced by intrinsic factors like knowledge and perceptions about that particular issues.

Knowledge

Knowledge per se is usually said not to lead to environmental action. However, it should still be acknowledged as an important precondition for the development of competence leading to action and behavioural adjustments in relation to the environment [7]. Environmental knowledge can be seen as informing environmental action both 'conceptually' and 'instrumentally'. Conceptually because it provides the grounds for thinking and acting, it provides the knowledge about the difference between the present and the ideal situation, while instrumentally it provides the tools used to think actions. These two are not compartmentalised, but rather inter-relate so that any applied knowledge is like a tool and can help assess the outcome as to whether it will be successful or not [8].

However, these effects depend upon the type of knowledge that is disseminated amongst people. If knowledge is related to effects of human activity on the environment, like for example, the damages caused to the water table by leachate leaking out of a landfill, it is useful to rouse concern and attention thereby creating a starting point for a willingness to act. However, since this knowledge is mainly of a scientific nature, it may contribute to an 'action paralyses' since no explanation is provided about why these problems are present, let alone how it is possible for people to contribute to solve them [8].

Another form of knowledge deals with the 'causal' dimension of environmental problems including the associated social and economical factors that influence our behaviour. For example, an increase in affluence is often accompanied by an increase in waste production. This knowledge belongs mainly to the sociological, cultural and economic spheres [8].

Knowledge can also provide strategy possibilities for change, that is, knowledge about how to control one's own life and how to contribute to changing the living conditions in society at large. This embraces both direct and indirect possibilities of action [8]. Finally, knowledge can also deal with alternatives and visions. Here the possibility of developing one's ideas for the future in relation to one's life, society etc is given. This type of knowledge is an important pre requisite for motivation and ability to act and change. In this dimension includes knowledge about how other people go about similar things in different countries, since knowledge about other possibilities can be a powerful source of inspiration for developing one's visions [8].

Stage of the issue

Awareness about environmental problems goes through stages. During these stages, the public perception of an environmental issue influences the level of attention paid to the environment within a society. This fluctuating attention is described in the issue-attention cycle, which describes the fluctuating attention paid to issues of crucial importance. An important influence on the attention paid is the media and the way that it operates [5].

Stages are normally divided into three. The first stage is when the problem is still a 'non-issue', that is, the phase before the problem itself and the resulting contamination become publicly identified [9]. At this stage only experts and interest groups have noticed the issue and may be alarmed. From this stage, often as a result of a disaster or a series of dramatic events, the awareness of the public suddenly arises. This discovery and the knowledge of hazards this problems entails causes alarm and interest about the possible solutions that might exist [5]. This is known as the 'public issue' phase. At this stage, the issue has moved from an individual to a group level and it involves institutions trying to contain the scope of the public issue, and the victims trying to expand its scope though collective action for example organisation and protest. The authorities cannot easily avoid a problem that has developed to the stage of public issue anymore [9]. A public issue becomes a 'political issue' when it expands into the political domain and involves other organisations – governments, companies, political parties, social movements and the mass media which aim to come up with solutions for this problem. After this stage, public interest gradually declines. This might be influenced by the media whereby people become bored by the attention given to the problem. Another possible damping of interest arises from the realisation of the costs that the solution for this problem involves. These costs often also involve changes in the arrangements found within society that might be benefiting certain groups [5].

Environmental action will take place mostly when the issue is either at the second (public-issue) or third (political issue) stage. In the first stage, action is likely to arise from organised environmental groups, while in the second and third stage, action is more likely to come from the general public. In the second stage, when people are alarmed about the situation and looking around for a solution, they are likely to resort to collective action like protest. When the issue becomes a political one and there are particular organisations and/or institutions dealing with it, people will have a term of reference with whom to file complaints. Therefore, at this stage individual action is more possible.

Health concerns and other influencing factors

When an environmental problem causes health concerns people demonstrate a certain level of interest because they are worried about their well-being and that of their family. For example, interest in waste separation and sustainable waste management is of particular concern to families because they are concerned about the well-being of their children [9]. Therefore, environmental degradation, especially for those that suffer the immediate consequences of pollution, can be a source of motivation for individual environmental action. A higher motivating factor for people to resort to environmental action is when emissions, coming, for example, from the burning of waste, can be easily observed or smelled and are therefore a cause for alarm [10]. For the people concerned, the potential costs of not taking action are greater than the costs involved (like time and money) of taking action. The fear resulting from perceived health hazards may override barriers to environmental activism that might exist. To involve themselves into such action people must also be convinced that individual actions can make an impact for the greater benefit [11].

Potential effects on health and well being interact with mediating factors that relate to the individual person like the social setting, the contaminant itself, intervention programmes and policies and other significant events. Other influences on environmental action and also methods for acting upon it, arise from different aspects like outdoor experiences and close contact with natural areas, involvement in environmental organisations and the loss and degradation of a valued place. Influence can also arise from family members who directed attention to the value of the environment or the importance of social justice [12] and the evolving responses of institutions to the problem [9]. When people's concern for the environment came due to a connection with a special place, this is usually part of the regular rhythm of life, like for example regular visits to the local reserved areas [12]. Influences come also from individual factors like personal characteristics such as a person's age, gender, occupation, length of residence or other demographic information, together with a person's general beliefs about self-efficacy setting and political, economic and cultural influences [12].

METHODOLOGY

This paper aims to examine the factors that motivate people to resort to environmental action, which in this case refers to the pressure exerted on Local Councils to introduce waste separation at source in their respective localities. In order to establish a theoretical background a literature review was performed, with the aim of establishing the main factors that instigate individual environmental action.

Following this, empirical research is then carried out in order to substantiate the literature review. A preliminary part of this research was 'observational' research that was made during a fair (Environment and Gardening Fair, Malta April 24th to 27th, 2003), when people visiting the fair came and put forward their complaints or sought information about waste separation themselves. In addition, email communication with a person that has done this type of environmental pressure was made. Finally, questionnaires (See Appendix 1 and 2) were compiled and used to obtain information from households residing in the locality where waste separation has been in place for eight months.

Initially data was collected to establish the level of participation of each household per day, with the aim of establishing the level of participation of each households, thus enabling the *Progett Skart Team* to determine how households are adapting to the system. Based on this information the questionnaires were performed in order to determine which factors encourage them [not] to participate and their willingness to engage in environmental action and encourage other people to do likewise.

The information collected about the level of participation of households was segmented in order to determine which households were participating and which were not. From these segments, a sample of twenty households was randomly chosen from the higher and lower participating households and out of these, the first fifteen that accepted were interviewed. This means that this is a non-probability quota sample. Because it's a non-probability sample it does not allow for sampling error to be measured, while since it is a quota sample, it means that a number of prescribed people/households out of several categories are interviewed. The interviews were arranged, that is, particular households were approached at their homes or offices and asked for an interview [13], while the questionnaires were schedule-structured and therefore the number of questions, their wording and sequence were identical for all respondents [14].

Personal interviews were carried out. This is mainly due to their flexibility – it gives the possibility to explain to the interviewee if he/she does not understand something, you can easily observe whether the respondent is taking time to answer or is answering in a hurry and because it gives the opportunity to collect interviewee comments which may be of interest for the research. A disadvantage of personal interviewing is that it is subject to interviewer bias and distortion [13]. Questionnaires consisted of ten/eleven main questions, which had a number of sub-questions. The questions were mainly closed but there were also some open questions. The total population is finite and consists of three hundred households. The size of the sample was of thirty, fifteen coming from the households that were participating eagerly, while fifteen coming from the households that were not participating.

Households from these two segments were chosen in order to examine the factors that motivate people to participate and those that were discouraging people to participate in waste separation. Moreover, it was also useful to examine why and what type of environmental action people engage into. Respondents were mainly of an average (post-secondary) and higher education background. Participating households were, by an absolute majority, professionally employed while non-participating households were mainly housewives, self-employed or professionals. In both cases, the average size of households was of 3.5 people and the majority of households had children. While in participating households half of the children were below the age of three and half above, in the non-participating households, children were mainly above three years of age.

As mentioned above, the sample size was thirty, therefore ten per cent (10%) of the total population were questioned. Although a larger sample would have increased the reliability of the results, it is usually accepted that five per cent (5%) of the population provides a reliable result [13]. This sample size was chosen because it covered the requirements for this research and due to time constraints. People co-operated well with the research. In fact only three households said that they were not interested in participating. Furthermore, other commitments for the householders caused some disturbances (e.g. small children) to the flow of the interviews and also increased time consumption. When preliminary tests were carried out, it was determined that these questionnaires would take between ten and fifteen minutes. However, when the real interviews started respondents took between twenty-five and forty minutes per interview. This discrepancy arose because the preliminary tests were carried out with people that were not separating waste while the real interviewees obviously were participating in waste separation and therefore had more opinions and suggestions to make. Respondents also complained about other people who were not participating as much as they were.

ANALYSIS

There are various aspects that cause an individual to act. An important precondition for action is knowledge because it provides the necessary grounds and tools for thinking and acting. The knowledge provided should help to arouse concern but also inform about other strategy possibilities for change and different solutions that can be used to arrive to a better waste management system than what is currently practised in Malta. This type of knowledge helps to breed 'frustration' because an individual is aware and feels that there are other techniques of dealing with particular environmental aspects but it is not possible to practise them because no infrastructure is present. The potential for action is enhanced if there are health hazards associated with the problem or if the effects of these problems are easily observable. Motivation for environmental action is also increased when the environmental problem involves the degradation of a valued place. Finally, if the issue is at the height of public concern, action is more likely to take place. The stage of concern reached influences knowledge and perception about that particular problem and if people believe that this problem is of concern to a number of them then there is likely to be both individual and collective action. If the issue is still a 'non-issue', it most likely that action would come from interest groups or experts that are aware of this problem and might be becoming alarmed about it.

Although various research projects have concluded that the Maltese do not show much interest in environmental issues [15], there have been some occasions when an environmental activity was organised and people responded well. For example, in an activity that took place in August 2002 to collect plastic bags and reuse them people responded very well and the activity was described as 'picking up momentum'. Those who handed in these plastic bags were paid Lm0.01 (0.024euros) for every two clean bags [16]. This shows that, when given the opportunity, a positive response is achieved and in some cases the lack of interest and negative environmental actions might be because no alternative behavioural possibility exists.

Obviously, lack of action is not always due to lack of infrastructure. Often it is the result of negligence or of reluctance to change especially when people are used to a particular system, which even though it is environmentally unfriendly, it is convenient. The latter is one of the problems faced by the waste separation system. For a long time the task of dealing with waste was in the government's hands. Furthermore, the convenient system, the lack of contact with waste and the problems it causes has removed all responsibility from waste producers to the extent that waste is now considered to be someone else's [the government] problem. However, since the environment is a collective good, collective responsibility has to be taken for it and this aspect still has to be developed in the Maltese society.

Often the waste issue has to deal with the 'handicap' that it is an out-of sight out-of-mind problem. In the case of Malta, the extent of the problem, although causing environmental havoc, can be an advantage to increase awareness about it and acknowledge that something has to be done to solve it. Due to this, more and more people are demanding a solution and some are aware that their collaboration is required for the system to function properly. Obviously a lot of work remains to be done in order to instigate the correct attitudes, responsibility and actions. Up till now, interest is being shown in waste separation but for this to be significant it needs to be spread amongst more people.

LOCAL COUNCILS

Local Councils have an important role in the implementation of the waste management strategy implementation process. Local Councils are responsible for the provision of MSW collection within their respective localities, the issuing of bye-laws on waste management and informing residents about waste management issues within their locality [2]. Therefore, they form part of the final stage of waste collection, which is the stage that reaches households, on whom the success of the waste strategy depends.

However, Local Councils have shown little interest in introducing waste separation at source. Various reasons have been given for this – more pertinent issues for that locality, lack of facilities and financial resources etc. Obviously, through the amending of contract, Local Councils will now be obliged to introduce waste separation. However, this imposition cannot be seen as a motivator for Local Councils to increase their effort in waste separation especially because, since they are aware that it will eventually be imposed upon them they are just waiting for their turn. This laid back attitude, has sometimes acted as a form of frustration for those individuals who would like to separate their waste. Because of this, some have contacted the Local Council and asked them to introduce the waste separation system. Response from the Local Councils has varied from a positive reaction (Birkirkara and Pembroke) whereby the Local Council has sought information from *Progett Skart* itself, to a neglect of the call (Gharghur, Mosta and St. Paul's Bay). Furthermore, in certain cases, Local Councils have told these individuals that when waste is separated, once it is collected it is again mixed and dumped at Maghtab. Such a response is obviously undermining the efforts of *Progett Skart,* which aims to achieve a separation rate as high as possible. Usually the interest shown by Local Councils in waste separation when individuals contact them, because they feel pressurised to please the people of their own locality to whom it is liable, rather than complying with a newly introduced law which they probably see as a burden.

Since the majority of Local Councils in Malta have not introduced this system, there is a potential to put pressure through individuals and increase the amount of waste being separated. Therefore, identifying such individuals and encouraging them to resort to this type of pressure is a useful means to motivate Local Councils to start looking into the changes required by the waste separation system themselves, rather than waiting for command and control regulation to be implemented. When Local Councils come forward themselves more motivation is likely to be demonstrated and in addition to this, when they realise that there are certain benefits for the Local Council itself, they would look at these changes more positively rather than just like another burden. However, identifying these individuals is not an easy task. In the majority of cases these people stepped forward and called the Local Council themselves. Other possible motivators to make stepping forward an easy process is through the use of broadcasts, publications, freephone and stands in fairs where people come forward themselves.

RESULTS

In its initial two questions the questionnaire aims to introduce the respondent to the current waste scenario and examine their knowledge with regards to other waste management practices. As was shown in a survey carried out during July 2002 [17], people are still requesting information regarding waste problems etc. A certain level of awareness with regards to different waste management practices was demonstrated in these questionnaires. This was both in the case of participating and also non-participating households. The main response (66% for non-participating and 86% for participating households) was knowledge on recycling, while other answers included knowledge on incineration (6.5% for each sample), reuse (13% and 13% for each sample) and composting (6.5% for non-participating households and 46% for participating households). Some of the respondents mentioned foreign practices with regards to waste management. Knowledge of this kind is important because, as was mentioned above, people are aware of other practices and it helps to instigate environmental action. However, it is important to keep in mind that in this case, the respondents are already participating in a waste separation system and therefore have some first hands on experience.

The third and fourth questions seek to understand how people who are already separating their waste look at the responsible local and central authorities. In both segments, respondents demonstrate a certain level of scepticism towards central and local authorities although for participating households this scepticism is slightly higher (for participating households 60% and 66% feel that the central and local authorities respectively are not putting enough effort, while for non-participating households, it is 53% and 46% respectively who claim this). It is likely that for this reason, those participating show more interest because they feel that through their participation they compensate for this lack of effort. At this stage, it is important to note that the Maltese in general look at public authorities in a somewhat sceptical way. The reasons for this stem from the unprofessional and clientalistic practices (consisting of the allocation of favours to governing parties devotees) with which these institutions have operated. The politicisation of operations, together with favouritism and unsatisfactory performance resulted in low legitimacy and under these conditions trust in public authorities dwindled. The organisations responsible for waste management have not been freed of this impression and people are still wandering whether they can deliver what they promised and tackle the waste situation seriously. This lack of trust can also be a possible instigator of environmental action since people look at authorities sceptically and demand better solutions. However, in order to achieve such actions, awareness about better solutions must exist and due to this more information and knowledge must be available.

However, central authorities' trust seems to be with some confidence because when respondents were asked 'whether they think the system is effectively supported by central government?' they responded positively for both segments (73% of participating and 53% of non-participating households confirm this). This is obviously positive for the *Progett Skart Team,* because although people feel that the central authorities are not putting enough effort to solve the waste problem they are aware that the team responsible for this issue is demonstrating a certain level of commitment. This enhances the possibilities for environmental action because people realise that although Local Councils have the appropriate support, no interest is being shown. When people understand this they will realise where gaps are lying and will be more willing and able to take environmental action.

The questionnaire then sets to examine how people look at the system itself. While the majority of both segments of respondents think that the system is beneficial (90% for participating households and 86% for non-participating), there still exist some problems that discourage participation. The most frequent problem mentioned is the frequency of collection. Before this system was introduced, mixed waste was collected daily so it is not surprising that this complaint is the most frequently mentioned. Complaints referred mostly to the frequency of collection of recyclable material (due to its bulkiness) and biodegradable waste (due to odours and pests). Other complaints concern the number of bring-in sites and the fragility of the biodegradable bags. The action that can triggered from these aspects is likely to be in the form of complaints regarding the misgivings of the system. For participating households the factors that would encourage them to participate more are information, convenience and green wardens (to help them understand the system better). Non-participating households would be more willing to participate if penalties for not-separating waste are utilised and if more environmental awareness exists.

In the final part of the questionnaire an enquiry is made about the respondents' inclination towards environmental behaviour and action. In the first part, the aspects that would encourage their pro-environmental behaviour is examined. In the case of non-participating households all reasons suggested (reasons where listed under three factors namely, 'because something needs to be done, because our children deserve a better country and because it is everyone's duty) that induce environmental behaviour fall under moral responsibility. In the case of participating households, since they already show a certain level of concern and commitment (a confirmation of this is given by 73% of respondents who say that they would still separate waste if bags were not home delivered but had to be collected), they were asked what actions they would take to induce Local Councils to introduce/increase waste separation at source. The majority of respondents said that the best solution to encourage them is to write them an email/letter (33%) or to call them (26%), while others said that they prefer to report immediately to higher authorities. Respondents were also asked what they think they can do to encourage other people to participate in waste separation and the most common answer was to discuss the system and its benefits with them (20%). Some also believe that reporting these people to the Local Council would be a good idea (17%). For non-participants, quite surprisingly, 86% said that they are willing to encourage people from other localities to participate but they did not specify how. Finally, the questionnaire tries to determine what factors need to be included to increase waste separation. Non-participants said that they would be more willing to participate if there are fines (46%) which are a form of negative reinforcement. Fines, or rather fees (usually based on weight and/volume), have in general worked very well in boosting waste separation at source. In Malta, there are no plans to introduce such fines because many fear that it will cause an increase in littering so as to avoid paying the fee. In addition to this, it is also expected to cause a lot of opposition since people will see the waste management project as another means for collecting government revenue [17]. However, in this case, it seems that there are people who will definitely not be motivated by any information etc given but need to be compelled into this action. Participants also feel that more enforcement is needed and again the majority of them (73%) point to fines as the best solutions. In this case, respondents mention also green wardens (40%) in order to increase enforcement and also to provide reinforcing incentives for people who participate well.

CONCLUSIONS AND RECOMMENDATIONS

Local Councils have a crucial role in the implementation and success of the waste strategy and it is therefore important to ensure that they are motivated to perform their job well. From the central Government's side it is often difficult to motivate local authorities because of the command and control style that is introduced during these type of activities. Due to this Local Councils will see the waste separation system as another regulation which they have to comply with. On the other hand, when pressure to introduce these systems is exerted from individuals they are likely to show more motivation because Local Councils will be responding to a request coming from a person in their locality and to whom they are liable, rather than complying with yet another regulation.

However, this type of action needs to be increased in order to make it more significant. For this to occur a certain level of knowledge with regards to the waste problem should be present. Obviously this knowledge should relate to the waste problem itself and the effects it is having on the environment and on human health, for example, emissions coming out from uncontrolled landfills and the harm that these are causing. Besides this, it is also important to include information about the possible solutions of dealing with these problems, how other countries are dealing with it and what will be achieved when a waste separation system is implemented. Information based on these two issues helps to breed concern about the issue and its effects and to increase 'frustration' about the system presently in use and demand a change in practice.

An important aspect is to ensure that *Progett Skart* continues to build a positive image and ensure people of what is being achieved. This is important to increase trust and to help motivate action towards Local Councils because when people understand the benefits they will realise that there are solutions to the waste problem but they are not able to do so because their Local Council has a laid back attitude towards these issues.

In the questionnaire those who said that they are willing to pressure Local Councils to introduce waste separation said they prefer to do so through email, letters or telephone. Apart from this, some also mentioned that they preferred to contact central authorities. Unfortunately, very little control can be exerted upon Local Councils to increase their availability, however in the case of central authorities, these must ensure a continued flow of information by participating in fairs, free phone services, website, email and other media means. This will make it easier for the public to put forward this type pressure.

REFERENCES

1.   MINISTRY OF THE ENVIRONMENT. State of the environment – Summary report for Malta, 1998, 88pp.

2.   MINISTRY OF THE ENVIRONMENT. A solid waste management strategy for the Maltese Islands, 2001.

3.   MINISTRY OF HOME AFFAIRS AND THE ENVIRONMENT. State of the environment – Report for Malta, 2002.

4.   LAWS OF MALTA. Local Councils Act, Chapter 363, 1999.

5.   TELLEGEN E and WOLSINK M. Society and its environment, an introduction. Gordon and Breach Science Publishers. The Netherlands, 1998.

6.   TANNER C. Constraints on environmental behaviour. Journal of Environmental Psychology, 19, 1999, 145-157.

7.   JENSEN B. Knowledge, action and pro-environmental behaviour, Environmental education research, Vol. 8, 3, pp. 325-334.

8.   COLDICUTT S. Environmental knowing and action, The Environmentalist, 18, 1998, pp. 251-261.

9.   McGEE T.K. Private responses and individual action, Environment and behaviour, Vol. 31, 1, 1999, pp. 66-84.

10.   BANNISTER G.J. and BLACKMAN A. Community pressure and clean technology in the informal sector: an econometric analysis of the adoption of propane by traditional Mexican brick makers, Journal of Environmental Economics and Management, 35, 1998, pp. 1-21.

11.   WALL G. Barriers to individual environmental action: the influence of attitudes and social experiences, Canadian review of sociology and anthropology, Vol. 32, 4, 1995, pp. 465-493.

12.   CHAWLA L. Life paths into effective environmental action, Journal of Environmental Education, Vol. 31, 1, 1999, pp. 15-27.

13.   KOTLER P. Marketing Management, International Edition, 9th edition, Prentice Hall Publishers, 1997, pp. 123.

14.   FRANKFORT_NACHMIAS C. and NACHMIAS D. Research methods in social sciences, 5th edition, Arnold Publishers, 2002.

15.   GRECH C.F. Society: Is environmentalism foreign to the Maltese society?, University of Malta, 1997.

16.   THE TIMES OF MALTA. Money used for plastic bags, August 6, 2002.

17.   FENECH M. Understanding public participation in source separation of waste. Implications for the implementation of waste management policies with particular focus on Malta and Sweden, MSc thesis, The International Institute for Industrial Environmental Economics, Lund University, Sweden, 2002.

THE WASTE PREVENTION STRATEGY IN HELSINKI METROPOLITAN AREA COUNCIL (YTV) FINLAND

R-L Hahtala
YTV Waste Management
Finland

ABSTRACT. In January 2002, the Board of Directors of the Helsinki Metropolitan Area Council (YTV) accepted the Waste Prevention Strategy. The target of the strategy is to utilise regional and national advice and guidance, so as to motivate the residents, enterprises and the public sector to avoid waste production, so that less waste will be produced per resident and per workplace in 2007 than in 2000. The strategy was organised as a project with the main parts to include (i) waste prevention in companies, concentrating on co-operation networks to be formed for different sectors and on information acquired and distributed with them, on the use of the waste benchmarking system maintained by YTV and on the development of waste management consulting for enterprises by a trainer network, (ii) the waste prevention in public administration covers offices, acquisitions and social and health care. The process starts by ecologising YTV's own operations, and proceeds towards waste reduction models to be prepared and introduced in co-operation with the municipalities of the region, (iii) the information service and awareness education is directed towards households and schools. An awareness campaign will be arranged for households in order to spread information on the reduction of waste and to develop more positive attitudes towards this idea. Education material and methods will be produced for preschool, primary school, senior high schools and vocational institutions together with the authorities in order to promote waste prevention (iv) through social influence and international co-operation, YTV takes an active part in the political discussion on waste at municipal, national and EU level by bringing up the theme of waste prevention. The development and measures related to waste avoidance both in Finland and abroad are also monitored closely.

Keywords: Waste reduction, Co-operation networks, Best practice models, Awareness campaign, Enterprises, Public administration

R-L Hahtala, as a member of the advisory team of YTV Waste Management, has participated in the development of the Waste Prevention Strategy, and she was nominated to lead the project introduced in order to implement the Strategy. Hahtala has drawn up different projects in waste advice for eight years.

INTRODUCTION

Helsinki Metropolitan Area Council (YTV) Waste Management

YTV Waste Management is responsible for waste management in the Finnish metropolitan area, i.e. in Helsinki, Espoo, Kauniainen and Vantaa. There are a little less than one million inhabitants in the metropolitan area. The area produces annually approximately 1.1 million tons of waste.

The waste management strategy aims at reducing the amount of waste, at recycling and reusing waste and at a safe final handling of mixed waste on landfill. In the near future, a waste handling plant will be built, after which some mixed waste will still be delivered for reuse, and only small amounts of treated waste will be placed on so called cell-landfill. With its environmental and quality systems, YTV Waste Management has committed itself to continuously improving the level of environmental protection.

The most important tasks of YTV Waste Management include the planning of waste management, the organisation of waste transport, waste handling, the post-handling of landfill areas that are no longer used, the collection of reusable waste, hazardous waste management and waste advice. YTV's tasks have been defined in a separate YTV Act.

YTV Waste Management is mainly responsible for the transportation of mixed waste from the residential properties in the region. Business, office and industrial properties, etc. arrange their own waste transportation based on YTV's waste management regulations. YTV also takes care of the separate collection of biowaste and for the collection of hazardous waste from households. Hazardous waste is received in a collection truck twice yearly and in approximately one hundred fixed reception sites.

Approximately 600,000 tons of the waste produced in the area are received in YTV's waste-handling centre at Ämmässuo in Espoo, which also hosts the only landfill area in the region. Annually, approximately 400,000 tons of landfill waste are produced. The separately collected biowaste, approximately 30,000 tons a year, is handled in a composting plant. Approximately 55 % of the waste produced in the metropolitan area is recycled and reused.

Waste Prevention

Waste prevention is on the highest level of the target hierarchy of EU waste legislation. It should be a primary measure, before the reuse of waste as material or energy or final disposal.

Prof. Dr. Gerhard Vogel from the Vienna University of Economics presented the future changes in the points of emphasis of waste management at the Avoiding Municipal Waste International Conference, between 2nd and 4th of October 2002, Figure 1.

The terminology of waste reduction has not yet established itself in daily speech. Generally speaking, waste reduction refers to everything that reduces the amount of waste falling into a waste container

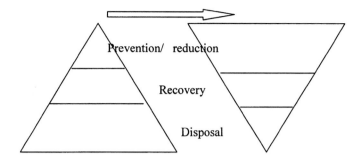

Figure 1 Prof. Dr. Gerhard Vogel, Vienna University of Economics, the future
changes in the points of emphasis of waste management. [1]

Thus, the recycling of waste for reuse is also waste reduction. In the political waste
discussion and in this strategy, the reduction of waste and, more precisely, the prevention of
waste production, have been considerably more restricted. Waste prevention includes all
measures preventing the production of waste in the first place. Waste reduction may be
quantitative or qualitative, such as a reduction in the amount of hazardous substances in
waste. [2]

Waste prevention measures should mainly be taken before the products have been
manufactured or have become waste or before the product is acquired (the consumer's
choices). Waste prevention thus takes place mainly in activities that are generally not
considered as part of waste management. Influences on waste prevention are already to be
found:

- in planning and development
- in manufacture
- in distribution and
- in choice and use of products.

Waste prevention can be observed through the reduction of material flow during the entire
life cycle of a product. In that case, waste prevention covers the measures aiming to reduce
the use of substances and materials as a whole, not just by recycling certain materials. This
can be achieved:

a) by not using a certain proportion of materials and products, i.e. by giving up certain
   material flows, and
b) by extending the life cycle of products and by a long-term reuse of products, i.e. by
   slowing down the amount of production. [2]

Thus, waste prevention often involves issues not directly linked to waste. We can, for
example, discuss sensible or sustainable consumption and material efficiency. Waste
prevention alone is not enough to solve all waste problems. The primary waste prevention
does not eliminate the need for material reuse, environmentally friendly waste handling and
safe final disposal. Therefore, waste prevention should be seen as a fixed and primary part of
waste management as a whole. [2]

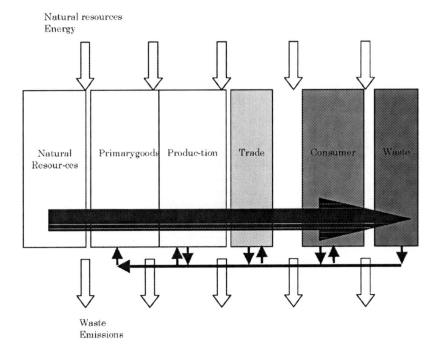

Figure 2 Waste prevention influences the waste and emission amounts in the whole
production chain from the utilisation of natural resources to the
waste management of a used product.

WASTE PREVENTION STRATEGY

Framework of the Strategy

The waste management section of YTV Strategy 2008 consists of three separate strategies: 1.
the Waste Reduction Strategy, aiming to reduce the amount of waste produced and to
increase recycling, 2. the Waste Handling and Final Disposal Strategy, aiming at the handling
of mixed waste in plants, at the introduction of the expansion area of Ämmässuo and at the
controlled closing of the current landfill area, and 3. the Strategy of Customer-Friendly and
Safe Waste Management Services, aiming to provide high-quality waste management
services that suit the customers' needs. The YTV Waste Prevention Strategy is part of the
Waste Reduction Strategy. It describes the target situation to be achieved by 2007, and
presents the central targets preventing waste and measures related to them, which will be the
focus of YTV's activities in the next few years. The strategy will be checked on an annual
basis. In a separate background research of waste prevention, the operating environment and
its changes have been described, as well as current waste amounts and estimates on changes
in them. It also includes orders affecting waste reduction and methods to promote waste
prevention. During the preparation, brainstorming sessions have been arranged among
different interest groups, such as environmental inspectors of various municipalities, the
environmental chiefs of municipalities, the representatives of environmental administration,
civil organisations, consumer groups and representatives of commerce.

The Target Situation of Waste Prevention

In YTV Strategy 2006, the target situation of waste prevention has been described as follows: "In order to prevent waste production, the waste management plant has a large co-operation network with institutions, organisations and enterprises acting in the region. In addition, regional and national advice and guidance have motivated the residents, companies and the public sector to avoid waste production so well that less waste is produced per resident and per workplace than in 2000". [2]   In the target situation of waste prevention for 2020, the point of emphasis of consumption has moved more towards quality, services and immaterial products as a result of changing values and ways of living. When making their buying decisions, the consumers follow closely the environmental statements marked on the products and their effects on the environment. Durable, repairable and reusable products that have conquered the market are given priority. The disposable culture which wastes natural resources is fading. Electronic media has replaced free papers in advertising, since a data network reaches every apartment. Enterprises consider their material flows closely and try to minimise the amount of waste produced. This is encouraged, besides the savings in material costs, by the heavily increased waste charges. Product planning aims at durable and repairable products, and the starting point of production is material efficiency. The use and development of information technology has enabled offices using small amounts of paper. [2]

Waste Prevention in Enterprises

According to the target situation of the Waste Prevention Strategy, in 2007 many sectors of enterprises will pay more attention to the efficiency of their use of material and to waste prevention and compare their achievements with waste benchmarking, at the same time competing for the nomination of the Year's Natural Resource Saver. Products will increasingly be replaced with services, and attention will be paid to their long life cycle. Shops will also offer the consumers alternatives producing less waste and pay attention to the maintenance and repair services for the products they sell.   In order to reach the target situation, waste prevention in enterprises will be promoted with three different sub projects: Enterprise networks and sector-specific waste prevention, Waste benchmarking and YTV's waste-sector trainer network.

Co-operation networks will be formed for different sectors of enterprises, in order to collect and spread "Best Practice" models for waste management and waste prevention. The waste benchmarking system maintained by YTV will be marketed at the same time. Commerce and construction will be the first sectors to be taken under consideration. Along with the progress of the project, other sectors with the most significant amounts of waste will be mapped. The waste management consultation in enterprises will be developed by creating a network of trainers, providing its members with training on material efficiency and on the projects and strategies of YTV Waste Management.

Enterprise networks and sector-specific waste prevention

The target of the project is to create co-operation networks for different sectors and invite experts of the sectors to join, such as representatives from the sector organisations, from the "model enterprises" in the sector, from the environmental inspectors of the municipalities in the metropolitan area, waste advice and the Finnish Environment Institute. Depending on the sector, the network can be expanded to include, for example, entrepreneur associations, the chamber of commerce and consulting enterprises, at the same time utilising existing projects and networks. The networks will also welcome representatives from enterprises and associations in the waste management sector. Co-operation opportunities with existing networks and different organisations, that examine, develop and promote waste prevention, will be sought.

The project will map the situation of waste management in enterprises, and produce and spread information on the opportunities for waste prevention in enterprises, among other things, with Best Practice models and BAT ("Best Available Technology") statements. Co-operation will also be sought with the Finnish Environment Institute and with the environmental authorities in using the environmental licence terms for the targets of waste prevention. The implementation of the environmental licence mapping starts as of the beginning of 2003 with the licences of the food industry.

The target of the project is to spread the Best Practice models created for different sectors through the co-operation networks to all enterprises of the sector in the metropolitan area, and at the same time to complete and update the models with the latest information in the sector. The model will encourage as many enterprises as possible to improve their actions to prevent waste and to join YTV's waste benchmarking system.

The co-operation networks will monitor the development of waste prevention, discuss and comment on the instructions and orders in their own sector and promote research and development in their own sector.

The benefit of the project is that contacts with enterprises will give us access to large potentials for reducing waste amounts. The interests of the enterprises are also closely connected to material efficiency and cost savings, which provides a good ground for co-operation. The project proceeds sector by sector. At the moment we are mapping the commercial sector by the end of 2003. The reports Waste Reduction in Wholesale Businesses [3] and Waste Reduction in Food Industry [4] were completed in 2002. The Best Practice model for the commercial sector was completed in spring 2003.

The studies regarding construction activity and enterprise offices will start in 2004. In 2003-04 it will be examined which other sectors of SMEs have large waste reduction potentials, and these sectors will be considered in 2005-07. Also, YTV will produce general waste management advice material for enterprises, such as waste mapping models.

Waste benchmarking

The target of the project is to develop and market a Waste Benchmarking system so as to ensure the usability of the system and the growth in the number of enterprises joining the system. The system as such will help enterprises to monitor the amounts of waste. The Benchmarking system will be available on the YTV website (http://www.ytv.fi/jateh/benchmark/index.html). It is available and free of charge for all enterprises. In spring 2003, the system covers approximately 200 enterprises. An enterprise that has joined the system can monitor its waste amounts with eight different reports. Below is an example of waste amounts in an enterprise compared to the average in the sector.

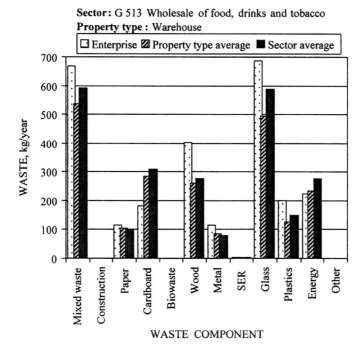

Figure 3   Reference result of waste amounts in an enterprise compared
to the average in the sector

The target is to create a continuously developing benchmarking system that corresponds to the users' needs. Also, a regular communication system with the users will be created, and sector-specific marketing campaigns will be implemented to acquire new users for the system. The system also collects information for the development of waste management and further improves YTV's waste management service. We also have the intention to give an annual award every autumn to the best enterprise in the area of YTV: The Year's Natural Resource Saver. The award has been granted annually since 1999.

The benefit of the project is that the more enterprises are involved the better the Benchmarking system functions. The granting of the annual Natural Resource Saver Award creates a positive public image and increases interest in waste prevention. The system can also be utilised to collect information for Best Practice models and to find BAT solutions for each sector.

Sectors are earmarked for inclusion in the system as the sub projects of waste prevention proceed: retail businesses in 2003, wholesale companies in 2004, public administration offices in 2004, construction in 2004-2005, enterprise offices in 2005, social and health care in 2005, vocational institutes in 2006, senior high schools in 2006, primary schools and day care in 2007.

YTV's trainer network for the waste sector

The waste advice provided by YTV will start allocating its resources more and more to the advice services for households and public administration, so that less time than before will be left for the advice for enterprises. To compensate for this, a consultancy network was established. The experts of the network receive training, among other things, on waste prevention and the strategies and projects of YTV Waste Management.

The target of the project is to develop purchased advice services as a new advice concept for waste advice for enterprises. Experience of similar services has already been collected by advising residents and schools as of 1993. The target is to increase sector-specific advice on material efficiency and waste prevention for enterprises and to activate enterprises to participate in waste benchmarking. The establishment of the network serves the project by increasing the number of waste advice experts. The training of the expert group chosen by YTV and further, the marketing and training feedback system, keeps YTV aware of the quality of the supply of experts.

Waste Prevention in Public Administration

According to the target situation of the Waste Prevention Strategy, in 2007 public administration, including YTV, will have included the Waste Reduction Strategy in its own activity. In order to reach the target situation, three sub projects have been started to promote the issue: Offices, Acquisitions, and Social and Health Care. By concentrating on these, the largest material flows in public administration are covered. The formation of the Best practice model base for offices and acquisitions of public administration will start by ecologising YTV's own operations. The waste management employees in charge of the environmental issues map the waste production of their own operations and the reduction of the amount of waste, and the same actions will be spread to all departments of YTV.

In the social and health care sector, YTV will start by forming a co-operation network consisting of the social and health care sectors of the member municipalities and of the Hospital District of Helsinki and Uusimaa (HUS). Common models for reducing waste will be produced with the network and spread to the different operations of special care, municipal health services, day care and geriatric care and social welfare of the disabled.

Offices and acquisitions of public administration

The purpose of the project is to create a co-operation network with the municipalities of the YTV region. The network will be used to prepare the Best practice model for public administration offices and to ensure the consideration of waste prevention in acquisitions. The target of the project is that less waste will be produced in the public administration of the metropolitan area in 2007 than in 2000. YTV and the municipalities in it are to monitor their waste amounts and they should have considered waste prevention in their acquisitions and office activities.

YTV's target is also to act as an example of waste avoidance in its own operations and to pay attention to waste prevention in acquisitions, operations and in the events it organises. The aim is to reduce the amount of copy and printing paper used by YTV by 20 % by the end of 2007 (compared to the 2001 level) and to train all personnel in waste avoidance. As a result of the waste prevention project that has already started at YTV Waste Management, a Best practice model base for offices will be created and processed further with the municipalities of the YTV region.

The target of the acquisition project is that YTV and its member towns consider waste prevention in their own acquisitions. YTV takes the initiative to introduce existing acquisition instructions and to improve their applicability. Furthermore, the need to collect criteria for different product groups to reduce the amount of waste due to acquisition is examined.

YTV is also part of co-operation networks discussing the consideration of environmental issues in acquisitions. The project also maintains co-operation with the Finnish Environment Institute, which is in charge of the marketing and training of environmental systems within state administration. Municipal acquisitions represent a large material flow in the area, and they benefit from an economically well-implemented project both through reduced acquisition costs and decreased waste management costs.

Social and health care

The purpose of the project is that YTV will establish a co-operation network consisting of the social and health care sectors in its member municipalities and of the Hospital District of Helsinki and Uusimaa. The co-operation network will discuss the opportunities to reduce waste in different operations and agree on how new solutions developed in different places can be spread to everyone for application. The target is to produce common models for reducing waste for the different operations of special care, municipal health services, day care and geriatric care and social welfare of the disabled. Models developed and tested in pilot targets are efficiently spread to be used in all units with similar operations in the metropolitan area. The total amount of waste produced in the sector in relation to the amount of operations can be reduced from the current level.

Social and health care is one of the greatest single municipal waste producers. The project will be utilised to make the instructions and practices of waste management function well, so that the amount of waste is reduced and as large a proportion of waste can be directed for reuse without affecting security. Thanks to these measures, the growth in waste management costs can be controlled.

Advice and Awareness Education

According to the target situation of the Waste Prevention Strategy, waste prevention and the intensifying of material economy will be commonly known concepts in 2007. The citizens' awareness of waste reduction opportunities will have increased, and the idea of sustainable consumption will be well known. In educational institutes, at schools and nurseries, waste prevention is emphasised both in teaching and in their own activities. In order to reach the target situation, three sub projects have been started to promote the issue: an Awareness Campaign for Households, General and Preschool Education, and Vocational Institutes.

The Awareness Campaign for Households aims at reaching the main target group: people in their forties and families with children living in apartment blocks. The awareness campaign creates a uniform external expression for the project and also markets waste prevention to possible co-operation partners of different sub projects. Co-operation partners are sought for the campaign both as investors and message spreaders. The campaign will be implemented on a national level. In the project of general and preschool education, learning material and methods are developed for comprehensive schools, senior high schools and preschools on waste prevention together with the municipal authorities in charge of education activities.

Similar co-operation is offered to vocational institutes in order to find teaching methods for different sectors, to spread fresh information related to the issue and to ecologise the institutes' own waste management.

Awareness campaign for households

The purpose of the campaign is to raise the interest of households and possible future co-operation partners in the subject and to provide tools for operations producing less waste. The target is to influence the households in the region so that a positive attitude is developed towards sustainable consumption and waste prevention, and the awareness of action opportunities and the willingness to operate with less waste increases. The main target group consists of city residents living in apartment blocks. The awareness education pays special attention to the parts of waste that households can particularly influence in order to reduce the amount of waste. These include packaging waste, the reduction of direct advertising and free papers, the reduction in the use of disposable products and the extension of the life cycle of old furniture and household appliances.

The general umbrella campaign aims to wake up many groups that can, with their own actions and in co-operation with YTV, promote the efficient spreading of the message even for other sub campaigns of the strategy. Such groups include municipal decision-makers and civil servants, such as consumer advisors and persons in charge of environmental issues, the employees of civil organisations, parishes and enterprise and employee organisations and those active in these. Journalists are also welcome to join, because all public discussion promotes the campaign.   The first part of the awareness campaign is planned for autumn 2003, weeks 39 and 40, and the campaign can then be seen, among other things, as a national TV campaign. The advertising of the campaign also aims at producing as little waste as possible. Newspaper advertising, for example, will not be used at all. The advertising will take place in mobile media, such as on the sides of buses and trams, in the subway advertisement places and on the radio. The main slogan of the campaign is "The smart produce less waste", Figure 4, and the message advertises a website with more information on the issue (www.fiksu.net). Other Finnish waste companies will simultaneously participate in the outdoor advertising in their own areas. The campaign will be repeated in spring 2004, and annually until 2007 with a different theme.

Figure 4 Proposal for the main poster for the awareness campaign for households, with the campaign slogan "The Smart Produce Less Waste".

General and preschool education

The purpose of the campaign is to spread information on waste prevention to comprehensive schools, senior high schools and preschools through different co-operation forms and networking. The target is that every schoolchild gets information on waste prevention once in preschool, in primary school and in senior high school. When the project ends in 2007, the co-operation network created will still be active and mediate and update information for the target group. The project will examine the waste flow information of the school and nursery properties and collect them in a waste benchmarking system. YTV will participate as an expert in the planning of different groups in projects and activities regarding material efficiency, such as the planning of teaching and the environmental systems at schools.

Children and young people are an important group as future and even current consumers and as future decision-makers. Information learned as a child and repeating itself as part of normal school work is a good basic support for applying the issue later, both in one's own household and as part of working life. Besides, teachers acting as reminders constitute an excellent network of co-operation partners.

The project started at the end of 2002 with the first phase where junior and senior high school teachers in the region, and other officials, were invited to a co-operation meeting to commit themselves to the project. The first 'idea seminar' was organised in spring 2003 to discuss the issue further and to decide on the proceeding lines. At the next stage, learning methods on waste prevention suitable for senior high schools will be developed.

Vocational institutes

The purpose of the project is to build a network of contact persons to forward regular updated information on waste prevention and material efficiency in different vocational sectors. Also, different kinds of teaching material and methods are developed for the teachers' use in co-operation. The monitoring of the waste amounts of educational institute properties is developed by utilising YTV's waste benchmarking system. At the same time, support is offered to the educational institutes to improve their waste management and to develop, implement and report on possible environmental programmes. The central target of the project is to familiarize educational institutes with sustainable consumption. The region has functioning co-operation networks, among other things, between secondary vocational institutes preparing for an occupation and different occupational sectors. The personnel and students of different institutes are aware of the opportunities of their own occupational sector to promote material efficiency and to prevent waste. They are annually provided with the latest information on the progress of material efficiency in their sector. The educational institutes also monitor their waste amounts and reduce waste according to the created models and instructions.

Youngsters studying for an occupation are an important group at future workplaces and as decision-makers and current consumers. Information included in studies preparing someone for an occupation spreads easily to working life together with the newly graduated youngsters. The project has proceeded so that in 2002 YTV's new advice and learning material, such as The Life Cycle Game, the Trouble Path and the Mobile Phone Quiz were marketed and delivered to educational institutes. YTV's exhibition Tietoterassi (Information Terrace) circulated in vocational institutes, Figure 5. "Material Efficiency – A Trainer's File" was also drawn up and handed over to the teachers in charge of environmental education in a few vocational institutes.

Figure 5   YTV's Information Terrace in square.

At the beginning of 2003, a co-operation meeting was arranged with principals of vocational institutes of technology and other civil servants in the sector, aiming to promote the addition of waste prevention to the study programmes of vocational institutes. Vocational institutes of other sectors will be included in the project as planned during the period between 2004 and 2007.

Civil organisation and association networks

The purpose is to form functioning networks among civil organisations and associations that are target groups of different sub projects. Along with networking, different small groups will be created (such as parents' activity), which also promote waste avoidance in their activities. The aim of the networks is to start different information campaigns related to waste prevention, to be implemented with the network members' own methods and to their own target groups. Along with the networks, the group promoting the issue will grow, and the message will be spread more thoroughly.

Social Influence And International Co-Operation

YTV takes an active part in the political discussion on waste at municipal, national and EU level by bringing up the theme of waste prevention.

YTV Waste Management and the environmental administration of municipalities will make suggestions e.g. to the Ministry of the Environment on issues related to waste prevention. These include, among other things, instructions for the implementation awareness and bookkeeping duty in accordance with the Waste Act and the principle of avoiding waste in public acquisitions.

YTV Waste Management follows closely the measures and developments related to waste avoidance both in Finland and abroad. The Waste Management Plant participates in possible Finnish and international co-operation and information exchange projects and familiarises itself with waste prevention measures in other cities and countries. The target is to maintain discussion on waste prevention and thus to influence the political attitude in order to reach the targets and to make use of others' experiences and study results so as to promote the issue. Another aim is to remain as one of the top actors in waste prevention both internationally and in Finland. A large communication plan was drawn up for the project, and communication action plans for the various sub projects. By implementing the action plans, it is possible to reach different media to a large extent, and thus also selected target groups.

Current projects

There has been International co-operation, among others, with the Eurocities organisation, as ISWA co-operation, for example, in the Vienna Conference in 2002, and in the development of waste management in Tallin, which is a co-operation project between the city of Tallin and the city of Helsinki. YTV is also involved in the newly launched BREF project through the Finnish Environment Institute. The project prepares BAT models for the EU. National co-operation has taken place with the Finnish waste advice co-operation body of the Solid Waste Association and in the advice network of ecoefficiency. YTV also co-operates with the Environmental School of Finland on the Life Environment 2001 – 2004 project, which aims at building an environmental estimation system for educational institutes. Furthermore, YTV also acts, among other things, in the Round Table of Environmental Educators, which is a national co-operation body of environmental educators and in the local agenda of Helsinki.

Other projects related to waste prevention include, among others, the Ecoevent Competition aiming at ecologising public events. In the competition, the public events of the area compete on which event produces the least waste per participant. Other criteria, such as successful sorting and the use of organic food, are also assessed. The award has been granted annually since 2002.

Separate studies

During the project, YTV has participated in other studies promoting waste prevention. One of the studies has examined the influence of a waste truck which weighs the collected waste on the amount of waste produced. The target is to find out what is the effect of the use of a weighing waste truck on the amounts of waste produced. Experiences on the existing collection systems and on their significance for waste prevention are gathered. During the project, YTV also participated in different life cycle studies related to waste prevention, and some such studies may possibly be made. In autumn 2002, a life cycle study on reusable nappies was completed: Vaipoilla on väliä (Nappies do matter) [5]

The influence of the Strategy

The assessment models for the influence of the strategy include pre and post models based on measurements before and after the measures, assessment of experts and participants on the effects of the measures. The influence is also examined with time sequence analyses. The assessments are included in the plans of the various sub projects. The effect of the Waste Prevention Strategy for households is mainly an increase in awareness and editing of attitudes. The effects and influence of the strategy are measured, in addition to the above, with an annual poll. The poll maps, among other things, the changes in residents' attitudes, the use of rental and repair services and the awareness of the reduction of waste amounts and of consumer rights. The last sample study was completed in January 2003. The questionnaire had a total of 28 questions, some of them with alternative answers and some that could be answered freely. The questionnaire was sent by mail to 3000 people, and a total of 1239 over 18-year-old people in the YTV region answered it. As questions related to waste prevention, for example the following questions were used:

Table 1   Question 8 of the YTV's Waste 2003 study. [6]

To which of the following characteristics do you pay attention when buying consumer durables? Please mark two (2) most important characteristics for each product.

	LONG LIFE CYCLE	REPAIRABILITY	TIMELESSNESS	PRICE	FASHION
Clothes	☐1	☐2	☐3	☐4	☐5
Household appliances	☐1	☐2	☐3	☐4	☐5
Furniture	☐1	☐2	☐3	☐4	☐5

According to the results of the year 2003 (Figure 6), a long life cycle is the most important criterion for choosing household appliances and furniture. The most important criterion for the purchase of clothes was the price. In the selection of furniture and household appliances, fashion was the least important criterion for the buying decision, whereas repairability was the least important criterion for clothes.

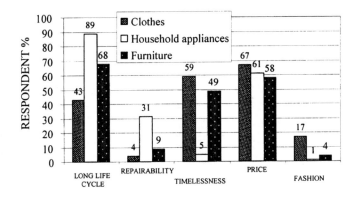

Figure 6   YTV's Waste 2003 study on the factors affecting the buying decision for consumer durables [6]

In order to monitor the waste amounts produced in the region, an annual large-scale waste flow study is needed that examines the waste flows in the region of YTV; the amount, quality and place of production, and a continuous monitoring system is created in order to realise any changes in waste amounts. The study is now underway.

Project Management And Resources

The Board of Directors approved the Waste Prevention Strategy in January 2002, and after this the project for implementing the Strategy was set up. The project is led by Project Manager R-L. Hahtala from the Communication Unit of Waste Management, who was nominated for the task. The project will be completed by the end of 2007.

The project is led by a management group including YTV's Waste Management Director J. Kaila as the chairperson, Development Manager J. Paavilainen from Strategic Projects, the chief of the Communication Unit, Quality Manager T. Tilli, and the leader of the waste advice team of the Communication Unit, T-R. Blauberg. The project group includes nine employees of the Communication Unit of Waste Management, five of them in addition to Blauberg (planner S. Kemppainen and waste advisors S. Huuhtanen, S. Kajaste, O. Linsiö, S. Stén) acting as leaders of the sub projects. Further, the implementation of the separate studies will be participated by employees from the various sectors of Waste Management.

The implementation of the Strategy has been assessed to require approximately 28 man years at YTV in the period between 2002 and 2007. YTV's costs due to the implementation of the projects, personnel costs excluded, amount to approximately 1.5 million euros. For the finance of the project, funds will be sought from different study funds, such as the Life finance of the EU and from national finance programs and from other project-specific co-operation partners.

REFERENCES

1.   PROF. DR. GERHARD VOGEL: Avoiding Municipal Waste International Conference, 2-4 October 2002, pp. 5-6

2.   PÄÄKAUPUNKISEUDUN JULKAISUSARJA C 2002:4, YTV:n jätteen synnyn ehkäisystrategia 2007 ja taustaselvitys (THE PUBLICATION SERIES OF THE METROPOLITAN AREA C 2002:4, YTV's Waste Prevention Strategy 2007 and background study), pp. 9-10

3.   HÄMÄLÄINEN HANNA, Jätteen vähentäminen tukkuliikkeissä, Pääkaupunkiseudun julkaisusarja C 2002:3 (Waste Reduction in Wholesale Businesses, the Publication Series of the Metropolitan Area C 2002:3)

4.   LEHTORANTA TIINA, Jätteen vähentäminen elintarviketeollisuudessa, Pääkaupunkiseudun julkaisusarja C 2002:12 (Waste Reduction in the Food Industry, the Publication Series of the Metropolitan Area C 2002:12)

5.   PÄÄKAUPUNKISEUDUN JULKAISUSARJA C 2002:14, Vaipolla on väliä
       (PUBLICATION SERIES OF THE METROPOLITAN AREA C 2002:14,
       Nappies Do Matter)

6.   TALOUSTUTKIMUS OY, YTV:n Jäte 2003, 2/2003 (TALOUSTUTKIMUS OY,
       YTV's Waste 2003, 2/2003)

FULL-SCALE DEMONSTRATION OF SUSTAINABLE AND COST EFFECTIVE ON-SITE LEACHATE MANAGEMENT BY WILLOW SHORT ROTATION COPPICE

G R Alker   AR Godley

R Marshall	J E Hallett	D Riddell-Black
RMC Aggregates	Water Research Centre	Independent Environmental
(Greater London) Ltd	WRc	Consultant
	United Kingdom	

ABSTRACT. A five year project has been carried out to demonstrate the management of landfill leachate by the application of willow short rotation coppice (SRC) plantations at two landfill sites in Hertfordshire. Following a one year establishment period, landfill leachate was applied to the plantations over three growing seasons (May to September). The yield and uptake of leachate components by the willows were monitored together with the impacts of leachate application on soil quality and aqueous emissions. The results were compared with control plots that were either irrigated with fresh water or left un-irrigated. Results indicated that there was significant attenuation of leachate components by willow uptake and soil microbial processes. There were no adverse changes in soil quality. Soil salinity increased slightly during the summer leachate application period but returned to initial levels after soluble ions were leached out during the winter. The concentration of leachate components in drainage water was consistently lower than the irrigated leachate due to dilution with rainwater in the soil. Methods of establishment, and operation, key results from the trials and an assessment of the costs are presented together with a discussion on the potential role willow SRC might have in leachate management for the waste management industry.

Keywords: Landfill leachate, Willow, Short rotation coppice.

G R Alker, is a Consultant Scientist at WRc, specialising in phytoremediation of wastewaters, waste and contaminated land.

A R Godley, is an Environmental Scientist at WRc, specialising in environmental microbiology.

R Marshall, is the mineral planning co-ordinator at RMC Aggregates (Greater London) Ltd.

D Riddell-Black, is an Independent Environmental Consultant. Her key areas of expertise include land application of organic wastes, bioenergy systems and plant uptake of heavy metals.

J E Hallett, is the agriculture and forestry field trials specialist at WRc.

INTRODUCTION

Landfill operators require cost effective methods for the management of landfill leachate. One option is to irrigate landfill leachate onto a fast growing non-food crop such as willow or poplar in short rotation coppice (SRC) systems. This is an intensive silvicultural technique where the trees are planted at high densities (up to 25,000 trees ha^{-1}) and repeatedly harvested on a much shorter rotation (3 to 5 years) than conventional forestry, without the requirement to replant after each harvest. Rapid biomass production uses large amounts of water and nutrients and these can be supplied by wastewaters, including landfill leachate. The SRC technique was originally developed because it offers a source of renewable energy (as a bioenergy crop) and in the UK there is currently a strong impetus to develop bioenergy crops due to the governments commitment to meet its renewable energy obligations.

TRIAL DESIGN AND OPERATION

Willow SRC plantations were planted at the Hatfield and Westmill landfill sites in 1998. These were maintained without leachate application in the first growing season and were then cut back (coppiced) to encourage more rapid growth in the spring of 1999. Leachate was applied to the plantations in the summer of 1999, 2000 and 2001 with different leachate application regimes practised at each site. The volume of leachate applied was increased in each subsequent irrigation season and the responses of the willows to the leachate were compared with un-irrigated and water irrigated control plots. Six varieties of willow, bred for bioenergy production were used to increase resistance to disease and pests and maximise the production of woody biomass.

The aqueous emissions from the plantations were monitored during the intervening winters. This monitoring included sampling and analysing soil pore water (at both sites), and at Hatfield the plantation field drains and surface water ditch into which the drains discharged.

Monitoring of soil quality and of the composition of the willow leaves and stems were carried out throughout the trials. In February 2002 the plantations were harvested and the yield and composition of the wood product determined. For a full description of the trials and detailed technical results see [1].

RESULTS

The trials successfully demonstrated the technical feasibility of managing landfill leachates by application to SRC in tandem with the production of a crop, which may be used to generate bioenergy.

The ammoniacal-N in the leachate was almost completely removed by a combination of microbial nitrification to nitrate in the soil and by the uptake of N (nitrate and ammonia) by the willows as nutrients. Some of the nitrate may also have been removed by microbial denitrification in the soil. During the leachate application periods there were no aqueous emissions from the plantations, indicating that all the water applied was used by the willows. This resulted in the transient accumulation of salts in the soils during the summer leachate application periods, although this was not detrimental to soil quality or plant growth.

The accumulated salts were flushed from the soil and diluted by winter rainfall and presented little environmental risk to receiving surface waters.

Metal concentrations in the leachates were low and loadings were insufficient to present a risk to the soil. Metal offtake in the harvested willow actually exceeded the inputs of metals into the soil from the leachate application, without tissue metal concentrations being elevated compared to the controls.

The nitrate emissions in winter leached water were low and presented little environmental risk to receiving waters. Leachate N removal by the willows was significant and dependent on the growth rate of the willows and the N loading rate. Nitrate emissions can be controlled and minimised by matching the leachate N application to willow N growth requirements.

The mass of wood harvested in 2002 was found to be greater in the leachate treated areas compared with the water irrigated and un-irrigated controls (Table 1). This is evidence that leachate provided nutrients and water to the willows. The yields from the trial plots irrigated with water were also higher than un-irrigated controls at Hatfield indicating that growth in the un-irrigated control plots was water limited. The composition of the harvested wood was unaffected by leachate application and typical for willow SRC coppice. This means that the quality of the wood as a bioenergy crop was not adversely affected.

Table 1   Machine harvested willow growth yields at Hatfield and Westmill (t dry wt ha^{-1} y^{-1})

		LEACHATE	WATER	UN-IRRIGATED
Hatfield	12000 trees ha^{-1}	6.65	5.36	3.38
Hatfield	24000 trees ha^{-1}	6.39	5.11	3.71
Westmill	15000 trees ha^{-1}	3.17	2.14	2.34

Loading Rate Limits

Leachate application rates may be limited by either the tolerance of willows to certain leachate components or by environmental emissions. Since landfill leachates are a complex mixture of components, the limiting component is likely to vary from leachate to leachate. Results from this study, suggest that either salinity (total dissolved salt content), ammoniacal-N, or components such as chloride and sodium which are not required in large quantities for plant growth are most likely to limit loading rates. At both Hatfield and Westmill the willows showed no symptoms of toxicity from the applied leachate.

The loading rates used were also not high enough to result in detrimentally elevated aquatic emissions, soil or plant tissue concentrations for most leachate components. Hence the absolute operating window for loading rates was not reached for most components in this study. The maximum loading rate for chloride was established, because in the final year of the study at both sites, the chloride concentration of the soil solution did not return to initial levels after the winter period.

The maximum leachate N loading that can be applied depends in part on the uptake of N by the willows. At Hatfield the N balance suggested that about 41% of the applied N was removed as offtake in the harvested wood and that there was little accumulation of excess nitrate in the soil. Only about 21% of the applied N was leached from the soil as nitrate. The balance of the N can probably be accounted for by that still present in soil, present in the willow roots, biologically removed by microbial denitrification, and volatilised as ammonia during irrigation.

Willow growth was poorer at Westmill and the nitrate accumulation in the soil greater. About 20% of the applied N was estimated as being taken up by the harvested wood and a greater percentage of the applied N was leached from the soil. The N loading rate is therefore dependent on the growth rate and N uptake of the trees and should be estimated on this basis.

Operational Recommendations

Differences between Hatfield and Westmill site performance identified that the establishment phase of the SRC is critical to develop the maximal potential leachate treatment in the shortest time. The key to this is good plantation establishment during the initial stages, particularly ground preparation and weed control [2].

Identification of the leachate component or components that restrict the leachate application rate needs to be on a case by case basis. This is so that the required plantation size for the leachate quantity and quality can be calculated during the design stage. It is then paramount that the hydraulic loading rate and chemical composition of the leachate is monitored and controlled within the specified design. Maintaining a regular supply of leachate and having a suitable leachate, monitoring, buffering or pre-treatment system to protect the plantation from atypical leachate is also recommended.

COST ASSESSMENT

A cost assessment (Table 2) indicated that willow SRC has low capital and operating costs, and given the same plant life as conventional treatment systems (20 – 25 years), may be considered as a favourable option in some cases. Willow SRC also has the potential benefit of providing an income from the land. Currently the market for willow wood chips as a bioenergy crop is limited. If it is assumed that such a demand would develop, as a result of UK government encouraging activities to meet its renewable energy obligations, then the potential income from willow SRC (£74 ha^{-1}) might be favourable compared with current agricultural incomes (£30 - £50 ha^{-1}).

APPLICABILITY OF SRC LEACHATE MANAGEMENT

Leachate management by willow SRC might be applied on existing operating sites and on old closed sites where leachate management is still required. These sites are regulated by a variety of legislative measures, including the Control of Pollution Act [3], Waste Management Licensing Regulations [4] and Pollution Prevention and Control Act [5]. It is anticipated that the application of willow SRC could be carried out for any site under a site licence or PPC permit, provided that the process is well designed and operated and the impact of the leachate management system on the local environment is well understood.

Table 2   Indicative cost of leachate management using SRC and alternative options.

TREATMENT SYSTEM	m³/day	Capex (£)	Opex (£/m³)
Tanker	Unlimited	-	10 - 20**
(Hatfield estimate 2002)	15		13
Methane stripping and sewer connection	60	100,000+	4
(Hatfield estimate 2002)	15	116,000	Approx. 1
Sequencing batch reactor & discharge to water course	250	450,000	1
Willow SRC (Hatfield)	40*	30,000	0.8

* Irrigation during summer
** Includes cost of treatment at sewage treatment works

The project recommended suitable loading rates for a number of leachate chemical components. The Hatfield and Westmill leachates used were of relatively low strength, so the hydraulic loading rate could be relatively high, without exceeding these loading recommendations. In order to not exceed chemical loading rates for extremely concentrated leachates, the hydraulic loading rate may need to be too low for the method to be economically or logistically feasible. In this case, land availability or establishment costs may result in the system being less attractive than alternative options. Therefore leachate management is probably best suited to lower strength leachates.

By only applying leachate during the growing season, full advantage can be taken of the plant uptake capacity of the system. This can be achieved by providing winter storage in lagoons or within the landfill or using an alternative method of treatment during winter.

CONCLUSIONS

The main conclusions from the trials are summarised as follows:-

- Willow growth was stimulated by the water and fertiliser value of the leachate, indicating beneficial reuse of leachate components.

- About 40% of the N applied in leachate was recovered in harvested willow stems and there was an overall 80% reduction in leachate N. Nitrogen in leaf litter and roots may subsequently be released from the system, potentially reducing the efficacy in the long term.

- Leachate treatment by willow SRC resulted in almost complete ammonia removal from leachate by a combination of plant uptake and nitrification.

- The heavy metals in the leachate tested were low and did not accumulate in soils to which leachate is applied.

- Water draining from the willow SRC, even at its peak concentration after the summer irrigation, was diluted compared with the applied leachate.

- The cost of leachate treatment by willow SRC is competitive with other leachate treatment management systems.

- Acceptance of the technology by regulators is likely to be on a case by case basis where the environmental impact can be assessed locally.

ACKNOWLEDGEMENTS

The project was funded by landfill tax credits from the RMC Environment Fund and by RMC Aggregates (Greater London) Ltd.

REFERENCES

1.   ALKER, G. R, GODLEY, A. R. AND HALLETT J. E., Landfill leachate management using short rotation coppice: Final Technical Report. WRc Report No. CO 5126, 2002, p199.

2.   HALLETT J. E., GODLEY, A. R. AND ALKER, G. R., Landfill leachate management using short rotation coppice: Operational Guide. WRc Report No. CO 5127, 2002, p72.

3.   HMSO (1974). Control of Pollution Act 1974. HMSO London.

4.   HMSO (1994). The Waste Management Licensing Regulations 1994. HMSO London.

5.   HMSO (2000) Pollution Prevention and Control Act 1999. HMSO London.

MECHANICAL BIOLOGICAL PRETREATMENT OF RESIDUAL WASTE IN AUSTRIA

E Binner

BOKU University

Austria

ABSTRACT: Decomposition processes in landfills usually take place under anaerobic conditions and are therefore fairly time-consuming (up to several decades). High amounts of gas (CH_4 and CO_2) and leachate containing high concentrations of pollutants are set free. By the way of a mechanical biological pretreatment (MBP), however, decomposition is accelerated and takes place within a few months under aerobic and controlled conditions. Thus, the landfilled waste no longer goes through acid fermentation and, depending on the duration and the quality of the pretreatment, methane formation is largely or almost entirely prevented. Gaseous emissions as well as organic pollutants (BOD, COD) in leachate will be reduced by more than 90 % compared to untreated residual waste.

Keywords: Mechanical biological pretreatment, Landfill behaviour, Reactivity of wastes, Respiration activity, Incubation test, Gas Generation sum, Methane generation,

Erwin Binner, Department of Waste Management at the University of Natural Resources and Applied Life Sciences (BOKU University), Vienna, Austria. His major area of expertise is the reactivity of mechanically biologically pretreated waste.

INTRODUCTION

When municipal solid waste is landfilled directly, anaerobic biological breakdown processes produce landfill gas and leachate. Over 90 % of the converted organic carbon is released as CO_2 and CH_4 as landfill gas; the remainder is released in the leachate. Globally, methane from landfills ranks third, after rice and meat production (from ruminates), of all anthropogenic methane emissions. Methane also contributes up to 35 times more to the greenhouse effect than CO_2.

Within the scope of the Austrian Waste Management Law [1], with the intent of minimizing environmental impacts from solid wastes, the Austrian Landfill Ordinance (DVO) [2], came into effect on January 1, 1997. The stipulations set forth in that ordinance apply to new landfills, as well as those already in existence. Transitional stipulations apply to old landfills, which, by 2004, must gradually become compliant [3].

One of the objectives of the Landfill Ordinance is to significantly reduce the content of organic substances (organic carbon) prior to landfilling. For "Massenabfalldeponien" (municipal wastes, similar bulky wastes, similar commercial wastes) the ordinance stipulates that carbon content, measured in Total Organic Carbon (TOC) is limited to 5 % of dry matter (DM). Compliance with the TOC limit is also achieved with an ignition loss of ≤ 8 % DM. With residential waste, it is only possible to comply with these limits using incineration.

The laws do, however, foresee that waste that has undergone mechanical biological pretreatment can be landfilled, if the calorific value (H_o) does not exceed 6,000 kJ/kg DM and compliance with the remaining limits is achieved (Tables 7 and 8, Addendum 1 of the Landfill Ordinance), even if the TOC limit is exceeded (Variance §5 Par. 7 lit. f). The limitation of calorific value is intended to prevent the landfilling of thermally valuable components, as well as to improve the emissions behavior of the landfill.

Because calorific value does not allow assessment of the reactivity of wastes, our department (Department of Waste Management, BOKU-University Vienna) was commissioned by the Austrian Ministry of Agriculture, Forestry, Environment and Water Management to conduct extensive research on the mechanical biological pretreatment (MBP) of solid wastes to determine parameters to characterize reactivity [4]. The results of these studies, particularly those concerned with air emissions from MBP landfills, were incorporated in the "Guidelines for the Mechanical Biological Pretreatment of Solid Wastes" [5]. Limit values of Respiration Activity (= "Atmungsaktivität") $AT_4 \leq 7$ mgO$_2$/g DM, Gas Generation Sum $GS_{21} \leq 20$ Nl/kg DM and alternatively Gas Evolution (= "Gasbildung") $GB_{21} \leq 20$ Nl/kg DM) were established. Using these as yet non-mandatory guidelines, more research will be done in the coming years, which will then be incorporated into a mechanical biological pretreatment ordinance.

WASTE REACTIVITY ASSESSMENT

Within the course of discussions of suitable parameters to describe waste reactivity, we developed a test cell to estimate gas production potential [6]. 1.7 litre waste samples are placed in glass cylinders and incubated under optimized anaerobic conditions (Figure 1). The gas produced is measured and analyzed. The evaluation then lets us estimate the anticipated gas emissions from a landfill containing the specific type of waste.

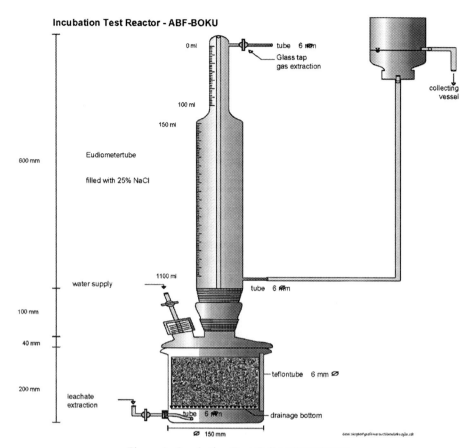

Figure 1   Incubation Test Cell (ABF-BOKU)

In Figure 2 results of diverse mechanical biologically pretreated wastes are shown. Tests were done with untreated wastes and MBP wastes after differing treatment periods (3 to 50 weeks) with differing treatment techniques (open windrow decomposition, using Wendelin technology, and in decomposition reactors). To estimate gas generation potential, 53 wastes with differing reactivity levels were tested for a period of 240 days. There is an excellent correlation between gas generation potential and gas generation sum after 90 days of testing ($r = 0.999$). This made it possible to reduce the test period for further experimentation to 90 days.

Because the limit value for the reactivity of MBP waste has been set at $GS_{21} = 20$ Nl/kg DM, further correlations are of interest for assessment purposes, particularly those correlations in the areas of low reactivity. The following figures show correlations for substances with a $GS_{90} < 50$ Nl/kg DM with the parameters AT_4 ($r = 0.873$) and GS_{21} ($r = 0.963$). When present, lag phases were compensated for in the calculations. Both parameters provide excellent opportunities for estimating gas generation potential. This means that a substance can be assessed in four to 21 days.

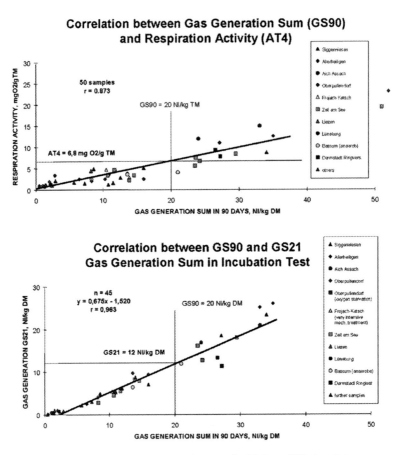

Figure 2   Correlation between gas generation sum in 90 days (GS_{90}) and the parameters respiratory activity in 4 days (AT_4) and gas generation in 21 days (GS_{21})

Because lag phases exceeding four days can occur during respiration activity determinations, aerobic and anaerobic testing must always be conducted in parallel, and compliance with both parameters must be assured.

The correlation coefficient between GS_{90} and dissolved organic carbons (DOC) in DEV-S4 eluate for the $GS_{90} < 50$ Nl/kg DM range was only $r = 0.638$ (23 samples), and for the entire reactivity range, was $r = 0.707$ (38 samples). On the one hand, there were repeated samples with high reactivity and very low DOC values, and on the other, there were substances with low reactivity with very high DOC concentrations.

EXPERIENCES IN AUSTRIA

Mechanical biological pretreatment (MBP) landfills

Even though the Austrian Landfill Ordinance has been in place since January 1, 1997, the technical specifications for MBP landfills were first defined in Mechanical Biological Waste Pretreatment Guidelines of 2002. This means that there are no landfills yet which fully comply with all stipulations of the ordinance. Many MBP plants have been halted during their planning stages due to the vagueness surrounding the law and technical specifications (eg, St. Pölten, Stockerau, Halbenrain), but now, rapid progress is being made.

That I am nevertheless able to discuss experiences in practice is due to Austria's special situation. The search for alternative waste disposal possibilities meant that as early as the 1970s, a good number of residential waste composting plants were established. Originally intended for composting, it was rapidly shown that the generated compost could not be used in agriculture due to the already existing high degree of contamination. In contrast to other countries, Austria continued to operate its 18 existing combined waste composting plants despite the problems with heavy metals. In the late 80s, up to 400,000 tons of communal wastes were processed into compost for recultivation measures and/or as a pretreatment option prior to landfilling.

The Austrian Ordinance for the Separate Collection of Bio Waste [7] went into effect on January 1, 1995. Out of the need to create plant capacities for composting the bio waste, five of the former household waste composting plants were converted to bio waste composting. Three plants were closed between 1993 and 1996. Two new plants were built and began initial operation at the end of 1996 and 1998, respectively.

Today, there are twelve plants in Austria in which around 275,000 t of residual waste and sludge from sewage treatment plants are mechanically and biologically treated. By intensifying the collection of recyclables and problematic materials, sorting lines preceding deposition, investments in preparation and conditioning, as well as converting to intensive decomposition processes, the plants are continuously adapting to the new technical specifications. Most of these plants, however, are not yet in compliance with the MBP Guidelines (calorific value < 6,000 kJ/kg DM, exhaust air treatment).

Because of Austria's low incineration capacity (there are only three household waste incineration plants), the fraction of residual waste which has a high calorific value is only incinerated at two sites, *Kirchdorf* and *Kufstein*. Sieve residues are transported to the waste incineration plant in *Wels* .

Impact of the mechanical biological pretreatment on landfill behavior

Database

As mentioned above, due to its high contaminant content, the majority of processed waste compost ended up in landfills. The evaluation of the data collected at these landfills by WURZ [8] made it possible to gain valuable knowledge of the long-term behavior of wastes that had been mechanically and biologically pretreated. The personal experiences made by landfill operators was also included in the interpretation of the data.

Because reliable baseline data are lacking, the conclusions drawn in this paper must be taken with the following reservations:

- Over the years, mechanical biologically pretreated municipal waste, rather than residual waste, was landfilled. But in Austria there is a very intensive separate collection of almost all compounds (biowaste, glass, paper, metals, plastics, ...). That is why residual waste has nearly the same composition as municipal waste. We must further observe that no special treatment was done that was specifically directed at compliance with calorific value (plastics, for instance, remained for the most part, in the landfilled material).
- Waste that had been treated mechanically biologically was not landfilled exclusively at any of the sites. Between 20 % and 80 % of the materials deposited were mechanical biologically pretreated. Almost always, bulky wastes, business and trade waste, residuals from composting, as well as sieve overflow from raw composting and harder materials from mature compost, as well as municipal sewage sludge, and even in individual cases, grass cuttings, were landfilled along with the waste. This means that there were certainly poorer conditions than those found in mono landfills. The following interpretations are thus quite conservative.
- The majority of the landfills were not built to today's technical standards. In some cases, a base liner is missing. Landfill gas extraction or removal systems, when they exist at all, merely consist of flares; older gas volume measurements are therefore not accessible.

In addition to landfill data, the results of decomposition studies, as well as those examining gas generation potential, are incorporated in this presentation.

Impacts on landfill volume

Waste composting plants were originally established to save landfill space by composting household waste, with subsequent sale of the compost. The idea of saving landfill space is also satisfied by biologically and mechanically pretreating residual waste. Even if the end product cannot be sold, the sorting of valuable materials (e.g., combustibles), loss of volume through composting, and improved compactability (measurements at the Siggerwiesen/Salzburg landfill showed that waste treated for only three weeks had a placement density of 1.3 t/m^3) saved 30 to 50% of landfill volume. If sieve residues are incinerated (at this time in Austria, this is practically nonexistent), calculations showed that it is possible to save up to 77% landfill space [9].

Leachate volumes

The water budget of a landfill is strongly influenced by local conditions (climate, precipitation), its geometry (surface, height of fill), and landfill engineering (placement technology, cover, leachate management system). It is therefore only somewhat possible to interpret collected data with regard to the impact of mechanical biological pretreatment on leachate volumes [10]. Even though evaluations show higher leachate volumes with shorter pretreatment periods (Siggerwiesen/Salzburg), the location studied has much higher precipitation rates of > 800 mm/a (in eastern Austria, rates are < 700 mm/a). In addition, leachate drainage was done by lowering the water table within the landfill. The higher leachate volume can thus, in part, be traced to encroaching groundwater.

FEHRER [11] observed extremely low permeabilitys in two different samples of pretreated wastes. The k_f-value of materials from Oberpullendorf (< 25 mm, 50 weeks treatment) was $k_f < 10^{-10}$ m/s, and that of the Allerheiligen landfill (< 12 mm, 5 weeks treatment) was $k_f < 10^{-11}$ m/s. With such low permeability, only very low amount of leachate can be expected.

Leachate composition

A clear indication that the quality of the leachate has been positively improved are the statements of several landfill operators with regard to incrustations in the drainage system. Inspections using video cameras showed only minimal incrustation. Almost all landfill operators who had the capacity to flush the leachate collection system determined that it was not absolutely necessary to flush the system annually. In contrast, conventional household waste landfills should be flushed two to three times per year.

Leachate analyses prove that the acidification phase is eliminated by mechanical biologically pretreated wastes (Table 1). Even newer mechanical biological landfills (< 4 years old) show pH values around pH = 8, while "untreated" MSW landfills indicate considerably lower pH values [8].

Table 1   Comparison of leachate concentrations from MBP landfills and "untreated" MSW landfills [8]

	MBP YOUNG < 4 years	MSW UNTREATED YOUNG < 4 years	MBP OLD > 4 years	MSW UNTREATED OLD > 4 years
MBP portion [%]	> 50	0	> 50	0
pH	7.9 to 8.5	6 to 8.3	7.5 to 8.4	7.2 to 8.5
COD [mg O_2/l]	up to 6,600	up to 60,000	200 to 2,300	up to 18,000
BOD$_5$ [mg O_2/l]	400 to 1,000	up to 30,000	10 to 350	up to 6,300
BOD$_5$ / COD	0.1 to 0.2	0.5 to 0.8	0.05 to 0.3	0.05 to 0.5
NH$_4$ [mg/l]	900 to 2,200	240 to 10.000	100 to 350	170 to 5,500
Zn [mg/l]	0.5 to 2.6	0.5 to14	0.02 to 0.2	0.1 to 0.7

Leachate concentrations are in the lower range of concentrations normally found in reactor-type landfills. Younger MBP landfills (< 4 years old) show maximum COD concentrations of 6,600 mg O_2/liter. A landfill with untreated waste of comparable age can have leachate peaks of up to COD = 60,000 mg O_2/liter. The COD concentrations in older landfills studied (> 4 years) lie between 200 and 2,300 mg O_2/liter, while those with untreated waste still show COD = 18,000 mg O_2/liter for the same period (Figure 3). The same trends can be seen for the BOD$_5$ parameter: 1,000 mg O_2/liter compared with 30,000 mg O_2/liter at younger landfills, and 10 to 350 mg O_2/liter compared with 6,300 mg O_2/liter. Thus landfills using mechanical biological pretreatment (MBP) show 90 - 95 % less organic burden than other landfills of the same age that were filled with untreated waste.

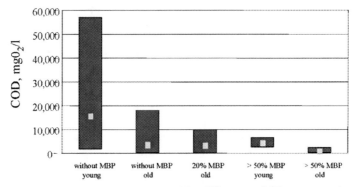

Figure 3   Impact of pretreatment and landfill age on COD concentration
in leachate (max, min, average)

Heavy metal concentrations were also reduced by mechanical biological pretreatment. To provide an example, zinc is shown in Table 1. With regard to ammonia-nitrogen concentrations, MBP landfills also show clear advantages (lower concentrations) as compared with untreated waste landfills [8].

Impacts on the geotechnical properties

To date, little attention has been given to the physical properties of MBP waste. One positive aspect is the improved compactability of the materials to be landfilled, which helps save landfill space. Leachate volumes are reduced due to its very low permeability. One of the reasons for this is the prescribed low particle size of landfilled material required for compliance with calorific value specifications. Despite efficient plastic separation and carefully executed composting processes, the incineration value of 6,000 kJ/kg DM can only be achieved within a particle size range of < 15 mm to < 35 mm. This fine particle size can, on the one hand, lead to difficulties in placement / sealing (particularly if there is precipitation), and on the other, it also reduces the stability of the landfill body [11]. For MBP waste < 80 mm he found a friction angle $\varphi = 40°$, whereas the MBP waste < 25 mm showed only $\varphi = 32°$. Because reinforcement is lacking (plastics, which could provide such reinforcement, must for the most part be removed), it is not possible to transfer tensile strength. If embankment slopes are not reduced accordingly, slides may occur.

Impacts on landfill gas emissions

The duration of pretreatment has been shown to have a significant impact on the generation of landfill gas. After a 3-week pretreatment period, landfill gas volumes of 5 m^3/t wet matter (WM) and year were suctioned off [9]. Following a 14-week decomposition period, values dropped below 1 m^3/t WM and year (with a 50 % MBP portion).

A particularly stunning example is provided by the impact of converting to biowaste composting at the Lustenau/Vorarlberg landfill. Since 1989, only untreated residual waste was landfilled here. This now 6-year old landfill has 7 m^3 gas/t WM and year; in the sections where primarily material that has been pretreated for 14 weeks is placed (even though the oldest material is already 20 years old and no landfilling has taken place for the last 5 years before sampling), 0.9 m^3 gas/t WM and year were extracted. Older measurements are not available, because the gas extraction system was first installed in 1989 [10].

Very little gas is generated at the Allerheiligen Landfill in Styria, where up until the conversion to MBP, residual wastes that had undergone 30 weeks of pretreatment were landfilled. The low volume of 0.25 m^3 gas/t WM and year, with a CH_4-portion of only 3 to 28 %, was problematic for flaring off. The flare (designed for $220 \text{ m}^3/\text{h}$) can barely be operated 1.5 hours per day [10].

These landfill estimates are confirmed by experimentation in the laboratory. Figure 4 shows the impact of different pretreatment times on the gas generation potential of residual waste. Compliance with the MBP Guideline limits (our tests show that a treatment duration of 8 to 12 weeks is necessary to achieve compliance) reduces gas generation by at least 90 to 95 % as compared to conventional landfilling of untreated waste [4].

Gas Generation Potential in Incubation Test - Impact of Mechanical Biological Pretreatment

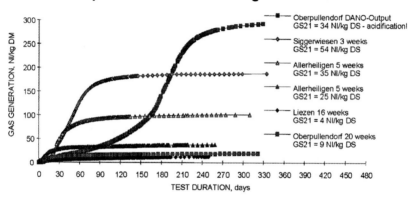

Figure 4   Impact of MBP on the gas generation potential of residual waste

SUMMARY

For more than two decades, Austria has been using mechanical biological pretreatment of wastes (MBP). At this time, twelve plants treat about 275,000 t annually. With the enactment of the Mechanical Biological Waste Pretreatment Guidelines in 2002, additional plants will be built. When waste is treated for 8 to 12 weeks, it is possible to comply with the limits set forth in the Guidelines with regard to reactivity ($AT_4 \leq 7 \text{ mgO}_2/\text{g DM}$ and $GS_{21} \leq 20 \text{ Nl}$ gas/kg DM). The limits established by the Austrian Landfill Ordinance for calorific value ($Ho \leq 6.000 \text{ kJ/kg DM}$) can, however, only be achieved when mechanical treatment is modified accordingly.

Tests at the Department of Waste Management, BOKU-University Vienna, show that the parameters GS_{21} (gas generation within 21 days of testing) and AT_4 (respiration activity within 4 days of testing) are very suitable for characterizing the reactivity of the waste.

Excessive lag phases due to the type of materials being treated can, however, lead to lower findings in both tests. In the gas generation test, lag phases occur primarily in waste that has not undergone any pretreatment, or subjected to only short biological treatment, due to acidification. In the respiration activity test, metabolic products generated during the biological treatment can hamper aerobic decomposition in the test, can also lead to lower findings. To avoid making erroneous interpretations when assessing MBP output, it is essential that both parameters be analyzed at all times.

Currently existing practical data show that stipulated landfill volume can be reduced by 35 % (when combustibles are landfilled) and 70 % (when combustibles are removed) by mechanical biological pretreatment of waste. Reliable practical data regarding leachate volumes are not available. Because the permeability of the landfilled material is significantly reduced, it can be expected that the occurrence of leachate is lower than with conventional landfills.

The observations of several landfill operators regarding incrustations in the drainage systems can be used as clear indicators of the positive impact MBP has on leachate quality. Inspections using video cameras showed little incrustation; one annual flushing of the system was viewed as adequate. Practical data from existing landfills (the landfills evaluated contain 20 - 80 % MBP wastes, with the remainder untreated waste), confirm the improvement in leachate quality. Concentrations of COD and BOD₅ drop by over 90 %. Heavy metal emissions are significantly reduced. Ammonia concentrations, however, only become better than those of landfills with untreated waste after longer landfill times.

The positive impact that biological pretreatment has on landfill gas emissions can also be verified using data from residual waste landfills. Primarily problems arise with flaring off landfill gas due to insufficient amounts indicates that, in the case of biological pretreatment, gas generation is significantly reduced. Laboratory tests show that with appropriate pretreatment (compliance with the reactivity limits established by the MBP Guidelines), gas generation can be reduced by 90 to 95 %. When the surface geometry of the landfill is optimized with regard to methane oxidation capacity and recultivation, both gas extraction measures and cover seal are no longer necessary.

REFERENCES

1. BMLFUW (Austrian Legislation): AWG- 325. Bundesgesetz vom 6.Juni 1990 über die Vermeidung und Behandlung von Abfällen (Abfallwirtschaftsgesetz), BGBl. 325/1990, Vienna, 1990.

2. BMLFUW (Austrian Legislation): DVO - Verordnung des Bundesministers für Umwelt über die Ablagerung von Abfällen (Deponieverordnung), BGBl. Nr. 164/96, Vienna, 1996.

3. BMLFUW (Austrian Legislation): WRG - Bundesgesetz vom 6.Juni 1990 über .... (Wasserrechtsgesetz), BGBl. ?/1997, Vienna, 1997.

4. BINNER E., ZACH A., LECHNER P.: Stabilitätskriterien zur Charakterisierung der Endprodukte aus MBA-Anlage, Forschungprojekt am ABF BOKU im Auftrag des BMLFUW, Vienna, 1999.

5. BMLFUW (Austrian Legislation): MBA-RL - Richtlinie für die mechanisch-biologische Behandlung von Abfällen, EU-Notice number 2001/423/A, Vienna, 2002.

6. BINNER, E.: Inkubationsversuche zur Beurteilung der Reaktivität von Abfällen, In: Waste Reports 02 „Emissionsverhalten von Restmüll", Arbeitsgespräch am ABF BOKU-Wien, Vienna, 1995, pp53-59.

7. BMLFUW (Austrian Legislation): Bioabfall-VO - Verordnung des Bundesministers für Umwelt, Jugend und Familie über die getrennte Sammlung biogener Abfälle, BGBl. 456/1994, Vienna, 1992.

8. WURZ H.: Auswirkungen mechanisch-biologischer Restabfallbehandlung auf das Deponieverhalten, dissertation at ABF-BOKU, 1999.

9. [9]RANINGER B.: Verfahren zur Vorbehandlung von Restabfall, In: Waste Reports 02 „Emissionsverhalten von Restmüll", Dokumentation eines Arbeitsgespräches am ABF BOKU-Wien, 1995, pp24-31.

10. BINNER E., LECHNER P.: Stand der mechanisch-biologischen Abfallbehandlung in Österreich, Präsentation at 2. Niedersächsischen Abfalltage, Oldenburg 1998.

11. FEHRER K.: Geotechnisches Verhalten von MBA-Material, dissertation at ABF-BOKU, 2002.

GOVERNMENT TARGETS VERSUS PUBLIC PARTICIPATION: BRIDGING THE GAP

P Tucker
University of Paisley
United Kingdom

ABSTRACT. Traditionally, public engagement in household waste minimisation has been driven mainly by strong altruistic motivations, environmental concern, dislike of waste, reinforced through social norms and persuasive campaign messages, but constrained by insufficient enabling infrastructure. This paper examines the relative roles of infrastructure, persuasion, demographics, and habit on the behavioural responses of householders. It specifically considers the effects of alternative kerbside collection regimes on dry recyclate yields, and tracks two Derbyshire districts though an episode of change. Conversion from paper-only kerbside collections to multi-material collections in those districts produced significant step changes in paper yields as well as total yields. Reducing residual (dustbin) waste collection frequency at the same time produced the highest increases. The paper reviews key psychological models of recycling behaviour and behavioural change, and examines the psychological context of the Derbyshire results. Increased convenience and heightened saliency are considered to be the major triggers of behavioural change.

Keywords: Recycling behaviour, Behavioural change, Kerbside collection, Recyclate yield, Demographic dependences.

P Tucker holds the Newspaper Industry Foundation Chair of Environmental Technology at the University of Paisley. His main research interests focus on waste and pollution management and the development of environmental decision support systems.

INTRODUCTION

The obligation to meet Government targets to recycle or compost 25% of household waste by 2005 (2006 in Scotland) from 2000/01 levels of 11.2% (England) and 7.0% (Wales) and 2001/02 levels of 7.4% in Scotland, imposes severe and immediate challenges on most local authorities. The major responses to the challenge have been (i) to increase the supporting infrastructure, which normally means increasing kerbside collections or increasing the number of materials they collect, and (ii) to raise public awareness through educational and promotional campaigns. Current advice additionally recommends the use of financial inducements or penalties, such as variable waste charging, as additional levers for participation [1]. Using a model of household waste management behaviours [2], Tucker and Speirs [3] forecasted the endpoints that could be achieved given a scenario of voluntary household participation operating without economic incentives or penalties (figure 1). The analysis showed the relative contribution that enhanced infrastructure might make, and how much would need to be addressed through allied promotional activities. The conclusions were that increasing convenience and engendering attitude change were both essential to achieving recycling rates of 25% and above. The infrastructure enhancements that are needed are clear (e.g. nation-wide household kerbside collection of recyclables and organic wastes, augmented by increased home composting and waste reduction and reuse measures, [1]). The details of the required 'education' elements, however, are much less clear. Normally, campaign strategies hinge on stimulating behavioural change, through conveying 'awareness' and giving 'procedural instruction', often backed up by persuasive messaging intended "to change attitudes".

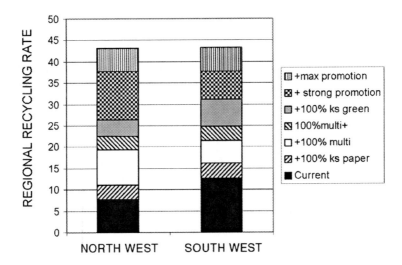

Figure 1 Predicted relative contributions of infrastructure provision and promotion to overall recycling rate. 100% multi assumes 100% coverage of kerbside collections of paper, glass, textiles and cans. Multi+ also recovers cardboard and plastic. Green means garden wastes. Strong promotion implies intensive, well-orchestrated awareness, information delivery and persuasive messaging. Max promotion assumes all personal costs are overcome.

Psychological models such as the theory of planned behaviour [4] and altruism [5] can help explain current household recycling performances. Attitudes relating to self-efficacy, awareness of consequences, and acceptance of responsibility, highlighted in those models, have all been found to be significant determinants of participation [6]. alongside negative perceptions [7] and the level of local facilitation. Recycling habits become set with early experience, and long-term behaviours are generally quite stable [6]. Drop-out is generally triggered through specific 'bad' experiences, including poor service, problems encountered, recognition of untenable personal cost in terms of time or effort, loss of saliency [6]. Recruitment normally depends on better facilitation and/or attitude change.

The launch of a new waste management scheme may be the crucial time to shape attitudes. At that time, the saliency of waste management issues to individual's life styles may become relatively high, increasing their susceptibility to change. Documented evidence shows that significant positive step changes in recycling performances have almost always been linked to changes in infrastructure provision, e.g. [8, 9]. In comparison, even the most intensive promotional campaign, e.g. [10], have only realised quite modest improvements in recycling rates.

EVALUATING THE EFFECTS OF RECYCLING PROVISION

The analysis by Tucker and Speirs [3] provided some guidance on the staged enhancements that would be needed to achieve the highest recycling levels, though did not go into the finer details of how individual local authorities might best progress towards their own local endpoints. Such solutions need to be worked out locally. Most local authorities have now made their strategic decisions on what to collect and how to collect it, by bag bin or box, co-mingled or source separated, fortnightly or weekly. But how much do these decisions matter? Does one system have advantages over another? These questions were addressed through significance testing the differences between alternative kerbside collection regimes. The results, based on CIPFA 1999-2000 statistics [11] showed that, on average, kerbside yields of paper and card were higher:

(a)   for schemes that collect other materials alongside that paper and card, with around 15% more paper and card being collected in a multi-material collection (Table 1). The difference was statistically significant (Mann-Whitney, $p = 0.0483$).

(b)   for schemes that use collection boxes (or bins) rather than bags or sacks (Table 2). On average, box collections were yielding some 15-20% more paper and card per household served than were bag collections. Again the difference was statistically significant (Mann-Whitney, $p = 0.0306$).

(c)   the frequency of collection was not a significant determinant of yield (Table 3; Kruskal-Wallace, $p = 0.499$; ANOVA, $p = 0.57$).

Table 1   Weights of paper and card collected according to collection type (Kg/serviced household/week)

SCHEME	N	MEAN	MEDIAN	STDEV	MIN	MAX
Paper-only	85	0.95	0.84	0.48	0.17	2.30
Multi-material	80	1.09	1.03	0.51	0.22	2.41

Table 2 Weights of paper and card collected according to container type
(Kg/serviced household/week)

CONTAINER	N	MEAN	MEDIAN	STDEV	MIN	MAX
Bag	86	0.91	0.87	0.36	0.17	2.21
Bin	4	1.31	-	-	1.05	1.69
Box	74	1.14	1.01	0.59	0.20	2.41

Table 3 Weights of paper and card collected according to collection frequency
(Kg/serviced household/week)

FREQUENCY	N	MEAN	MEDIAN	STDEV	MIN	MAX
Weekly	35	1.04	1.10	0.48	0.22	2.41
Fortnightly	120	1.01	0.88	0.50	0.17	2.30
>2 weekly	9	0.85	0.79	0.27	0.51	1.32

It can also be noted from these results that there are considerable variations amongst schemes of similar type in different communities. Spatial heterogeneities are also evident at more local levels, between individual neighbourhoods and streets [6]. So why do different localities respond differently to similar recycling provisions?. Traditionally, it has been accepted that differences may be linked to demographic factors such as social class, age, education, housing type, etc., however research has produced quite equivocal results on the significance of individual factors [6].

The research results now presented provide an evaluation of the significance of demographic differences within districts and the response of those districts to a step change in recycling provision, from paper-only bag collections to multi-material box + bag collections.

PERFORMANCE EVALUATIONS IN CHESTERFIELD AND NORTH-EAST DERBYSHIRE: A CASE STUDY

Kerbside recycling collections started in Chesterfield and North East Derbyshire in 1995 through the introduction of paper-only (blue bag) collections. From November 2001, both authorities commenced a rolling programme to convert those paper-only rounds to multi-material collections, recovering paper, glass, mixed cans, and textiles. In parallel, some of the areas covered by the new schemes in Chesterfield had their residual waste collection reorganised from a weekly to 2-weekly collection alternating with green waste.

The round conversions to multi-material produced significant step changes in material yield. Total kerbside yields in the new scheme averaged in excess of 2.3 Kg per household served per week in N.E. Derbyshire and nearly 1.8 Kg/household/w in Chesterfield (excluding the most recent conversions where yields of over 2.9 kg/household/w were recorded).

The recovered material comprised around 63-67% paper, 28-29% glass, and 4% and 3% cans and textiles respectively (Figure 2). It is also noticeable that recovered paper yields increased when paper was collected alongside other materials. Typically the paper yields increased by factors of order 1.1 to 1.4 across individual areas in N.E. Derbyshire, and by factors of 1.5 or more in Chesterfield. These outcomes correlate well with the national differences seen between the two types and means collection, that were noted above in the CIPFA statistics (Tables 1 and 2).

The change from paper-only collections to kerbside multi-material collections still maintained the historical differential in recycling performance between the two districts (except for the four latest conversions in Chesterfield where that differential was eroded). That is the historically poorer-performing areas remained the poorer performers in the new scheme. These results confirm the features predicted to occur at regional level (Figure 1). Model diagnoses of the Chesterfield and North East Derbyshire results showed that an equal step change in recycling attitudes might have occurred in each district.

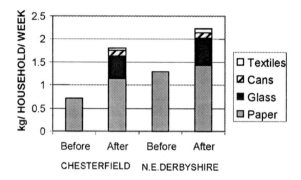

Figure 2   Kerbside yields before and after conversion.

The parallel alternate green waste and residual waste collections in Chesterfield had mixed effects on the dry recyclable yields. Four of the targeted rounds (corresponding to the latest conversions identified above) showed very high yields, whilst the yields of the other five rounds in the alternate week collection trial area were much closer to those of the normal collection regime (Figure 3). The high performing rounds occurred exclusively in the areas where the alternate week collections were launched at the same time as the multi-material kerbside conversions. In contrast, the poorer-performing dry-recyclate collections mapped onto the areas where alternate week green and residual waste collections had already been running six months prior to the multi-material conversion. Timing would therefore appear to be a critical factor.

The Influence of Demographics

Whilst the evidence presented in Figure 3 strongly suggests the importance of timing, it does not yet provide the statistical proof. The rounds from the three different collection regimes also had marked differences in their respective demographic profiles, with the better performing rounds all being drawn from the more affluent housing stock of the Borough. Could the differences be explained simply by demographics?

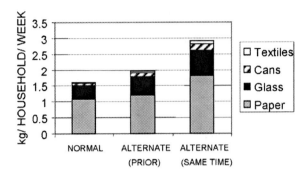

Figure 3   Effect of residual waste collection frequency on dry recyclate yield.

It proves very difficult to isolate demographic influences from operational influences in samples of mixed demographics. In this research, a separation was achieved by regressing the measured performance data against possible causal demographic, operational and situational factors. The regression equations used for modelling the round collection data were of the form:

$$Kg/hh/w = c1 \times \alpha + c2 \times \beta + c3 \times \gamma + \ldots..$$

where α, β, γ, etc. are percentages of households in a given group and c1, c2, c3 are scaling coefficients.

Stepwise regression was used in the analyses, allowing both forward selection and backward elimination to select the statistically significant predictors from the pool of possible predictors. In interpreting the results, it is stressed that the regression coefficients c1, c2, ... should not be interpreted in terms of the weights of material expected from each group. They are simply scaling factors that apply when all the other terms in the equation are also present.

The derived regression equation based solely on the Acorn demographic classification for Chesterfield and North East Derbyshire combined was:

$$Kg/hh/w \text{ (all materials)} = 0.0366 \times \% \textbf{ A} + 0.0231 \times \% \textbf{ B} + 0.0197 \times \% \textbf{ F} + 0.0185$$
$$+ 0.0166 \times \% \textbf{ E}$$

where A - F are the Acorn market research indicators for the following designations: A (thriving), B (expanding), D (settling), E (Aspiring), F (striving).

It is noted that the relative magnitudes of the coefficients by and large conform to the traditional wisdom that the more active recycling groups may be drawn preferentially from the more affluent sectors of the population. Regressions based on 'raw' census variables confirmed that the higher positive coefficients were associated with the percentages of detached housing and more mature residents, with private sector rentals having a negative effect. Regressions based on demographics alone had similar explanatory powers with R^2 values around 0.5. The goodness of fits were improved markedly ($R^2 \sim 0.7$) when operational and situational variables: **Alt_prior**, **Alt_sametime** and **In_chesterfield** (as opposed to in

N.E. Derbyshire) were introduced into the pool of potential predictor variables. **Alt_Prior** denotes those rounds where alternate week collections were introduced six months prior to the multimaterial conversion, and **Alt_sametime** denotes rounds where the two changes occurred together.

$$Kg/hh/w \text{ (all)} = 0.0319 \times \% \text{ A} + 0.0229 \times \% \text{ B} + 0.0207 \times \% \text{ D} + 0.0191 \times \% \text{ F} + 0.0186 \times$$
$$\% \text{ E} + 0.00890 \times \% \text{ Alt\_sametime} - 0.00280 \times \% \text{ In\_Chesterfield}$$

The key point in these results is that the '**Alt_sametime**' factor is identified as a statistically significant factor with a positive effect on yield, whereas '**Alt_prior**' was not identified as being significant in any of the regressions. The place of residence (in Chesterfield or N.E. Derbyshire) also proved to be a statistically significant determinant of recycling yield. Paper and glass yields remained higher within the traditional recycling bases, whereas can and textile yields appeared to be more positively correlated with the traditionally-poorer recycling groups (Table 4), though those correlations were relatively poor compared to the fits for paper and glass (Table 5).

Table 4 Ranking of coefficient magnitudes for the demographic (Acorn) groups

MATERIAL	RANKING (MOST POSITIVE → LEAST POSITIVE)
Paper	A B D E F
Glass	A B D F E
Cans	F B D
Textiles	F A D B

Table 5 R^2 values of regression equations for individual materials

ALL MATERIAL	PAPER	GLASS	CANS	TEXTILES
0.699	0.627	0.603	0.511	0.243

DISCUSSION AND CONCLUSIONS

In Chesterfield and N.E.Derbyshire, the conversion from paper only 'blue bag' kerbside collections to multi-material collections by box plus bag, significantly raised the tonnages of paper collected as well as the overall tonnages realised by the two schemes. This turned around low, relatively stable, but gradually declining, recyclate yields (Figure 4) into yields that were well above the national average (based on 1999-2000 CIPFA returns). It was not possible to deconvolute the respective contributions of single versus multi-material, or bag to box collections from the monitored data, nor was it possible to isolate the effects of those infrastructure changes from the influences of accompanying publicity. However, national analyses and model forecasts showed just how important those factors might be. The change from paper-only collections to kerbside multi-material collections still maintained the historical differential in recycling performance between the two districts (except for the latest conversions in Chesterfield).

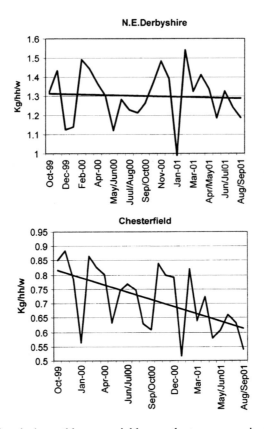

Figure 4   Historical monthly paper yields over the two years prior to conversion

Whilst that differential may, in part, relate to demographic differences between the two Districts, historical event chains may have contributed as well. This is witnessed by the stronger decline in paper-only yields in Chesterfield than in N.E.Derbyshire in the two-year period leading up to the conversion. The reasons for the higher decline are not yet known.

Another significant effect on yield was linked to how the residual (dustbin) waste was collected. Introduction of fortnightly collections of residuals, as opposed to weekly collections, increased the source separated dry recyclate yields. However, the effect was marginal in areas where fortnightly residual waste collections were already well-established prior to the start of the multi-material scheme. In contrast there were very large (four fold) increases in dry recyclate tonnages in the areas where the two changes were introduced at the same time. The imposition of a reduced frequency of residual waste collection effectively reduces the amount of 'dustbin space' available for that waste. The intended outcome is that householders will have to seek alternatives, preferably recycling, for disposing of their excesses. For some householders that would need a significant life-style change. Such events will significantly raise the saliency of the waste management issue to householders (in terms of potential personal costs).

Providing the solution (i.e. the multi-material recycling box) at the same time, leads householders to adopting that solution. In contrast, imposing the lifestyle change prior to providing the solution effectively renders that solution almost irrelevant when it eventually comes. Lifestyle adjustments have already been made.

Other authorities in the country, e.g. Wealden [9] have tried cutting residual waste collections in areas where multi-material kerbside provision was already well-established. Large step changes in dry recyclate yield were reported for Wealden [9], though they did not match the four-fold increases witnessed in Chesterfield.

The research has shown that whilst alternative strategies and mechanisms of kerbside collection do, on average, affect the tonnages of materials collected, there can be considerable variations amongst areas running similar schemes. Good and bad examples can be found for all types of scheme. The research has added further evidence that demographic differences and the chain of history may also be important in shaping today's recycling performances. Large step changes in recycling behaviour are almost always linked to changes in provision. This is considered to be due to the combination of increased convenience, higher salience, and accompanying education and persuasive messaging. Education and persuasion do not appear to work effectively in isolation. National and regional awareness campaigns such as "Doing your bit", "Do a little, change a lot", "Slim your bin", "Rethink rubbish" have so far failed to produce any evidence of achieving measurable behavioural changes.

ACKNOWLEDGEMENTS

The research in Chesterfield and North East Derbyshire was funded under the Landfill Tax Credit Scheme by WREN (Waste Recycling Environment). The ongoing research into recycling behaviours is sponsored by Sun Chemical, Stora Enso, Bridgewater Paper Co. Ltd and Cheshire Recycling.

REFERENCES

1.   STRATEGY UNIT. Waste Not Want Not. Cabinet Office: London, 2002.

2.   TUCKER, P., SMITH, D. Simulating Household Waste Management Behaviours Journal of Artificial Societies and Social Simulation (1999). Vol.2, No.3, http://www.soc.surrey.ac.uk/JASSS/2/3/3.html

3.   TUCKER, P., SPEIRS, D. Model Forecasts of Recycling Participation Rates and Material Capture Rates for Possible Future Recycling Scenarios. http://www.strategy.gov.uk/2002/waste/downloads/recycling_participation.pdf, 2002

4.   AZJEN, J. From Intentions to Actions: A Theory of Planned Behaviour. In Action Control: From Cognition to Behaviour, (Kuhl and Beckmann, eds.), Springer Verlag: New York, 1985.

5.   SCHWARTZ, S.H., Normative Influences on Altruism. Advances in Experimental Social Psychology, (Berkowitz, L. ed.), Vol.10, Academic Press: New York, 1977.

6. TUCKER, P. Understanding Recycling Behaviour, ISBN 1-903978-01-7, University of Paisley: Paisley, Scotland, 2001.

7. TUCKER, P. A Survey of Attitudes and Barriers to Kerbside Recycling. Environmental and Waste Management, Vol.2, No.1, 1998, pp. 55-63.

8. CLAMP, F. Kerbside Collection of Paper for Recycling. Hertsmere Borough Council, 2000.

9. WOODARD, R., HARDER, M.K., BENCH, M., PHILIP, M. Evaluating the Performance of a Fortnightly Collection of Household Waste separated into Compostables, Recyclates and Refuse in the South of England. Resources, Conservation and Recycling, Vol.31, 265-284, 2001.

10. READ, A.D. A Weekly doorstep Recycling Collection, I had no idea we could! Overcoming Local Narriers to Participation. Resources, Conservation and Recycling, Vol 26, 217-249, 1999.

11. CIPFA. Waste Collection and Disposal Statistics: 1999-2000 Actuals, CIPFA:London, 2001.

COMPOSTING INDUSTRY SITUATION
IN TURKEY AND EU STATES

R Külcü

O Yaldiz

Akdeniz University

Turkey

ABSTRACT. The volume of municipal and agricultural wastes produced in Turkey is increasing, because of the increase in the country's population and changes in lifestyle, from 15 million tonnes in 1991 to 23.7 tonnes in 2002. Turkey's municipal solid waste generally consists of wastes generated from residential and commercial areas, industries, parks and streets, and is not sorted at source, but collected in the same waste bins. Composting plants have been installed in some cities while in other centres disposal practice varies from landfilling to dumping in quarries, streams and even the sea. Adaptation to EU protocols has reinforced the importance of composting for Turkey. For example, Istanbul Metropolitan Municipality is the biggest composting plant in Europe and it has operated since 2001. Most European countries have developed their own composting industry because they are forced to by EU directives. This paper evaluates the current state of the composting industry in Turkey and also compares the performance and profile of different composting plants with that of EU countries.

Keywords: Composting, EU States, Turkey, Composting Industry.

R Külcü is a Research Assistant at Akdeniz University, Faculty of Agriculture, Department of Farm Machinery, Antalya Turkey.

O Yaldiz is a Professor at Akdeniz University, Faculty of Agriculture, Department of Farm Machinery, Antalya, Turkey.

INTRODUCTION

The disposal of organic wastes is an increasing problem both in Turkey and throughout the world. Landfilling, composting, incineration, and recycling constitute the alternatives to solid waste removal. Decreasing volume and mass, and the removal cost of solid wastes are the main purposes of solid waste elimination studies. Composting, by which the organic materials are processed biologically, is one of the methods that is used for this target. Compost is used in agriculture, forestry and horticulture.

A phenomenal increase has occurred in biological waste treatment in Europe in recent years. Investigations indicate that at least 35 percent of urban wastes and a large proportion of industrial residuals could be biologically treated via composting and/or anaerobic digestion. This recoverable organic fraction is about 49 million metric tonnes per year - equal to 40% of the total waste production in Europe [1].

Production of municipal and agricultural waste is strongly related to economic development of a country. Composting seems to be an important element in sustainable waste management for Turkey, because only 0.45% of the residential waste had been composted in 1996 [2]. There are only three large-scale composting plants. One of them is the Istanbul Metropolitan Municipality Plant, operated since 2001. This plant is the biggest composting plant in Europe and it shows that the importance of composting is increasing in Turkey.

COMPOSTING IN EUROPE

The Directive 99/31/CE on Landfills

The Directive 99/31/CE on landfills requires that the landfilled biowaste be sharply reduced over the next few years. This directive is aimed at reducing the production of biogas at landfilling sites and improving the conditions at which landfills get operated. The targets sets by the landfill directive are set out in Article 5 of the directive and require the following [3]:

- Not later than 16 July 2006, biodegradable municipal waste going to landfill must be reduced to 75% of the total amount by weight of biodegradable municipal solid waste produced in 1995.
- Not later than 16 July 2009, biodegradable municipal waste going to landfill must be reduced to 50% of the total amount by weight of biodegradable municipal solid waste produced in 1995.
- Not later than 16 July 2016, biodegradable municipal waste going to landfill must be reduced to 35% of the total amount by weight of biodegradable municipal solid waste produced in 1995.

In order to achieve this, biological treatment and composting are playing a major role; composting is also the most "natural" way to manage biowaste.

Composting Industry Situation in EU States

The collected and treated amounts of organic matter are different in the EU countries. Approximately 17 million tonnes of bio and green wastes are separately collected and treated of a total recoverable potential of 49 million tonnes biowaste [1]. Total food and green waste compost production within the EU is approximately 5,000,000 t/y. Germany, Austria and the Netherlands are the leading countries in compost production as well as in the enhancement of strategies to promote it [4].

The organic waste activities in Europe can be divided into 4 categories. The policy in Austria, Belgium (Flanders), Germany, Switzerland, Luxembourg, Italy, Spain (Catalonia), Sweden and the Netherlands is implemented countrywide. These countries recover around 80% of the separately collected and treated (mostly by composting) organic waste fractions in the EU. Digestion also plays a minor part at the moment. Denmark, UK and Norway form the second category of the implementing states. These countries have built up the parts of the political, quality and organizing framework for separate collection and composting. Finland and France form the third category, and have developed strategies at the starting point. In the forth category we find countries where there is no effort on composting of source separated organic waste like Spain, Greece, Ireland and Portugal. These countries landfill mixed urban wastes [6].

Table 1   Compost production in the European Union [4].

EU COUNTRİES	COMPOST PRODUCTION, 10^3 t/year	COMPOST PRODUCTION, 10^3 m³/year	L/İNH/YEAR
Austria	500	909	113.2
Denmark	250	454	87.1
The Netherlands	650	1181	76.9
Germany	2400	4363	53.5
Belgium (Flanders)	200 (180)	363 (327)	35.8 (55.7)
Sweden	100	181	20.5
Luxembourg	3	5.45	14.2
Finland	30	54	10.6
Italy (Northern Italy)	250 (200)	454 (363)	8.1 (14.7)
France	240	436	7.7
United Kingdom	159	289	4.9
Greece	/	/	< 1
Ireland	/	/	< 1
Spain	/	/	< 1
Portugal	/	/	< 1
Total	4782	8694	23.5

Norway

Norway has 33 composting plants with capacities between 300t/y and 25,000t/y. Simple windrow composting plant is used by farmers (1,500t/yr). This approach is expected to become more common for farmers beginning to commit themselves to composting. Three more similar facilities are planned in southern Norway [6].

Sweden

In Sweden a decision made by the Parliament in 1998 showed that landfilling of organic wastes will not be allowed after the year 2005. Sweden compost production is about 100,000t/y from biowaste and 100,000 t/y from greenwaste. Trends are towards high technology standards in composting. Composting plants in Sweden are typically built in densely populated areas. These technologies range from forced aeration of membrane covered windrows to modern tunnel composting plants [6]. Sweden has 100 greenwaste composting plants near landfills, based on windrow technology [6].

Finland

Approximately 80,000 tonnes of biowaste are treated in centralized treatment plants in Finland. This is 10% of the total amount of organic waste. According to the Waste Management Plan (1998) and its recycling target of 70% of the organic part in MSW, the total number of centralized biowaste treatment plants is going to be raised from about 5-10 to 40-50 by the year 2005 [7]). Organic wastes are generally treated by composting in Finland. There is only one anaerobic digestion plant, which treats about 25,000 tonnes of biowaste and 1,000 tonnes of sludge annually [7].

Denmark

Denmark has 5 million inhabitants, generating 12 million tonnes of waste a year [8]. Denmark has a low reliance on landfill and employs a range of treatment options for the management of BMW (Biodegredable Municipal Waste). In 1998 5.3% of BMW was consigned to landfill, 54.3% to incineration with energy recovery, 29.6% to composting, 10.4% to recycling and 0.4% to anaerobic digestion [9]. Denmark has 5 biogas and 134 compost plants. These plants treat 37,000 tonnes of organic household waste and 615,000 tonnes of garden and park waste. The composting plant in Denmark produced 388,000 tonnes of compost in 1999 [7].

Great Britain

50 compost plants operate in Great Britain, of which two have a production of more than 25,000t/y. A tunnel composting system started with the capacity of 16,000t/y of garden waste, dewatered sewage sludge and sugar beet residues in Suffolk. There is also a prototype container composting system. Most other facilities in the UK utilise open windrows [6]. Tunnel and box composting systems are becoming the favoured technology. This new trend is based mostly on bad odour experiences from central Europe [6]

Netherlands

About 92% of households are connected to a separate waste collection system in the Netherlands (16 million people in approximately 6 million households). The separate collection system for organic wastes results in 1.57 million tonnes of biowaste being processed into 0.6 million tonnes of compost [10]. Netherlands has 26 composting plants for domestic biowaste with a total capacity of 1.57 million tonnes [6]. There are only two anaerobic digestion plants in Netherlands with a total capacity of 88,000t/y. A further anaerobic plant processes the organic fraction of residual waste; another for so-called biological mechanical pre-treatment is in the planning phase [7].

Belgium

Seven composting plants and one anaerobic digestion plant are operational in Flanders, with treatment capacities of between 35,000 and 60,000t/y. While five of the composting plants operate on the closed hall method, the others are based on the box and tunnel method. Flanders has only one digestion plant with a capacity of 35,000t/y. A second digestion plant will become operational in 2003. Separately collected biowaste (340,000 t/inhabitant/y or almost 60 kg/inhabitant/y) and green waste (460,000t/inh/y or some 78 kg/inhabitant/y) accounted for about 25% of the total amount of collected MSW (2001) [7]. Greenwaste is composted in open air windrows, on a hardened floor, with collection and re-use of leachate. Aeration is obtained by regular turning of the compost material. In some cases, aeration tubes are installed in the floor to allow active aeration of compost piles [6].

Germany

Seven to 8 million tonnes of waste are collected separately in Germany at the moment. Total potential organic raw material amounts to up to 9 million tonnes. About 60-75% of all the inhabitants used an in source biowaste separation system in 2001 [7]. Approximately 700-900 composting plants were operating in Germany between 1990 and 2001. These plants produce 4 million tonnes of compost products. While 80 percent of the composting plants work with heaps or windrows, 10% use boxes, 3% compost in enclosed, aerated metal containers; 3 % utilise continuously revolving decomposition drums; another 3% use a channel system with aerated side-by-side rows, separated by concrete walls, or tunnels (basically covered channels) [11].

There are 500-800 anaerobic digestion plants operating in Germany. While most of these anaerobic digestion plants use agricultural wastes, a few use co-digested biowaste. Around 37 large industrial digestion plants treat pure biowaste, with a capacity of 500,000 tonnes; 20 of them are members of a voluntary quality assurance system of the Quality Assurance Organization BGK [11].

Luxembourg

41% of the 430,000 inhabitants in three large inter-municipal associations are connected to source separation system at the moment in Luxembourg. Luxembourg has three composting plants with a total treatment capacity of around 27,000t/y. There are another three in project phase. Anaerobic digestion and mechanical-biological-pretreatment activities are not used in Luxembourg [7].

France

Seventy composting plants treat mixed MSW and the average treatment capacity of these composting plants is around 3.1 million tonnes. They produce about 1.5 million tonnes of compost products [7]. In France approximately 40 greenwaste windrow composting plants are processing 400,000t/y. There is an unknown number of smaller plants (2,000 to 3,000t/y) [6].

Austria

80% of the households were involved in the separate collection system, generating 1.1 million t/y of biowastes [5]. Only about 11% may be addressed as organic wastes in a narrower sense for biological waste treatment [12]. More than 50% of this is composted in small facilities (<5,000t/y) owned by farmers and municipalities. For greenwaste, generation of around 620,000t/y can be assumed [7].

There are only three industrial biogas plants for organic household waste with a total capacity of 45,000t/y (12,000, 13,000 and 20,000 tonnes respectively) [7].

Austria has more than 500 organic waste composting plants with an overall capacity of 760,000t/y. In 2000 the actual quantity of processed biowaste amounted to 610,000 tonnes [13]. 50% of composting plants have a capacity of less than 500t/y. As with another 100 plants with an annual capacity of less than 2,000t/y, composting mostly is done in windrows. Twenty-one other plants processe between 2,000 and 5,000t/y. Only 14 of Austria's composting plants have a capacity of over 5,000t/y. Static windrow composting, forced aeration, box composting and rotating drum systems are used in addition to turned windrows [6].

Switzerland

Around 80% of source separated organic waste is composted in open windrows with capacities between 100 and 20,000t/y. The larger plants operate on the forced aeration method. A large enterprise (ROM) with mobile technology supervises many decentralised windrow composting plants and is thus responsible for processing 50,000 to 60,000t/y of organic materials [6].

Italy

Municipal solid waste are generally landfilled in Italy, 74.4% of total MSW was landfilled in 1999, mainly without any pre-treatment, while source separation and recycling averaged 13.1% of national MSW production [7]. However, a law passed in February 1997 sets a target of 35% source separated collection of organics by 2003 [6].

The number of composting plants in Italy increased from 10 in 1993 to 114 in 1999 (135 plants considering also sites with a capacity of less than 1000 tonnes per year) [5]. These plants treat principally organic waste from source separation of MSW (food and yard waste). The average size of the plants is relatively small, around 10,000 tonnes (75 plants); whereas 37 plants treat between 10,000 and 30,000t/y. Recent facilities most frequently have a capacity above 30,000 t/y (18 plants); Region Veneto (which has the highest capacity, more than 500,000 tonnes, ie. more than 100 kg/inhabitant) shows most frequently a different pattern, with some facilities between 50,000 and 100,000 tonnes [7].

Portugal

No activities regarding separate collection and composting of organic wastes are known in Portugal.

Nonetheless, some facilities for mixed MSW have long been running (5 plants with 400,000 tonnes capacity, with capacities to be enlarged to 480,000t/y in 2002). Lately, some facilities to tackle source separated organic waste (e.g. from big producers) have been planned and/or are being constructed. Also, some of the old facilities are planned to be partially upgraded into quality composting sites for source separated organic waste [7].

Ireland

Almost all the organic components collected with MSW (92%) and commercial waste are landfilled in Ireland at the moment. A number of home composting schemes have been set up by local authorities, with eight such projects receiving funding until 1999. Currently Ireland has no significant biological waste treatment capacity. Two of the four centralised composting plants treat only greenwaste, the others treat kitchen waste. Further development of centralised composting for organic waste is strongly recommended by the EPA (2000) [7].

Spain

Spain has a windrow composting plant with forced aeration and a capacity of 25,000t/y located in the centre of Madrid, 500m from the Government Palace. It is expected to treat green residues from public parks. There is work on very big plant (260,000t/y) in La Coruna, which will be in operation by 1999. It will utilize mechanical-biological treatment, composting and digestion [6].

Greece

The overall production of MSW in Greece is estimated at around 4,000,000 tonnes for the year 2000. Until 1994, waste disposal was characterised by thousands of dumpsites (4,850 were recorded officially), 70% of which were uncontrolled (corresponding to 35% of the total waste quantities). The proportion of the population served by a regular collection system was around 70%, while in numerous small islands and isolated villages collection was poorly organized [7].

In Greece, apart from some pilot projects, composting activities are limited to the construction of an MSW composting plant for the city of Athens. It is sponsored by the EU and will have a capacity of 100,000t/y [6].

COMPOSTING INDUSTRY SITUATION IN TURKEY

Organic Waste Potential in Turkey

Turkey's Municipal waste potential is about 65,000 tonnes per day [14]. Organic components collected with MSW and commercial wastes are generally landfilled in Turkey. In 1996 [2] 31.75% of the total amount of daily collected garbage was disposed of at the main city dump, 42.83% at the municipality dump, 2.35% at the other municipality's dump, 12.76% at the sanitary landfilling (only six municipalities have sanitary landfills), 0.45% at the compost plant, 1.96% to open combustion, 0.02% to rivers, seas and lakes). Turkey has over 2,000 open dumps. Open dumping means that solid wastes are dumped without any precautions being taken. They are neither compacted nor covered, they have no systems to deal with leachate or methane gas collection, and they pollute the environment continuously. They encourage both insects and vermin to breed, and endanger public health and safety. Serious accidents, such as the methane explosion at the Ümraniye Open Dump, Istanbul, in April 1995, which killed 39 people, or the slippage of a huge mass of solid waste from the Kemerburgaz Open Dump (also Istanbul) onto the neighbouring road in May 1996, demonstrated the significant threat of this method. Such open dumps should be closed immediately and/or rehabilitated, in order to avoid yet more severe accidents in the future.

Approximately 32,000 tonnes of dry matter animal manure, 110,000 tonnes of dry matter plant (vegetable) wastes are produced per day, while 486,000 tonnes of slaughterhouse wastes, 1,410,000 tonnes of milk processing wastes and 86,000 tonnes of olive oil mill wastes were produced annually in Turkey. Agricultural wastes are generally burned or buried by farmers so agricultural wastes have an important role in environmental pollution.

By the time the Solid Waste Control Regulation of Turkey was published in 1991, the municipalities had started to stop using and/or rehabilitate the existing open dumps, and construct sanitary landfills according to the new standards. This regulation permitted only three waste disposal methods. These methods are sanitary landfill, composting and incineration for energy production.

Composting in Turkey

Composting seems to be an important method in sustainable waste management for Turkey, because only 0.45% of residential waste was composted in 1996 and also there are only three large-scale composting plants in 2002.

The first composting plant was built in Izmir and has operated since 1969. Its capacity is 250 tonnes MSW/day and it produces about 100 tonnes of compost per day at the moment. Composting facilities are processed by the technologies with DANO-Biostabilisation, which transferred from Denmark. There are two DANO Composting units with a batch capacity of 75 tonnes. It is possible to increase to 150 tonnes per day for 8 hours retention time. In DANO–Drums the temperature rises to 55-65°C in a short time period and aerobic degradation begins rapidly. The compost heaped for 2-3 months period is ripe and can be sold as humus material for soils [15].

The second composting plant was built along the western coastal zone of Antalya in Kemer and has operated since 1999. Most of the 4 million foreign tourists spend their vacation in this region. Therefore, Kemer region has a higher degree of importance for the Turkish economy. This plant has a capacity of 50,000t/y and a brokolare composting system. It also produces 1,200 tonnes of compost per year [16].

In order to solve the solid waste disposal problem, which is one of the main scopes of the Southern Antalya Tourism Development Project, a composting plant has been constructed in the near vicinity of Kemer city. The plant was built on an area of 12,000m^2. Kemer composting plant is fully equipped with an air-conditioning unit with a capacity of 12,000m^3 air. The source separated wastes are also separated manually in the composting plant. Separated organic wastes are mixed with green material and then pressed to form briquettes. The briquettes are placed in maturation rooms and kept in for 2-3 weeks to achieve the decomposition of the wastes. At the end of the maturation period, the compost product is screened by rotary sieves and packed for marketing [16].

The third composting plant, which has a capacity of 1,000 tonnes per day, is located at Kemerburgaz region, on the European side of Istanbul. Istanbul Metropolitan Municipality has been operating this plant since 2001. It produces 250 tonnes of compost per day. This plant is the biggest composting plant in Europe. Approximately 10,000 tonnes of MSW per day are produced in Istanbul Metropolitan city. MSW is mainly treated by landfilling and composting in Istanbul now. About 10% of the total amount of MSW is being disposed by composting, the rest to sanitary landfills [17]. The capacity of Istanbul Solid Waste Composting and Recycling Plant was planned according to the values in Table 2

Table 2   Waste Characteristics for Istanbul

Average moisture content	55 %
Organic matter	45 %
Total solid waste quantity	300000 tonnes/year
Solid waste quantity of fermentation unit	150000 tonnes/year
Composting time	56 days
Compost quality	According to 4 M-10[*] standard compost temperature is 30-40°C

[*]LAGA (Landerarbeitsgemeinschaft Abfall) Merkblatt M10
(German compost quality standard 1997)

The composting unit in the plant is 230m in length and 35m in width. This unit is separated into two sections: the primary and secondary decomposition sections. The primary section is completely covered by stainless steel and retention time is 3 weeks, compared to 5 weeks in the secondary section. Air passes through the piles of the second decomposition part, and then this air is used for aerating the piles of the primary section. Aeration time and aeration degree can be adjusted automatically. Technical information about the decomposition unit are given in Table 3 [18].

Table 3   Technical information of Istanbul Solid Waste Composting and Recycling Plant

PARAMETER	CHARACTERISTIC
Operation time	8 weeks (7 transfer operations)
Number of decomposition areas	2 lines (each line consists of 8 areas)
Pile height	2.5m in regular operation (peak height is 3m)
Mixing width	31m (maximum)
Aeration of 1-3 area	Compressed aeration
Aeration of 4-8 area	Absorbed aeration
Aeration of 2-8 area	Moisture addition during mixing

CONCLUSIONS

Management of solid waste still continues to be a problem around the world. Since handling and disposal of solid waste is an expensive process, most countries are trying to minimise the generation of solid waste. It is also possible to evaluate this valuable component as compost for energy production.

European countries are making great efforts to treat organic waste streams as far as possible with biological means. The trend generally favours separate collection and composting or digestion of organic wastes from households, gardens and public parks. Increasing environmental protection requirements, market forces and laws do not permit other alternatives.

Composting plants are installed in some cities of Turkey while in others disposal practice varies from landfilling to dumping in quarries, streams and even the sea. There is not enough compost production in Turkey and new composting plants should be installed. But composting plants in Turkey operate at a loss because Turkish farmers do not use compost as a fertilizer and/or soil conditioner.

In conclusion:
- Turkish farmers should be educated about the benefits of compost as a fertilizer and/or soil conditioner.
- Composting of solid wastes should be encouraged in rural areas.
- New composting facilities should be installed.
- Cost-benefit analyses should be made before deciding to construct composting plants.
- Household wastes should be connected to a separate collection system.
- The Directive 99/31/CE on landfill target should be addressed to Turkey's solid waste management.

REFERENCES

1. BIONET, 2002. Biological Waste Treatment in Europe Technical and Merket Developments.
    http://www.bionet.net/de/waste/index.htm

2. DIE (The national Institute of statistics of Turkey), 2000. Results of municipal surwey of
    environment in Turkey in 1996. www.die.gov.tr/TURKISH/SONIST/CEVRE/cevre.html

3. COUNCİL DİRECTİVE 1999/31/EC of 26 April on the landfill of waste. Offical Journal L.182,
    16/07/99 p 0001-0019.

4. CENTEMORE M, RAGAZZİ R., FAVOİNO E, 1999. Label Policies, Marketing Strategies and
    Technical Developements of Compost Market in The European Countries.Internaional
    conference of biological treatment and environment 2nd-4th September 1999 Welmar, Germany,
    p. 355-363.

5. FAVOINO E, 2003 Composting Across Europe.
    http://www.hua.gr/compost.net/Favoino%202,%20composting%20across%20Europe.pdf

6. BARTH J, KROGER B, 1998. Composting Progress in Europe. Biocycle International, April 98,
    V.39, issue 4.

7. COUNTRY REPORTS, 2003. http://www.compostnetwork.info/countries/index.htm.

8. VELTZE S A.,2002. Managing without landfill - The Danish example. Waste Management
    World. V.2, N.1.

9. CROWE M, NOLA, K, COLLİNS C, CARTY G, DONLON B, KRİSTOFFERSEN M, 2002.
    Biodegradable Municipal Management in Europe. European Environmental Agency report.

10. OUWERKERK H 1999. Waste management in the Netherlands. European Conference on Waste
    Management Planning,

11. HOGG D, BARTH J, FAVIONIO E, CENTEMERO M, CAIMI V, AMLINGER W,
    DEVLIEGHER W, BRINTON W, ANTLER S 2002. Review of Compost Standards in
    Germany. The Waste and Resources Action Programme (WRAP) report, ISBN:1-84405-011-4.

12. AMLINGER F, PEYR S, GESZTI J 2001. Compopst Management in Austria-Quantities and
    Qualities. 15th-17th May Biowaste Conferance, Pt.Pölten, Vienna.

13. COUNTRY REPORTS, 2001. Organic waste treatment plants in Austria 15th-17th May
    Biowaste Conferance, Pt.Pölten, Vienna.

14. MINISTRY OF ENVIRONMENT OF REPUBLIC OF TURKEY, 2002. Solid waste.
    www.cevre.gov.tr.

15. ERDIN, E 2002. Composting Factory in İzmir. http://erdin.deu.edu.tr/pubs.htm.

16. KURT P B, TOPKAYA M. B, ÖZDEN T, GÜNE, N. 2002. Composting experience and practice
    in Antalya Province. Appropriate Environmental and Solid Waste Management and
    Technologies for Developing Countries Sysmposium, Istanbul, p. 1337-1344.

17. DEMIR A, TOSUN I, ÖZKAYA B, BILGILI M. S, GÜNA, A, AVŞAR F, KARAASLAN Y.
    2002. Aerobic composting of municipal solid waste in Istanbul: Start up and operational
    experiences. Appropriate Environmental and Solid Waste Management and Technologies for
    Developing Countries Sysmposium, Istanbul, p. 1329-1336.

18. DEMIR, A, BAŞTÜRK, A, KARAASLAN, Y, ÖZKAYA, B, DEBIK E, BAYHAN, H,
    KANAT G 2002. Composting process and the product quality in Istanbul composting and
    recycling plant. Appropriate Environmental and Solid Waste Management and Technologies for
    Developing Countries Sysmposium, Istanbul, p. 1321-1328.

THEME FIVE:

ASSESSING ENVIRONMENTAL IMPACT

GETTING WASTE WISE:
REFORMING PRODUCTION AND TAMING CONSUMPTION

P Modak

Environmental Management Centre

India

ABSTRACT. Sustainable production and consumption are two key aspects that we need to focus *simultaneously* while seeking the goal of sustainable development. Wastes are produced not just because of inefficient production but also because of over consumption. While the production systems are getting reformed towards *sustainable thinking,* these changes are rather limited to the developed world. Much work still needs to be done in the small and medium enterprises of the developing countries that are expected to play a critical role in influencing the global markets. The world is still divided when it comes to production reforms and consumption patterns. Disparity between consumption patterns in the developed and developing economies is rather high. Life cycle oriented policies, environmental and social stimulation of global trade and responsible advertising will play an important role here. There is a need to bring in a behavior change in the Governments, businesses and the communities in this direction. This paper makes several observations and recommendations to this effect.

Keywords: Waste management, Sustainable production, Consumption, Trade, Responsible advertising

Prasad Modak is a Partner at the Environmental Management Centre (EMC), Mumbai, India. His field of interest has been Environmental Assessment and Sustainable Production and Consumption. In these areas, he has been working closely as a Consultant to the World Bank, United Nations Environment Programme and United Nations Industrial Development Organization. Modak has worked extensively in several countries, with national governments and with international donor agencies and has numerous publications to his credit. Earlier to the establishment of EMC, Modak served as a Faculty at the Centre for Environmental Science and Engg at Indian Institute of Technology, Mumbai, India.

INTRODUCTION

Waste is an outcome when we seek materials or energy. For manufacturing operations, the nature and quantum of waste depends on the raw materials, processes used, product design and extent of production. For service industry, e.g. hospitality sector, waste is generated depending on the nature of services made available, and the response from customer depending on his/her consumptive behavior.

Waste is something that is not desirable – from both economic and environmental point of view. Higher waste results into less productivity and increased costs of waste treatment, storage and disposal. More generation of waste thus leads to lesser profits, reduced competitiveness and even poorer public image. Today, many have realized that cutting down of wastes is an important business strategy.

CHANGING TRENDS IN PRODUCTION

Over the last two decades, much work has been done to substitute/improve equipment/technologies that will generate less waste and provide a better quality of products. We call these categories of processes as low or non-waste technologies, or clean/cleaner technologies, environmentally friendly technologies or zero waste technologies. Some of these technologies indeed cost more than conventional technologies in terms of capital. However, for green field projects, these technologies are now preferred due to rising costs and difficulties of securing raw materials, demand on quality products, and increasing costs of waste management.

In many manufacturing sectors, we see today improved processing configurations (e.g. multiple loop, counter-current, fluid bed systems), efficient process design and operating conditions (e.g. better raw materials, improved catalysts, optimum sequencing). Double contact double absorption process in sulphuric acid plants has become a default design and black liquor recovery is integrated in most paper and pulp industries.

As a result, many key industry sectors are operating at substantially improved benchmarks of specific resource consumption and waste production. Water consumption benchmark in textile wet processing sector for instance is around 60 to80 liters/kg today as against the 160 to 180 liters/kg benchmark that was thirty years ago.

Today, the early technology limited focus has widened somewhat – as human factor engineering and management systems are understood to be as critical. In the last decade especially, we therefore see an emergence of management systems and a suit of techniques such as, Total Quality Management, Total Productive Maintenance, Quality/Environmental Management Systems, Japanese 5S, Six Sigma etc. These systems and techniques ensure that the technologies are well adapted, used and continuously improved and at every stage of manufacturing operations the defects and wastes are meticulously tracked and consciously reduced as a group responsibility. In many ways therefore, we see a confluence of waste related concerns and productivity interests in the senior management of the industries.

The process limited interest has now expanded to design of products. Here, leadership has been taken by consumer products industries that have been constantly looking for "cutting costs" (e.g. through simplified packaging!) "branding" (in terms of codes of environmental and social conduct) and by "offering something different" (e.g. organic). An inquiry into product design has led to much holistic thinking – that includes not just at the optimization of waste management systems but at optimization of raw materials and minimization of life cycle impacts of product's use, abuse and disposal. You see today products that boast slogans such as chlorine free, ozone friendly, organic, eco-friendly etc influencing the choice of the consumers.

Such a progression has over the years brought changes in the environmental consulting and engineering market as well as environmental education. Earlier most environmental consulting was limited to permitting, designing, building and operating waste treatment systems. With the interest expanding from waste treatment to waste prevention, consulting opportunities have widened to include newer areas such as waste audits, waste minimization, environmental management systems, impact assessment, product design etc.

This has led to new entrants to the conventional environmental consulting that include chemical technologies, economists, product designers, management and communication specialists etc. The university curriculums have adapted to this change, offering more multidisciplinary programmes on environmental management –broadening from initial civil engineering bias. Many business management schools have started offering courses on environmental economics, business and environment etc. Product design schools have introduced courses such as life cycle analyses that expand the student's understanding on generation of wastes/residues and their environmental impacts in the entire life cycle of the product.

PRODUCTION IN THE DEVELOPING WORLD

This thinking has been limited to large industries and corporates with multinational presence and indeed so in the developed countries. The very scale of operations and need to remain competitive in changing global markets, have compelled these business houses to follow a strategic approach to mainstream environment.

The small and medium industries (SMI), especially in the developing world, have not yet reformed their production practices in this direction. The SMI sector in developing world today takes a nearly 50% share of its economy and employment and hence cannot be neglected. These SMIs use outdated processes, second-hand equipment (generally sold from developed countries), produce products that could be unsafe and of poor quality. Most of the times, the SMIs are seen to operate in clusters and close to resident populations causing thereby severe environmental and health impacts. The business operations in SMI typically change over time and are many times on the verge of "life and death". There is hardly any interest to reform the production processes.

Increased competition and globalization of trade have triggered however pressures on the SMI in developing countries today. These pressures include requirements on phasing out of environmentally harmful substances, substitution of hazardous processes by safer alternatives and satisfactory treatment of wastes to ensure meeting of the local law and reduced impacts on the neighbourhood. To be a member of the supply chain, some retailers (especially in the textile, leather tanning and food segments) have imposed requirements of social and ethical codes of conduct that ensure fair treatment and compensation to the workers.

This sure will be a process of gradual change – and the gap will slowly reduce between the manufacturers that insist and meet international manufacturing/sourcing requirements and manufacturers that limit to local markets where such demands are not generally made.

Today, we see SMIs in developing countries struggling to organize, survive and restructure their operations in this direction. Restructuring is seen to happen in terms of closures, amalgamations, modernization/expansions. In some cases, SMI's have indeed turned around to converting market pressures (or threats) into opportunities by taking advantage of cheaper resources, especially labour.

There are already signs of reverse product flow between developing world and developing countries because of tactical modernization, building technology absorption capacities and export orientation of the government policies. China and India have already becoming SMI powerhouses to the large textile, leather and electroplating retail houses and manufacturers of the developed countries and countries like Taiwan, Malaysia and Vietnam have already established their export niche in the electronic sector.

Unfortunately, these changes are rather left to the market and not steered well through efforts of local and national governments by devising a right mix of policies and strategies and by promoting financing and capacity building mechanisms.

UNSUSTAINABLE CONSUMPTION

Markets have always been undoubtedly the critical factors and are intimately tied up with the manufacturers/service providers. Although, the manufacturers/service providers are reforming the production practices on counts of competition, cost-concerns and branding, the consumption patterns in the word market have remained untamed.

Consumption patterns depend on economy, culture and advertising. With the propagation of television in rural areas and mobile handsets in urban areas, advertising is becoming a great stimulus to drive consumption patterns. Consequently, the average consumption patterns on per capita basis have doubled over the last two decades, leading to more intensive utilization of resources and waste generation. Some believe that as production will reform and roll more money in the pockets, the average consumption patterns will even triple!

The term ecological footprint is expressed as an area of land and water required to support a defined economy or population at a specified standard of living. Developed economies are considered to require far more land than they have, thus, through trade, impacting on resources in other countries.

Figure 1 shows distribution of average regional ecological footprints. It is evident that ecological footprints of the world are strongly divided – e.g. North Atlantic countries have footprints four times of the ecological footprints of the African world! Consumption needs to be looked at both "over consumption" and "under consumption" related perspectives. The Figure also indicates the importance of understanding product flows on a global scale and a need to consider the World Trade in a life cycle perspective.

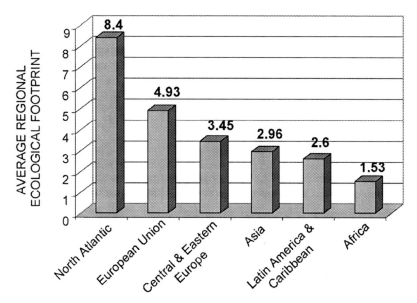

Figure 1 Average regional ecological footprint. (Data sourced from Global Status Report on Sustainable Consumption, UNEP DTIE 2002).

What you need is a change in the attitude that leads to avoiding, correctly choosing, sharing or using less followed by reusing, recovering the products and exchange wastes. But here we are talking about changing the behavioural pattern.

We need to expand life cycle thinking therefore to predict the behavioural changes instigated by innovations in products and technologies. The question is what changes and what behaviours are desired? What behaviours, techniques and devices can help us achieve them? To answer these questions, economists and technologists must collaborate with behavioural scientists (e.g. psychologists, sociologists, cultural anthropologists). There is also a need that the Governments develop a long-term policy based on insights into behavioural science. To do so, the Governments must take steps on their own to demonstrate their will and commitment.

The inclusion of sustainable development principles in procurement practices is already a reality in a number of countries such as Canada, the Netherlands, Norway, United States and South Africa. The experiences in these countries indicate that incorporating sustainable production and consumption considerations into public purchasing is not only a viable option, but also helps to develop sustainable markets.

Influencing consumption patterns in a responsible manner should become an interest of not just governments but that of the businesses as well who are driving the markets and are responsible for production. Industry should stop caring only about economic feasibility, and start paying attention to ecological applicability and social responsibility (the 'triple bottom line').

Industry should create new demands on "sustainable products" or pass on environmental costs in the price of the product to sensitize the consumers. The interaction between the market, the community of designers and the consumers is critical here.

Producers produce goods and services for consumption. Advertising enables the information on these goods and services to reach consumers. This would then translate into buying decisions if the consumer can afford or need it. However, often the need is created by enticing the consumer to purchase and consume. As advertising influences consumption it can be one important area of intervention to correct the worldwide unsustainable consumption trends.

In villages across India we can find TV sets that run on generators (as there is no regular electricity supply) but no such arrangement that village children may read or write in the evenings. Even the very poor of the villages get to see (and enjoy) the programmes and the advertisements on these TV sets and make slow changes in their consumption patterns [1].

For developing countries the problem is twofold as there is both over-consumption and under-consumption. Advertising and marketing not only makes people consume more but is also responsible for the consumption of luxuries while forsaking necessities. In India, it is estimated that advertising has an influence on the consumption habits of this group of about 431mn people. Advertising also, to an extent, influences the 'aspirants' who number another 275mn. Indeed time has come to observe 'Buy Nothing Today', or 'Watch no TV week.'

Advertising often leads to a switch from traditional time-tested products and services to costlier sophisticated alternatives. According to a McKinsey report on India's consumption patterns, Indians have given up many of their traditional items used as soaps, shampoos or medicinal remedies.

Advertising often results in overspending and over-consumption by creating artificial needs. And often, if not always, such over-consumption is unsustainable in the long run.

The urban poor are more affected as they are more open to the influence of unsustainable lifestyles through advertisements and also because of the contiguity to more affluent lifestyles. Also satellite television is more common in cities. Satellite TV channels being less regulated and having huge footprints cutting across rich and poor nations airs large number of advertisements of potentially harmful products and unsustainable lifestyles.

While there is the issue of producer responsibility as to what should be produced and how and in what amounts, there is a need to move towards more responsible advertising to reorient and tame consumption. The Paris-based International Chamber of Commerce (ICC), the biggest membership body of business, has a 23-article code of advertising practices. The code aimed to promote high ethics by self-regulation. However, this code is only a guideline and not a regulation. What may be better is the option of co-regulation of voluntary codes or their national adoption.

CONCLUDING OBSERVATIONS

Sustainable production and consumption are two key aspects that we need to focus simultaneously while seeking the goal of sustainable development. Mere focus on process oriented waste minimization is not enough. A life-cycle thinking is necessary in the commercial and regulatory world that examines alternatives and impacts right from extraction of raw materials to disposal of products, in a holistic manner. This will need that our databases on impact assessment are improved and the environmental modelling skills become more exhaustive.

To achieve sustainable production and consumption, the Governments, Businesses and Communities will need to act together as what we are taking here basically changing of the behaviour patterns. Sustainable product design and responsible advertising will play an important role here.

We must realize that world is still divided when it comes to production reforms and consumption patterns. Mainstreaming environmental and social thinking in the international trade may perhaps provide an important mechanism to address this gap. Here, the governments in the developing world will need to understand this broader dimension of trade, and provide support and capacities to enable the businesses (especially the SMI) to remain competitive.

REFERENCE

1. *Advertising and consumption: The Unholy Nexus,* Source – http://cuts.org/sc99-1.htm#top

Many of the suggestions made on responsible advertising have been based on this article

WASTE MINIMIZATION IN MINERAL AND ROCKS EXPLOITATION BY INTEGRATED MANAGEMENT AND PRODUCTION OF VALUE ADDED MATERIALS

M Grossou-Valta

J Mavroyiannis

F Chalkiopoulou

National Technical University of Athens

Greece

ABSTRACT. In the framework of new technological developments and international competition, the European industrial mineral wealth needs to be re-evaluated and new research targets set. Re-evaluation should take into account the requirements of modern society and industry, in relation with environmental protection and legislation aspects. The majority of current research projects related to the exploitation of industrial minerals' wastes aim to develop methodologies that contribute towards preserving mineral resources; extending reserves, water and/or energy; increasing recycling rates; and upgrading existing production procedures in parallel with the production of improved quality or new products. Very often these projects address specific market needs such as: (i) production of standard high quality, tailored raw mineral materials with enhanced properties that increase the value of the final products, (ii) production of ready mixes and (iii) development of new / improved products in order to substitute conventional materials or enter new marketplaces. Within this paper, five cases will be presented as examples of integrated industrial mineral and rock waste management. These are: (a) alternative uses of marble wastes, (b) use of aggregates rejects for the production of low quality fillers, (c) exploitation of run-of-mine magnesite fines, (d) treatment of feldspar wastes and by-products for the production of high quality feldspar concentrates and (e) use of olivine-dunite rocks for environmental applications.

Keywords: Wastes, Industrial minerals, Environment, Marble, Fillers, Aggregates, Magnesite, Feldspar, Dunite, Olivine, Composites.

M Grossou-Valta M.Sc., is a Mining Engineer and Metallurgist, Director in the Mineral Processing Dpt, ex-lecturer of N.T.U.A., with 30 years experience in mineral processing. She has vast experience in coordinating and participating in R&D national and European projects. She is also a panel evaluator for E.C. projects.

J Mavroyiannis Diploma in Mining Engineering National Technical University of Athens 1982. 18 years experience in IGME in Mineral Exploration / Evaluation, experience in participating in R&D projects as Project Leader.

F Chalkiopoulou is a Mining Engineer and Metallurgist. She is Deputy of the Industrial Minerals Section with 20 years experience in Mineral Processing and extensive experience in R&D projects' management. She is involved in a variety of local and international projects.

INTRODUCTION

At the beginning of the 21st century, the mining and quarrying industries, amongst others, are more and more pressed and are becoming more conscious of the need for enhanced environmental performance. Besides other impacts, they are also faced with the acute problem of the excess of wastes produced by their activities, and of their disposal. Rapidly rising public concern for the environment, the expansion of human activities over increasingly larger areas of land, the diversification of land-use, restrictive legislation adopted by states, all apply severe pressure on mining firms to: (a) confine themselves to smaller and remote areas, (b) reduce the amount of wastes released to the environment and (c) strictly control their pollution effect.

At the same time, physical depletion of rich mineral resources forces them to produce more and more waste per unit of useful material. As a result, the enterprises involved in the sector have concentrated effort on R&D projects in order to develop alternative uses for their wastes. The potential use of by-products and wastes by client industries could alleviate the difficulties experienced by mines and quarries in this respect, providing at the same time a basis for a better environment and a better quality of life.

Industrial minerals and rocks are no exception to the above situation. Even though such materials are considered inert in most of the cases, huge quantities of wastes and disposal area issues comprise a major environmental problem in many cases. Some example cases are examined within the framework of this article.

GENERAL APPRAISAL OF THE CURRENT STATUS IN EUROPE

The minerals' industry's contribution to the European GDP is about ten times larger than that of metal mining. The European minerals industry is world class and many producers are amongst the world's largest [1]. The statistics show a reasonably good economic outlook for several European minerals. For some of these commodities, the EU holds a significant part of the world supply – well above 20% of the total world output. The EU is also a significant producer of industrial rocks and the mining sub sector is by far the most important and active of all [2]. Depending on the mineral or rock type, as well as the mine/quarry size, various materials may stem from the exploitation (excavation and processing) that are usually stockpiled in the surrounding area. These wastes may account several million of tonnes. A sound example of this are the ornamental stones' derived from marble exploitation. Although these materials are inert there are impacts for the environment, such as severe landscape alteration.

In general, extractive operations display features that require a thorough, and sometimes delicate, balance between economic, environmental and social concerns. Extractive operations also raise the question of the depletion of non-renewable resources [3]. The priority issues expressed by the Commission for Sustainable Development of the extractive industry refer to increasing the level of environmental protection by improving the overall environmental performance of the industry and the sound management of mining waste, including recycling. Economic aspects, social performance and employment are the two other pillars of sustainable development for the extractive industry today. Also, the concept of a European Research Area (ERA) for research and technological development is now a reality within FP6.

The European need for better management and partial or full exploitation of wastes stemming from the exploitation of minerals and rock is a need well recognized within the European industry and the scientific community. It is reflected in the proceedings of several European networks, such as EUROTHEN, E-CORE, OSNET, where a significant number of projects address related problems.

Process technology has played a key role in the development of the minerals sector, particularly in the production of high-value products, and also can provide the means to develop specific properties so as to introduce improved or new products into the market. Also process technologies can utilize previously unusable minerals resources, and thus expand the reserve base [4, 5].

The development of mineral exploitation in respect to the environment has been directed towards addressing four issues:

1. New routes for the processing of conventional minerals and the production of conventional products with the application of more environmentally friendly methods/techniques.

2. Development of innovative processing methodologies for common, abundant rocks and minerals in order to produce materials appropriate for various uses.

3. Development of processing methodologies of industrial minerals and rocks, including wastes, in order to produce materials appropriate for new fields of application.

4. Development of innovative processing methodologies for wastes and by-products in order to produce high added-value materials.

Besides innovative technological findings and developments, other specific 'tools' are also currently available to support and orientate such efforts, namely IPPC, BAT documents and environmental legislation as well as Life Cycle concepts. New instruments such EMAS may also be adopted by the industry.

CURRENT STATUS IN GREECE

Mining activity in Greece is as ancient as the Greek civilisation and besides metals, the exploitation of certain non- metallics were known to the Greeks since prehistoric times. The Industrial Minerals' sector is quite dynamic in Greece and most of the mineral commodities produced are export oriented. The entry of Greece to the EU enriched several sectors of the EU minerals industry. In addition to the world famous Greek marbles, Greece produces and processes a wide variety of industrial minerals and rocks. It is a leading producer of bentonite, second only to the USA, the world's third largest producer of perlite, the chief global exporter of pumice, a major magnesite trader and the sole global producer of huntite [6, 7].

The use of Greek industrial minerals in traditional fields of application is well established within Greece, Europe and internationally (Figure 1). Innovative products, applications and considerable technological progress have been witnessed during the last decade. In many cases these are outcomes of European R&D projects. It is also noted that a significant number of projects, externally or internally funded, concern waste management, treatment and re-use for the production of high added-value products, new construction products and new applications.

Mineral Resources - Production Volume Share

Mineral Resources - Export Value Share

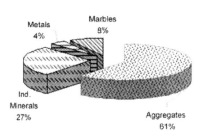
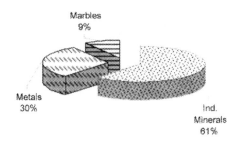

Figure 1   Production volume share and export value share
of Greek Non-Metallics compared to metals

Thus, in response to technological developments, international competition and environmental legislation, the Greek Industrial Minerals' sector has been activated towards: a) incorporating new production technologies, b) developing methodologies to process lower mineral grades and reduce stockpiling, c) promoting use of wastes in innovative combinations with other materials, d) producing high-added value products from low-cost wastes and by-products. A number of typical examples of such projects are presented in the following paragraphs.

EXAMPLE CASES OF INDUSTRIAL MINERALS WASTES TREATMENT

General

The research team of the Greek Institute of Geology and Mineral Exploration (IGME) has played a vital role in mineral sector and especially in the processing of wastes. For this purpose, the scientists of IGME have successfully collaborated with scientists from many European correspondent Institutes, as well as people from the Industry. The significant experience gained is reflected in the examples that follow. These were selected on the grounds that they represent cases of common interest within many countries of the European territory, as well as materials of significant quantities.

More specifically they concern: (i) exploitation of wastes and rejects stemming from the extraction and treatment of white and/or off-white carbonate rocks (either marbles or aggregates), (ii) processing of residues from feldspar exploitation, (iii) processing of wastes and off-grade ores from magnesite excavation and treatment (dunite - harzburgite gangue, run-of-mine fines). According to the procedures followed in all the above cases, a generic flow sheet may be devised (Figure 2), describing the basic steps necessary in order to develop a feasible plan for the exploitation of wastes.

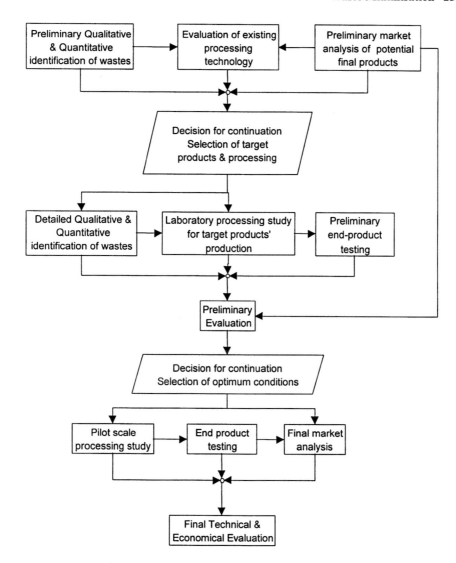

Figure 2   Generic Flow Diagram to Evaluate the Potential of Wastes Exploitation

Marble Waste Exploitation

Greece is one of the major European marble producers. According to recent data and calculations, the annual production is at least 400,000 m^3, comprising mainly calcitic and dolomitic marbles, while a minor portion are limestone ornamental stones and travertine. Considering that the average recovery ratio in a marble quarry does not exceed 15%, it can be estimated that the correspondent waste quantity generated annually approximates 7.000.000 tonnes.

These materials are stockpiled in the surrounding areas of quarries affecting landscape aesthetics, while non-renewable natural resources remain unexploited. The path to the exploitation of ornamental stone wastes is a complex one. Very often one has to investigate the perspectives of a vast spectrum of uses in relation to local and foreign market needs whilst keeping in mind that the more processed a product is so to is its value and marketability. Thus, production of aggregates may be the easier alternative for such wastes. However, the feasibility of this specific use depends on the quarry location in relation to the consumption centers and the appropriateness of the raw materials' properties. In order to achieve higher prices, the possibility for the production of high-added value products must be investigated.

The most promising use for a significant number of types of these wastes is the production of fillers, in the case, of course, of white materials. Calcium carbonate is by far the most widely used mineral filler in paper, paint and plastics. Chalk, limestone and marble are the main raw materials used for the production of ground calcium carbonate (GCC) in Europe. Depending on the type of paint or coating different filler properties are examined. GCC loadings in paint generally fall within the 10-35% range. The loading levels of the filler in resins vary according to the resin type and may be 20-60%. General requirements for GCC as filler in plastics are low cost, low oil absorption, dispersibility, low abrasion and high brightness [8].

Minerals are used in paper as fillers substituting minerals for fibre or as part of the coating in coated paper. Among the different paper mineral filler types in Greece, limestone and marble are the prevailing ones. For the establishment of a new type of filler, important characteristics are the price, the top cut size, the standard size distribution, the whiteness and the chemical composition and content in elements affecting the final performance of the product (e.g. iron content etc).

The main problem addressed in the Greek market of carbonates is the luck of standard quality. Due to this fact many industries using carbonates in their production procedure are importing the material. On the other hand, there is a wide range of applications (adhesives, asphalt membranes, detergents, etc.) that require lower filler grades. In such a case the price of the material would be decisive for its marketability. It is noted that materials with size distributions between 10-40μm cover about the 40% of the market and prices can vary from 0,020 – 0,22 euro/ kg depending mainly on the top cut size and the surface treatment.

IGME has carried out a number of relevant projects, on both a national and European level, in collaboration with organisations, universities and enterprises. Typical examples include: (i) a regional study for waste exploitation from the Epirus prefecture ornamental stone quarries, (ii) the development of low cost modular dwelling components based on the combined use of marbles waste raw materials and an innovative polymeric formulation.

Epirus prefecture marble wastes [9]

The aim of the project was to conduct a technological evaluation of the existing marble waste of the Epirus prefecture (NW Greece) for filler production and animal feed. The project included geological mapping, processing tests, measurement of specific properties, application tests, local market research, development of a GIS data base, and a final technological evaluation. It is noted that the limestone waste quantity is estimated at 10,000,000m^3. Three major uses were examined, resulting to the following conclusions:

- The material may be used as filler combined with polymers for *composite prefabricated products* (Figure 3). Ground and classified material (<250µm) was used in combination with polyester resins up to 30%. The results were promising. Further research is needed in order to reduce the portion of the resin.

- The material may also be used as powder filler in the *paint industry*. Laboratory testing showed that selected materials may be used for the production of filler having Dr Lange whiteness >90%, applicable to a variety of products, even for white emulsions and acrylic paints.

- The material may finally be used as an additive in *animal feed*. It was successfully tested as an additive to the feed of ruminants and poultry.

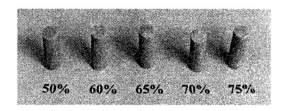

Figure 3   Composite specimens for testing (aggregate weight portion is presented)

Development of modular dwelling components [10]

This case study refers to a CRAFT project with the main objective of developing innovative, entirely factory made, easy to erect, flat pack, modular, interlocking composite construction panels, out of which low cost housing could be easily assembled by unskilled local labour. Fundamental to the project is the use of waste raw materials, consolidated within an innovative polymeric matrix. The use of waste raw materials from the quarrying industry is expected to provide a positive influence on cost effectiveness, and a reduction of environmental impacts. Within the project, many combinations of different marble wastes (from Portugal, U.K., Greece) with innovative polymeric matrices were studied in order to:

- Increase the raw material proportion within the matrix, as compared to the usually applications,

- Import excellent properties (iso-metric strength to weight ratio, flame-proof, hydrophobic and wear resistant facings) to the end products, far better than the conventional concrete prefabricated panels.

The technical concept includes also the incorporation of an insulation material (like polyurethane foam), which is expected to provide excellent thermal and noise insulation. Finally, ease of use is anticipated for these products, making them suitable for conventional building, and ideal for constructions in cases of emergency, where the need for quick building is required (e.g. refugees' accommodation, post-earthquake housing).

Aggregate Quarry Rejects Exploitation [11]

This case was thoroughly examined within the framework of a Brite EuRam project. One aim of the project was to develop filler applications for the fine residues from a typical Greek aggregate quarry. Many laboratory and pilot plant tests were performed on the material, including ultra-fine grinding with a laboratory jet mill. Many properties of the powders were also measured such as whiteness, oil absorption and abrasiveness in order to appraise their appropriateness for various uses. Extensive application tests were also performed by a Greek paint producer and a Greek paper producer.

According to the results obtained within the project from the laboratory test work, specific by-products stemming from the production of limestone aggregates and mainly fines from dust capture circuits (~5% of the production) may be further processed with fine and ultra fine grinding. The products obtained are fine materials (top cut ~10μm) with $CaCO_3$~97-98% and Dr Lange whiteness ~88-92% (Figure 4). Such carbonate materials may be used as lower grade fillers by the plastics and paints industry and potentially by the paper industry. It must be noted that the production cost for such grades will be decisive for further exploitation.

It was also concluded sand from the aggregate production may be used as feed in a potential future circuit for filler production. It must be noted that the "fillers" produced under these conditions are medium grade appropriate for less demanding applications.

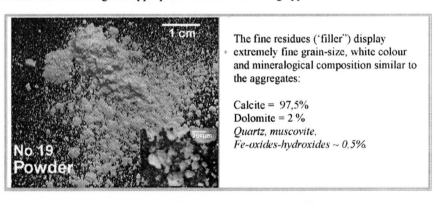

The fine residues ('filler") display extremely fine grain-size, white colour and mineralogical composition similar to the aggregates:

Calcite = 97,5%
Dolomite = 2 %
Quartz, muscovite,
Fe-oxides-hydroxides ~ 0,5%

Figure 4   Calcitic aggregate residues as fillers

Feldspars' Exploitation Residues Processing [12]

World Market analysis shows an increasing demand for high quality feldspar concentrates. The Greek reserves account several million tonnes, mainly sodium feldspar and secondary potassium feldspar (Central Macedonia, Thrache). Large reserves of potassium feldspar exist in the tuff rocks of the island of Milos. The perspectives of the domestic market are envisaged to be increasing in future years mainly for tile manufacture with expected consumption growth to around 35 million m^2 [12].

A specific case was studied concerning feldspar deposits in N. Greece within the frameworks of both a Brite EuRam and a national project. In this case, the host rocks are pegmatites with the main impurities being iron minerals ~1% and occasionally titanium minerals up to 1.5%.

Mining of deposits in this area is conducted selectively and thus unexploited lower grade deposits are left behind. More specifically, 10% of the totally exploited ore is of high quality, 30% is of lower quality, but saleable at a lower price, and 60% is a non marketable waste material that still contains a significant recoverable quantity of minerals.

Taking into consideration environmental aspects for non-fluoride treatment of the material, a detailed laboratory and pilot plant investigation was performed in order to develop efficient methodologies for the processing of the lower grade deposits. It was shown that:

- The proposed technologies eliminate completely the use of hydrofluoric acid which creates tremendous environmental and health problems.

- According to application tests, the end products are feldspar concentrates of high quality and mixed feldspar-quartz concentrates of low iron content suitable for the ceramic or glass industry.

- There is economic interest for industrial exploitation of the project's results

Magnesite Exploitation and Processing

The problem of residues stemming from the exploitation of magnesite has been addressed several times in Greece. The residues vary and are derived from several steps of the excavation-processing procedure. Their chemical and mineralogical composition and magnitute depends on the ore characteristics and the beneficiation applied. Many different types of wastes may stem from the exploitation of the same ore. Two example cases are referred to below: one is relevant to magnesite run-of-mine fines, and the other is relevant to the overburden wastes from the excavation of magnesite ore at the same areas in N. Greece.

Run-of-mine (R.O.M.) magnesite fines [13]

This case study refers to a magnesite ore deposit that has been mined and processed since the beginning of the century. About 4.3 million tonnes of crude magnesite ore were annually treated in the existing beneficiation plant in the period 1990-1995, to produce over 300,000 tonnes of raw magnesite. This product is further fired in a plant with a total capacity of 200,000 tonnes, to produce different types of dead burned and caustic calcined magnesia.

The processing routes that were applied included optical sorting effective (at that time) in the size range of +12mm. Thus, from the treatment of the aforementioned magnesite ore deposit, a significant percentage of the run-of-mine feed (26.1 % to 37.5 %) with grain size <12mm, containing about 30% $MgCO_3$, has being disposed and stockpiled as waste material. It is noted that 50×10^6 t. of this R.O.M. had already been stockpiled (1992) and this quantity had annually increased by 1.6x10*1.

This fact has a number of serious impacts: a) high disposal area requirements, b) valuable mineral ($MgCO_3$) losses and c) transportation and other operating cost burdening. During raw material appraisal it was discovered that it is very difficult to process, mainly for three reasons: increased content of 'fines' (<5mm), many different kinds of impurities, and wetness/stickiness. The solution was applied "upstream", i.e. the pre-separation of fines by drums/rolls before scrubbing, rather than by dry screening or by washing/scrubbing the entire

0-12mm run-of-mine material, proved simple and quite effective (Figure 5). It was also established that further downstream in the flowsheet, no single method could provide the required results, but rather a combination would be necessary.

Magnesite R.O.M. wastes - Stockpiled

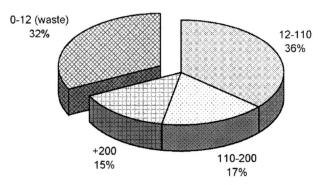

Figure 5   Participation of 0–12mm material in the excavated ore

Intensive research was performed at laboratory and pilot plant scale, including modeling and application of innovative technological achievements (TRIFLO heavy media separator, cryogenic magnetic separator, drum 'sieving'). New equipment was also designed and commissioned by the owner Company, namely a new reticon camera sorter for the 6-12mm material.

According to the project results, the problem of the recovery of magnesite from run-of-mine fines came close to a solution and that, in the process, significant independently-viable hardware and software "by-products" were generated. Productivity gains of the order of 60% were foreseen for the magnesite production circuit, with consequential lowering of overall costs and reduction of waste disposal area requirements. Furthermore, it is envisaged that the beneficiated magnesite and the derived products could result in an increased market share into the EC at the expense of outside competitors.

Dunite - Harzburgite waste from magnesite quarrying

Abundant, large in size, dunite bodies occurring in ophiolite complexes in Northern Greece that may host mineral deposits such as magnesite and chromite. Therefore, during the exploitation of magnesite, considerable quantities of dunite may stem as overburden waste or gangue that may be used for olivine production. Olivine sand is appropriate for the substitution of the hazardous quartz sand used in foundries. Olivine is also suitable for the neutralization of industrial waste acids. The potential for these applications was investigated within a Brite EuRam project. The work programme included extensive industrial steel casting tests. The use of several binders and resin-bonded processes was also studied. The project results were evaluated from technical and economical points of view. More specifically the above study concerned a number of deposits. An important case examined was that of the dunite - harzburgite wastes produced from a magnesite ore open pit. The scope of the specific research was to investigate the possibility for production of olivine foundry sand. It was concluded that from the overall rock extraction quantities (800,000 tpa) about 140,000 tpa consist of gangue material that may be used for the production of approximately 72,000 tpa olivine foundry sand (<0.6 >0.2mm). The industrial tests performed verified the potential for this certain application.

CONCLUSIONS

The problem of mineral and rock exploitation wastes extends over all of Europe, with the main environmental impacts being landscape alteration and land use problems. Besides this, these materials represent non-renewable natural resources which, under the view of current concepts for competitive and sustainable growth, should be efficiently exploited. Solution to the problem through partial or full use of such wastes requires the combined effort of both mining enterprises and hi-tech manufacturers as well as qualified and experienced laboratories. Altogether they seek to introduce raw materials, currently considered as redundant by- products of the extractive and processing industry, for the needs of the construction, cement, chemical and agricultural sectors [14].

Most of the projects designed and run nowadays for the exploitation of mineral waste aim to develop methodologies that contribute in: (i) mineral resources preservation/extension of reserves, (ii) water/energy preserving, (ii) increasing of recycling rates, (iv) upgrading of existing production procedures and generally an enhanced environmental performance of their operations.

The creation of markets is also an aim of such projects, including: (i) production of standard high quality, tailored raw mineral materials with enhanced properties that increase the price of the final products, (ii) production of ready mixes, (iii) development of new / improved products in order to substitute conventional materials or enter new market places.

Decrease of waste produced on site in mines, quarries or plants will minimize the environmental problems caused by their disposal. Areas affected by deposition will be smaller, more land will become available to other social activities and human recreation, and damage to the landscape from waste piles will be less, as will the danger of soil, surface and ground water contamination.

REFERENCES

1.   EUROMINES Annual Report 2002

2.   E.G.S., Opinion 9 "Minerals in Europe: The risk of outsourcing", 27 March 2000

3.   COM (2000) 265 Final "Promoting sustainable development in the EU non-energy extractive industry"

4.   M. GROSSOU-VALTA, F. CHALKIOPOULOU: "The value added to Industrial Minerals after Processing with New Technologies", Workshop with the title: "New techniques in the aggre-gates' quarrying industry of Cyprous", Nicosia - Cyprous 21/11/1998.

5.   S.E.VALJ: "Process technology a key to increasing profits in industrial minerals" Industrial Minerals Sept. 1999

6.   M. GROSSOU-VALTA, F. CHALKIOPOULOU: "The Greek Industrial Minerals and their applications in local and foreign markets", Greek Ceramic Company Conference, Athens 1998.

7.   K. HATJILAZARIDOU, M. GROSSOU-VALTA, F. CHALKIOPOULOU: "Greek industrial minerals, current status and trends", Industrial Minerals June 1998.

8.   F. CHALKIOPOULOU, M. GROSSOU-VALTA: "The prospects for the exploitation of the European White Marbles as Raw Materials for Fillers' Production", under edition, volume devoted to the NTUA Professor, A. Frangiskos, April 2001.

9.   N. KAKLAMANIS, F. CHALKIOPOULOU: "Technological Study of Industrial Minerals, By-products and Quarry Wastes in the Epirus Prefecture for their Suitability as Fillers", national funded project, internal report, PEP 1999-2001.

10.   J. MAVROYIANNIS "Development of Low Cost Modular Dwelling Components Based on the Combined Use of Waste Raw Materials and Innovative Polymeric Formulation" CRAFT-1999-70263, 2001-2003.

11.   M. GROSSOU-VALTA, F. CHALKIOPOULOU, "Development of novel processing for the production of low cost by-products fillers as a replacement for high-cost fillers" BE97-5078.

12.   P. CHARALAMBIDES, M. GROSSOU-VALTA, AND N. ARVANITIDES, "Environment Friendly Treatment of Feldspar Waste and By-products for the Production of High Quality Feldspar Concentrates", Conference "Recycling and Waste Treatment in Mineral and Metal Processing-Technical and Economic Aspects", Volume 2 p.377-388, Lulea Sweden 16-20 June 2002.

13.   V. NIKOLETOPOULOS, F. CHALKIOPOULOU ET AL: "New process routes for the recovery of magnesite run-of-mine fines", Synthesis Report, BRE2-CT92-0388.

14.   M. GROSSOU-VALTA, F. CHALKIOPOULOU,: «Technology, a tool for the increase of the mineral ore reserves and the elimination of the environmental impacts», 3rd Congress of Mineral Wealth, Greece, Athens, November 2000.

THE USE OF GIS FOR OPTIMISING WASTE COLLECTION

G Wassermann

BOKU University

Austria

ABSTRACT. Waste collection and waste transport account for approximately half of the costs for all the processes under waste management. This makes waste collection a very important process for optimisation in waste management. The ecological impact of waste collection and transportation within the whole system is dependent upon the transported waste material, but it can rise to a very high extent. Under waste collection there are several possibilities for making improvements such as: the collection system, collection trucks, mode of transportation or routing. The main focus in the presented paper has been placed on the route planning and optimisation of collection routes. On the one hand many of the current collection routes came into being due to their history and do not have any economic or ecological foundation. On the other hand in order to bill each municipality properly, waste is collected separately. Using the example of two case studies, we will discuss to what extent geographic information systems are suitable for optimisation in waste collection and whether it is possible to find guidance levels for the shortest general route to use as a sort of benchmark for waste disposal companies or as a tool to control expenditures for municipalities.

Keywords: GIS, Geographic information system, Routing, Route optimisation in waste management, Waste logistics, Benchmarking.

Gudrun Wassermann has been a research assistant at the Department of Waste Management at BOKU University, Vienna, since 2001. She works in the field of waste management and environmental management, waste collection systems, waste prevention, recycling technologies and methods for decision-making in waste management. Her special interests include the application of LCA and Geographic Information Systems in waste management systems.

INTRODUCTION

Modern waste management aims to minimise the impacts on natural resources and protect the environment from the negative influences associated with waste processing. Apart from the aspect of ecological optimisation, private waste disposal enterprises as well as municipalities are also particularly keen to minimise their expenditures for municipal waste collection.

Most municipal waste is carried to disposal or recycling sites by collection trucks. This road transport is expensive and causes considerable amounts of air pollutants [1]. WOLFBAUER & LORBER [2] asserted that waste collection and waste transport account for approximately half of the costs of all the processes in waste management. An Austrian case study showed that the ecological impact of waste collection and transportation within the whole system depends upon the transported waste material, but can rise up to 90% [3, 4]. This makes waste collection a very important part for optimisation in waste management in terms of both the cost aspect and the ecological impact. As one can imagine, this ecological and economic optimisation often go hand in hand with one another.

Under waste collection there are several possibilities for making ecological and economic improvements such as:

- Optimising the collection system (bring system or kerbside collection);
- Optimising the collection trucks (smaller or bigger trucks, private cars); or
- Optimising the mode of transportation or routing.

Often it is difficult to estimate the impact of changing the collection system or the collection route, because it is not possible to test the different alternatives. To this end geographic information systems are a well-known decision support tool.

A Geographic Information System (GIS) is a computer-based set of tools for the study, analysis and visualisation of objects and processes by integrating various sources of data on the basis of a common geography. With the aid of GIS, it is possible to analyse data in a very efficient way. Using analytical methods, one can get new information. In this paper we shall examine:

- to what extent geographic information systems are suitable for optimisation in waste collection; and
- whether it is possible to find guidance levels for the shortest general route to be used as a sort of benchmark for waste disposal enterprises or as a tool to control expenditures for municipalities.

THE USE OF GIS FOR OPTIMISATION IN WASTE COLLECTION

Depending upon the collected waste type and the spatial spread of the collection area, one has to take different aspects of optimisation into consideration. In particular, the required data (waste data as well as spatial data) for the analyses have to be taken into account. Depending upon the extent of the investigated area and the settlement structure, different qualities of all spatial data are needed.

Different spatial units can be defined as:

- every single household (simplified as furthermost regions that must be reached);
- special parts of cities (e.g. recycling banks, civic amenity sites); and
- boroughs, cities and villages (only relevant for superior routing).

To begin with a proper GIS analysis for the optimised routing, the following data are needed as a precondition:

Spatial data: First a digital database is needed for the street courses for the area under investigation. Normally it is easy to get good data from the superior street network, but it is difficult or expensive to get data on the lowest street level. The street data have to reach every point of waste collection.

Second the locations of the collection points are needed. Depending on the investigated areas, there may be problems pinpointing all the collection points exactly. If only the furthermost points are considered, this should lead to similar results. At least for Austria it is difficult to get more detailed digital spatial information than on the city level. Another possibility is to digitise the information oneself.

Waste data: For proper analysis it is necessary to know at least the amount of waste per year, the capacities of the collection trucks and the minimum collection frequency. The following two case studies shall show the use of GIS for the optimisation of two different collection systems.

Collection of Municipal Waste with Kerbside Collection in Rural Areas

The collection of household as well as commercial waste (similar to household waste) requires a continuous pickup of waste at each single household. The collection of municipal residual waste was investigated for the Austrian district of Hartberg in order to assess the optimisation potential. Hartberg has an area of approximately 958 square kilometres. Two waste disposal enterprises are responsible for waste collection among 50 municipalities.

The current collection routes came into being as a result of their history and have no real economic or ecological foundation. In order to bill each municipality properly, the waste is collected separately.

A digital street network as well as waste collection data (waste quantities, number of bags and bins etc.) were available, so it was possible to investigate the optimisation potential in waste collection to a reasonable degree. The desktop GIS application ArcView 3.1 with the network - extension was used to find optimised routes for waste collection within the whole district.

Firstly, the existing collection systems (collection frequency, assignment of a collection area to a certain collector, collection routes) were left unchanged in order to simulate the current situation. Secondly, the optimisation potential through optimising the collection frequency was investigated. Thirdly, the possibility of extending the collection routes beyond the municipal border was investigated and the collection area was assigned to the closer of the two waste disposal enterprises.

The top part of Figure 1 shows an example of the existing waste collection, and the bottom part shows an example of the optimised routing and the optimised division of the collection areas.

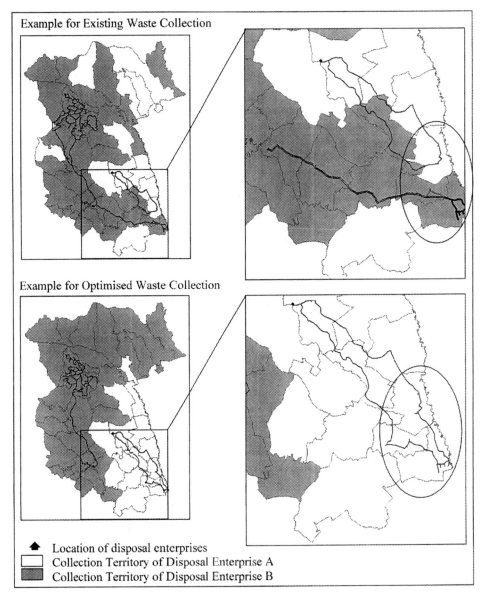

Figure 1 Example of the existing and the optimised waste collection for six municipalities in Hartberg. The right section shows the zoom for three municipalities.

The top example shows that the same collection truck has to drive the same way twice and that the collection truck from the second disposal enterprise has to drive another route. In the optimised example, only one route has to be driven to achieve the same result. The new arrangements for the collection areas help to shorten the routes.

The comparison of the calculated driving kilometres for one year showed that by optimising the collection frequency, 25% of the distance could be saved. By optimising routing across municipal borders, 32% could be saved (Figure 2).

The real optimisation potential with the use of geographic information systems seems to be low at about 10% (26,445 km after the frequency optimisation and 23,665 km after the cross border routing optimisation). This is due to the fact that in the investigated area 15 out of the 50 municipalities already have a nearly optimised collection system. Combining those municipalities with others is not possible due to the high amount of waste.

If one compares the effect for only those regions where optimisation is possible, another 20% of the existing collection kilometres can be saved. For the three municipalities shown in Figure 1 after changing the allocation of the collection area to one waste disposal enterprise and after optimising the routing across municipal borders, the collection distance could be reduced from 742 km to 329 km. This means an optimisation potential of more than 50%.

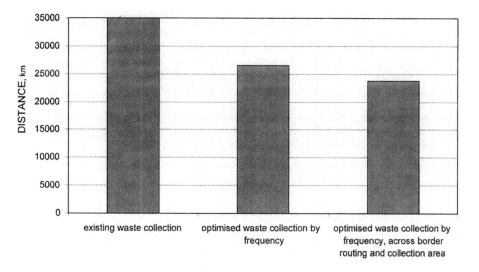

Figure 2   Comparison of results for existing and optimised waste collection

Collection of Municipal Waste in the Bring System at Collection Points and Collection of Commercial Waste

Waste collection with the bring system causes a higher amount of waste at each collection point but fewer collection points than with kerbside collection.

Under the context of an investigation into the ecological relevance of transport logistics in waste management [3, 4] the optimisation potential for the transportation of old refrigerators was investigated [5]. A polygon scheme incorporating the areas of the different communities and the grid of primary and secondary roads along with pertinent statistical and demographic data (such as population and area dedication for specified uses) served as the basis for establishing a GIS model.

To estimate the transportation distances similar to the aforementioned example for residual waste collection, a route planning system using three different scenarios was investigated and a comparison of these cases was performed.

The first scenario starts from the operational base of the waste disposal enterprise from which point trucks with a maximum load capacity of 50 units pick up the old refrigerators at the collection locations and haul them back to the operational base in direct journeys. Under the second scenario, circulating trips, which head from one collection station to the next were completed. Based on a given number of refrigerators in the specific area per year, a certain number of such trips would result. No time restrictions were taken into consideration. This case would result in the maximum transportation distances per unit. For the last option, planned tours were assumed, taking into consideration restrictions in load capacity and available time. The required segmentation of tour planning was done interactively.

The investigation presented a result of nearly equal distances for direct journeys and planned routes. The reason for this surprising result could be found in the fact that at most of the collection stations more than 50 refrigerator units accumulate and thus represent the full load capacity of the transport vehicle; no optimising is thus possible. The advantages of route planning, however, can be put to good use in integrating these various collection points when only small quantities of obsolete refrigerators are the rule.

In comparing the calculated values with a value achieved using the practical operation of one specific waste disposal enterprise, it could be shown that optimised planning of transportation routes could reduce transportation expenditures by more than 50%.

BENCHMARKS

Apart from the use of GIS for route planning as a direct use, it can also be interesting to find guidance levels for the shortest general route as a sort of benchmark for waste disposal enterprises or as a tool to control expenditures for municipalities. In the above mentioned study on the ecological relevance of waste transports [3], the attempt was made to establish such criteria for disposal systems in terms of transport expenditures. A research project to classify Austrian settlement patterns [6] served as the basis for the study. To classify waste-related structures, the following additional input parameters were taken into consideration: general spatial information, street graphs on the basis of communities, waste generation on the basis of communities, number of pick-ups per month, and the location of bags and bins.

In a first step, optimised routes for kerbside collection were acquired and the distance covered in kilometres was noted. In a second step, the investigation tried to find a correlation between collection output (km per tonne) and the settlement structure. The analyses for the collection transport showed that for waste-related questions it is possible to establish five categories as shown in Table 1.

For the category "waste management co-operatives, concentrated, small" the investigations amount to an extensively harmonised picture for all the examined regions. It was thus possible to define an average collection transport distance of 11 km per tonne for plastics and compounds. This figure seems to be independent of other influencing factors.

Table 1   Established categories for collection transports.

	CATEGORY				
	1	2	3	4	5
Organisation	sep. munic. waste coll.	waste management co-operative			
Settlement structure	urban	concentrated		dispersed	
Catchment area		small	large	small	large

For the dispersed areas no uniform image could be found. An average value of 32 km/t has to be defined. The measured values of the GIS were between 15 and 47 km per ton of collected material. Altogether the values for the collection areas in the dispersed regions are explicitly higher than for the concentrated regions.

In addition to the real collection transport, the so-called distance transport between single collection regions is also relevant. It can be assumed that the collection areas of the waste management co-operatives approximately match the canton borders. With the use of GIS, the distances were calculated for the reference cantons, which reflect the Austrian reality from large (more than 2,000 square kilometer), to small (less than 1,000 square kilometer). The investigations resulted in a coefficient for the distance transport of 5/3 of the radius with a large conformity for all the catchment areas. The transport output was considered for the two extremes, with 10 km/t for small waste management co-operatives and 34 km/t for large waste management co-operatives.

CONCLUSIONS

Geographic information systems seem to be a limited but useful tool for optimisation in waste collection. Two case studies in Austria showed that in many cases, in spite of the fact that collection routes are often the result of a historic evolution, the optimisation potential with the use of GIS often is smaller than the optimisation potential for example of optimising the collection frequency.

A very good field of application for GIS can be found in particular for smallish collection regions, where because of the low amount of waste per region and year there is a possibility of pooling such regions together. Here the use of GIS can lead to an optimisation potential of about 50 %.
Furthermore it could be shown that sometimes the actual collection route arbitrarily chosen by waste disposal enterprises differs very much from the optimal route. In one case study an optimisation potential of 50% was calculated. The use of GIS as tool to control expenditures seems justifiable despite potential expenditures for purchasing the software and digital databases.

Another aspect to using GIS as tool to control expenditures is the creation of general benchmarks to determine the shortest route depending on different settlement structures, trading areas and the organisation of waste disposal. In an initial investigation five different categories for waste management problems could be established. For only one category with a concentrated settlement structure and small trading area was it feasible to find a sort of benchmark. For urban regions it should be possible to find another correlation, but at the time of the investigation the available database was too small to achieve clear results. For this aspect, further investigations should be made.

REFERENCES

1. BURGER, A., HARTMANN, K., Optimierte Kehrichttransporte belasten die Umwelt weniger (Opimised waste transports pollute the environment less), Umwelt Aargau, Nr. 8, Januar 2000, pp 21-24.

2. WOLFBAUER J.; LORBER K., Benchmarks for Municipal Waste Collection. In: VIIth International Symposium Waste Management Zagreb, pp 301-322, 2002.

3. SALHOFER S., SCHNEIDER F., WASSERMANN G., Ökologische Relevanz der Transportlogistik in der Abfallwirtschaft (The ecological relevance of transport logistics in waste management). Waste Reports Nr. 10, Wien 2002.

4. WASSERMANN G., SALHOFER S., SCHNEIDER F., Ecological relevance of transport logistics in the separate collection of LDPE films and EPS. In: VIIth International Symposium Waste Management Zagreb, pp 285-299, 2002.

5. ODIC, M., Anwendung von GIS zur Abschätzung von Abfalltransporten am Beispiel von Altkühlgeräten (Application of GIS to estimate waste transports using the example of old refrigerators). Project report, unpublished, Neudörfl, 2002.

6. FEILMAYR, W., und KALASEK, R., Klassifikation des österreichischen Siedlungsraumes (Classification of the Austrian settlement area). Unpublished project report i.A. der TELEKOM CONTROL Ges.m.b.H.. 2000.

SUSTAINABLE WASTE MANAGEMENT IN THE CATALAN WINE INDUSTRY

M T Gea Leiva   **A Artola Casacuberta**   **A Sánchez Ferrer**

Escola Universitària Politècnica del Medi Ambient

X Sort Camañes

Miguel Torres SA

Spain

ABSTRACT. The wine industry in Catalonia (Spain) plays an important role in the economy of the region. Miguel Torres S.A. is a well-known industry specialised in the production of high-quality wines and brandy. Miguel Torres possesses its own vineyard.

Two of the main solid wastes produced in this kind of industry are:

- Stalk, waste from grape harvesting: a mixture of leaves and branches from vineyard collected during grape harvest. This waste is only produced during September and October.
- Wastewater sludge from the operations of wastewater treatment, mainly biological sludge. They are steadily produced during the year.

These two wastes are usually treated outside the wine industry by sanitary landfilling or incineration. Miguel Torres would like to integrate the management of these wastes into its waste management system. In this work, a composting process is proposed to treat these two organic wastes generated in the wine-producing industry. The wood-based wastes (mainly branches) have the porosity required to act as a bulking agent for the biodegradable wastes (leaves and wastewater sludge). The composting process produces a stabilized pathogen-free fertilizer (compost) that will be recycled in the vineyard crop, as Spanish soils are poor in organic matter. The process is proposed as a sustainable waste management strategy for the wine-processing industry.

Keywords: Composting, Wine industry, Wastewater sludge, Stalk, Waste management.

M T Gea Leiva is a PhD student at the Escola Universitària Politècnica del Medi Ambient in Barcelona, Spain.

A Artola Casacuberta is studying at the Escola Universitària Politècnica del Medi Ambient in Barcelona, Spain.

A Sánchez Ferrer is a Professor and Senior Researcher at the Escola Universitària Politècnica del Medi Ambient in Barcelona, Spain.

X Sort Camañes works with Miguel Torres SA in Barcelona, Spain.

INTRODUCTION

The wine industry in Catalonia (Spain) plays an important role in the economy of the region. Spain is a European leader with respect to innovations in this area. Big companies' investments have been characteristic in the last years, which have supposed the improvement of the quality of wines and spirits and the image of the cellars [1]. Miguel Torres S.A. is a well-known industry specialized in the production of high-quality wines and brandy. The main operations in the production process of Miguel Torres S.A. are grape harvest; grape pressing; grape juice fermentation; wine clarification; wine filtration; wine bottling; product distribution. During the production process significant amounts of liquid and solid wastes are generated. Miguel Torres has a wastewater treatment plant in its installations for the treatment of all the wastewater and vinasse generated in the industry.

The 80% of the total generated wastes are organic wastes or by-products. The main organic wastes produced in Miguel Torres S.A. are grape pomace (50%), lees (20%), stalk (7.6%) and dewatered sludge from the wastewater treatment plant (7%).

Recycling of grape pomace (grape peels and seeds) which is produced during the grape pressing, and lees, consisting of the sediments from the fermentation and the clarification operations, is well established. They are used as by-products in the alcohol-producing industries for the production of alcohols and colorants, which are used in the elaboration of spirits and in the painting industry, and tartaric acids, which are used in the pharmaceutical, food and wine industries. The remaining materials after alcohol extraction of grape pomace are peels and seeds. Peels are composted to obtain a high quality fertilizer, or used for production of animal feed. Grapeseed oil is produced and the remaining material after the seed pressing is used for energy recuperation.

The management of the other two important organic solid wastes, stalk and sludge, is by means of an external company. Miguel Torres possesses its own vineyard and given that Spanish soils are poor in organic matter, vineyard crops require organic fertilization.

At present, there is a tendency towards the agricultural use of sludge. The unique legal restriction to direct land application of the sludge is given by its content on heavy metals and potentially toxic compounds. Clean sludges coming from agricultural industries may be suitable for direct land application. European Commission is preparing a new Directive on the biological treatment of organic wastes [2]. According to this Directive the hygienisation of sludge before its application to land will be mandatory. Composting is one of the alternatives proposed by the Directive in order to achieve sludge hygienisation.

Composting is the biological decomposition and stabilization of organic substrates, under conditions that allow development of thermophilic temperatures as a result of biologically produced heat, to produce a final product that is stable, free of pathogens and plant seeds, and can be beneficially applied to land [3]. Main factors influencing the composting process are temperature, water content, oxygen concentration in the composting matrix, porosity and free air space (FAS). Temperature is both a consequence of the composting process (microbial metabolism) and a control parameter. Temperatures providing the maximum degradation velocity are in the range 40-70°C [3]. The water content of sewage sludge is usually too high for direct composting even if a dewatering treatment is applied to this material. In a majority of cases a bulking agent is needed to reach optimum water content, porosity and free air space (FAS) [3]. A lot of experiences are reflected in the literature referring the research of optimum sludge-bulking agent mixtures for urban or industrial sludge composting. In addition to the type of bulking agent used, its particle size has been also empathized as an important factor in sludge composting. Large bulking agent particles will provide an excess of FAS that will result in an oscillating temperature profile during composting [4]. The proportion of bulking agent in the final mixture has also been highlighted as a factor influencing the composting process [4,5].

An excess of bulking agent will lead to a low content on biodegradable organic matter in the final mixture. Sludge composting experiences are presented in some cases at different scales, laboratory and pilot plant [4,6]. The main difference observed at different scales is temperature evolution. The amount of heat generated and the heat retention capacity of the material increase as the volume of the treated material increases [7].

Stalk appears to be an ideal bulking agent because of its branched structure, which provides a high porosity. Co-composting sludge from the wastewater treatment plant with stalk acting as bulking agent would generate a stabilized fertilizer suitable for its application to the vineyard crops. Moreover the composting process allows treating different wood wastes that are generated sporadically, as ground old oak barrels [8].

Sludge from the wastewater treatment plant is generated steadily during the year and stalk is produced during the grape harvest, a short period of 4-6 weeks around September every year. Stalk storage must be provided so it can be used for sludge composting during the whole year. Due to its high biodegradable organic matter content, a natural fermentation of the stalk occurs during the storage time, reducing its biodegradable organic matter content, and also changing structural properties as bulk density and porosity.

Applying the composting process to the organic wastes inside the wine industry where they are generated, and using the resulting compost in the vineyard crops will close the organic matter cycle. Co-composting these two organic wastes, sludge and stalk, is proposed as a sustainable waste management strategy for the wine-processing industry. Applying this strategy, Miguel Torres S.A. would take the responsibility of the waste management, assume the cost, and obtain the economical, environmental and agronomic benefits.

The specific objective of this study was to determine how the properties of ground stalk change with the storage time and how this influences the optimum ground stalk:sludge volumetric ratio for composting.

MATERIALS AND METHODS

Materials

The waste materials under study are dewatered sludge from the Miguel Torres S.A. wastewater treatment plant; and stalk from the wine production process after grape harvest. Sludge has a steady generation during the year and was sampled just after dewatering. Stalk is generated only during the grape harvest in September and October. Stalk was ground after the grape harvest and stored in a pile in the open air. The materials were sampled once a month for analysis and composting experiments through all the year after grape harvest in September.

Pine wood chips were used as alternative bulking agent. Sawdust was used as drying agent. Both materials were obtained from a local carpentry firm. All mixtures for the composting experiments were handmade.

Chemical and Physical Analyses

Parameters such as water content, total solids, organic matter content, bulk density, free air space (FAS), pore space (PS) and water holding capacity (WHC) were determined according to the standard procedures recommended by the TMCC [9].

Composting Experiments

The reported experiments were undertaken using 4.5-L Dewar® vessels conditioned for composting with a stopper and a rigid wire net near the bottom to separate the material from possible leachate. The stopper was perforated in two points for temperature monitoring and for air supply. Those vessels have been validated for their use in composting research and have been used for several composting experiments with different materials (municipal solid wastes, sludge from wastewater treatment plants, and sludge from paper mills) [10,11].

Temperature Monitoring and Air Supply

Pt-100 sensors were used for temperature monitoring connected to a data acquisition system (DAS-8000, Desin, Spain) which is connected to a standard PC. The system allows, by means of the proper software (Proasis® Das-Win 2.1, Desin, Spain), the continuous on-line visualisation and logging of different parameters (i.e. temperature and oxygen). Pt-100 sensors were placed in the material such that the measuring point was positioned at a point half the height of the material in the Dewar vessel. Oxygen concentration in interstitial air was monitored with an oxygen sensor (Sensox, Sensotran, Spain) and air was supplied into the vessels in aeration cycles where total time of cycle, aeration time and air flow were programmed and changed on the basis of the measured oxygen concentration.

RESULTS AND DISCUSSION

Previous Experiments

Stalk appears to be an ideal bulking agent for sludge composting, due to its physical properties, such as its porosity, resistance to biodegradation, and its chemical properties (a high C/N ratio, around 39) that compensates for the low C/N ratio of sludge [5]. Previous composting experiments were carried out in piles on a pilot scale using mixtures of stalk and sludge at different volumetric ratios [8]. These preliminary experiments were carried out just after grape harvest, with fresh stalk. It was observed that, due to the stalk's branched structure, it was difficult to achieve the integration of this material with wastewater sludge and several pile turnings were required before obtaining a suitable mixture. As a consequence of that fact a delay in the activation of the composting process of several days was observed.

If the stalk material was ground to reduce its particle size, the integration of the materials was easier and the activation of the process faster. The experiments were carried out in spring, 6 months after grape harvest, and stalk had been biodegraded during the storage time. Typical temperature profiles for the composting process were obtained. The experiments showed that a mixture of 2:1 ground stalk:sludge volumetric ratio was suitable for the composting process, reaching the thermophilic range of temperatures which ensured the hygienization of the material and yielding to high reduction of organic matter content and high nitrogen retention. It was also observed that the process was influenced by climatic conditions in maintaining adequate water content in the mixture and in the temperatures reached by the material.

Evolution of the Materials

Figures 1 and 2 show the water and organic matter content of the materials under study, reflecting the evolution of those properties for stalk with storage time. The sludge obtained after centrifugation in the wastewater treatment plant presents a high water content (around 88%), and an organic matter content about 70%.

Ground stalk, stored outdoors, presents a high water content due to the rainfall in winter and spring and the humidity of the area (there is a river near the emplacement). A water content within the recommended range of 40 – 60% [3,12] cannot be achieved mixing the two materials. Therefore it was decided to cover a part of the ground stalk storage pile to prevent the contact of the material with water (dry stalk). As Figure 1 shows, this strategy led to an important reduction in the water content of the stalk.

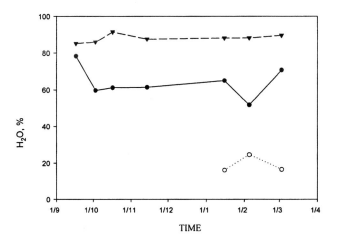

Figure 1 Water Content Evolution of the Materials Under Study:
—•— Stalk, ····o···· Dry Stalk —▾-- and Sludge.

Although some biodegradation has taken place, leading to physical and structural changes that can be visually recognised, no significant reduction of total organic matter content has been observed on stalk with increasing storage time. This is because stalk is composed mainly of branches (hard-wood fraction) and the biodegradation takes place only on the soft-wood fraction of stalk as leaves. The changes in the soft-wood fraction do not represent a significant total change for the whole material.

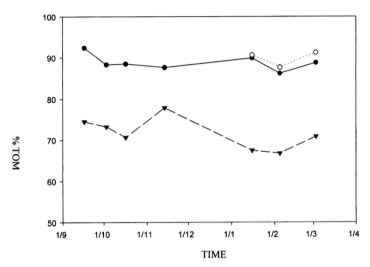

Figure 2 Total Organic Matter Content Evolution of the Materials Under Study:
—●— Stalk, ····o···· Dry Stalk —▼-- and Sludge.

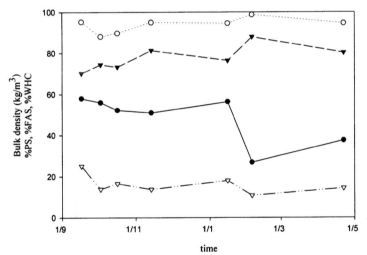

Figure 3. Evolution of physical properties of Stalk through storage time.
—●— Bulk density (kg/m³), ····o···· %Pore Space, —▼-- % Free Air Space,
—▽— % Water Holding Capacity, volume basis.

Figure 3 shows that there is a tendency towards a reduction in bulk density and water holding capacity and an increase in FAS and porosity with storage time. This is the result of the biodegradation of the soft-wood fraction of stalk, which consists of small size particles that fill the spaces in the structure of the branches. This reduction in stalk bulk density or the increased FAS may lead to a worse integration of this material with sludge when acting as bulking agent, and to an excessive FAS in the resulting mixture. As a result, an increment of the required stalk:sludge volumetric ratio must be expected as the storage time increases. The water holding capacity also decreases with time, so the stalk loses its capacity for absorbing the excess of water.

Composting Experiences at Laboratory Scale

To evaluate the composting potential of the wastes under study, different ground stalk:sludge mixtures with volumetric ratios ranging from 1:1 to 3:1 were tested after sampling in various occasions. Table 1 shows the initial water content and the maximum temperature reached in these experiments

Table 1   Maximum temperature achieved and initial water content in the undertaken composting experiments.

DATE	MIXTURE*	INITIAL WATER CONTENT, %	MAXIMUM TEMPERATURE ACHIEVED, °C
17/9	StS32	83.2	39.1
17/9	StS11	83.1	46.5
17/9	StS21	82.1	43.9
17/9	StS52	82.0	38.9
17/9	StS31	81.9	42.9
3/10	StS11	83.2	30.3
3/10	StS21	79.3	38.1
3/10	StS31	76.8	36.7
17/10	StS21	83.6	34.0
17/10	StS31	80.5	30.5
14/11	StS21	83.4	29.4
14/11	StS31	80.8	31.2
14/11	WS11	76.6	38.6
16/1	StS	83.3	27.7
16/1	StSaS	82.2	27.9
6/2	WS21	80.2	32.5
6/2	DS21	79.8	37.7
13/2	SSt12	73.2	38.5
13/2	SStSa111	70.5	38.7
13/2	SWSa111	68.6	44.2
24/2	SStSa313	73.3	49.5
24/2	SStSa316	70.6	38.8
6/3	DS21	77.6	33.3
6/3	DS31	77.1	36.4
6/3	DSaS613	75.5	28.0
6/3	DSaS814	70.8	31.9

* The letters in the name of the mixtures indicate the material used, where: Stalk – St; Dry Stalk – D; Sludge – S; Wood Chips – W; Sawdust – Sa. The numbers indicate the volumetric ratio. Thus, StS32 means a mixture of stalk and sludge in a volumetric ratio of 3:2.

Due to the high water content of the original materials, the mixtures resulted in water contents over 75%, most of them near 80%. The mixtures presented poor activity and only in a few cases temperatures over 40°C were achieved. In any of the undertaken experiments the thermophilic range of temperatures was reached, necessary to obtain a hygienizated material suitable for land application. Figure 4 shows the temperature profile obtained for three of the different undertaken experiments.

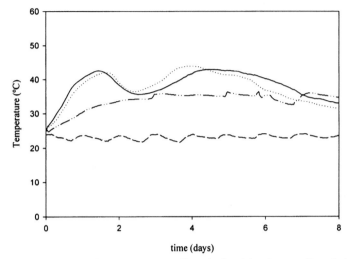

Figure 4. Example of temperature profiles obtained for three stalk and sludge
mixtures in different dates and for different volumetric ratios:
——— 17 Sept, 3:1········ 17 Sept, 2:1 —·— 3 Oct, 2:1——— Room Temperature

When covering the stored stalk to avoid humidification, its water content was reduced to about 20% (Figure 1). Different experiments were carried out with this dry stalk, but the resulting mixtures still presented high water content, because of the small volumetric ratios used and the low bulk density of stalk. It was decided to substitute stalk by wood chips in the sludge mixtures as an alternative bulking agent with a lower FAS. However, the resulting mixtures were too wet and the material did not reach temperatures above 40°C.

The studied sludge is dewatered by centrifugation. Adding polyelectrolyte to the sludge before centrifugation leads to lower water contents in the final product. To emulate the water content that could be reached improving the yield of centrifugation, it was decided to add sawdust to the sludge as absorbent material before mixing. This strategy improved the results of the composting process, the temperature reached around 50°C and this temperature was maintained for more than a week, as can be seen in Figure 5. Adding sawdust in moderate amounts has a positive effect in the composting process because, although the reduction in the total water content of the mixture is low, sawdust absorbs excess water from the sludge. In addition and due to its small particle size, sawdust produces a very homogeneous mixture with sludge breaking the typical compact structure of a dewatered sludge and making easier its integration with stalk. This effect could be observed when mixing the materials for the experiments.

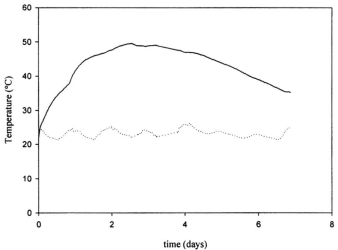

time (days)
Figure 5. Temperature profile obtained for the mixture of Sludge:Sawdust:Sta
in 3:1:6 volumetric ratio
—— Temperature profile ········· Room Temperature

Proposed Strategies

The poor results obtained in the laboratory contrast with the good results of the preliminary experiments on a pilot scale. This contradiction can be explained because of the different operation scale and the different composting system used (dynamic or static). The three variables - operation scale, bulking agent particle size, and bulking agent:sludge volumetric ratio are interrelated and affect the composting process in the same order of magnitude [11].

Thus, the particle size of ground stalk, which was successfully applied for composting on a pilot scale, may be too big for composting in vessels of only 4.5L operating volume. On the other hand, as the stalk bulk density decreases, the amount of stalk required to compensate some of the properties of the mixture with sludge (i.e. water content) increases, leading to high volumetric ratio requirements and a high mixture FAS. This effect of FAS over the recommended value is reduced when operating at higher scales because of the compaction phenomena of the composting material. Moreover, the larger the organic matrix is, the higher the rate of heat generation and the extent of heat retention, which also explains the different temperature profiles obtained [7].

When composting with dynamic systems the rate of frequent turning of the material helps to homogenise the mixture and integrate the materials. Turning and the exposure of surface area to the air promote the evaporation of water, which becomes very important when wet mixtures are treated. In closed static systems the surface area available is lower and the evaporation of water may be only achieved by aeration of the material. Excessive airflow may lead to the cooling of the mixture, preventing the high temperatures required for material composting may being achieved.

It can be concluded that if mixtures with appropriate water content are formulated and integration of the materials is achieved through turning or by reducing the individual water contents of each component, composting can be successfully applied to the wastes under study. Thus, two strategies are proposed: provide cover storage for the stalk and add polyelectrolytes to the sludge before centrifugation to achieve enhanced dewatering.

CONCLUSIONS

Co-composting of stalk and sludge generated in the wine industry is a suitable waste management strategy, which will produce a suitable fertilizer for application in the vineyard, closing the organic matter cycle.

Bulk density and water holding capacity of ground stalk decreases during storage as biodegradation of the soft-wood fraction takes place, while FAS and porosity increase. Thus, the stalk:sludge optimum volumetric ratio for composting should be increased during the year. Stalk should be stored covered to avoid humidification. It is strongly recommended that reduction of the water content of the dewatered sludge should be achieved by adding adsorbent material or improving the dewatering system, to get mixtures with appropriate moisture content for composting.

ACKNOWLEDGEMENTS

The authors wish to thank Quim Mach for his basic and always smiling support in the sampling of the materials, and the interest and help of Francesc Aguilera in the development of this work. The Spanish Ministerio de Medio Ambiente and Ministerio de Ciencia y Tecnología (Project 2000/074) provided financial support.

REFERENCES

1. Comunidad Autónoma de La Rioja. http://www.larioja.com. April 10, 2003

2. EUROPEAN COMMISSION. 2001. Working document. Biological treatment of biowaste. 2nd draft. http://europa.eu.int/comm/environtment/waste/facts_en.htm.

3. HAUG, R.T. 1993. The Practical Handbook of Compost Engineering. Lewis Publishers. Boca Raton, Florida.

4. JOKELA, J., RINTALA, J., OIKARI, A., REINIKAINEN, O., MUTUA, K. AND NYRÖNEN, T. 1997. Aerobic composting and anaerobic digestión of pulp and paper mill sludges. Water Science and Technology, 36(11):181-188

5. MORISAKI, N., PHAE, C.H., NAKASAKI, K., SHODA, M. AND KUBOTA, H. Nitrogen transformation during thermophilic composting. J. Fermentation Bioengineering, 67(1):57-61.

6. SESAY, A.A., LASARIDI, K., STENTIFORD, E. AND BUDD, T. 1997. Controlled composting of paper pulp sludge using the aerated static pile method. Compost Science and Utilization, 5(1): 82-96.

7. SHERMAN-HUTTON, R. 2000. Latest developments in mit-to-large-scale vermicomposting. Biocycle, 41(11):51.

8. BERTRAN I CASANELLAS E. Compostatge de residus vitivinícoles: rapa, roure i fangs de depuradora. 2002. Final year project. Universitat de Barcelona.

9. TMCC – Test Methods for Compost and Composting – U.S. Department of Agriculture & U.S. Composting Council (2001). Test methods for the examination of composting and compost. W.H. Thompson (ed.).

10. GEA LEIVA, M.T. 2001. Diseño y validación de sistemas de compostaje a escala laboratorio. Aplicación al tratamiento de residuos sólidos urbanos y lodos de depuradora. M. Sc. Dissertation, Universitat Autònoma de Barcelona.

11. GEA LEIVA, M.T., ARTOLA CASACUBERTA, A., SÁNCHEZ FERRER, A. Application of Experimental Design Technique to the Optimization of Bench-scale Composting Conditions of Municipal Raw Sludge. Compost Science and Utilization. In press.

12. SAÑA, J. Y SOLIVA M. 1987. El Compostatge. Procés, sistemes i aplicacions. Servei del Medi Ambient. Diputació de Barcelona.

AN ALTERNATIVE TO FURNANCE RECYCLING IN THE REUSE OF ALUMINIUM WASTES

G Di Bella   C Paradiso
F Corigliano   I Arrigo
University of Messian
L Mavilia
University of Reggio
Italy

ABSTRACT. The paper presents the results of the redox reaction in aqueous solution:

$$2AC^\circ + 2OH^- + 6H_2O \rightarrow 3H_2 + 2AC(OH)_4^-$$

when applied to aluminium wastes with the aim of finding an alternative to their recycling to the primary production where the latter is impractical for technical or economic reasons. The best chemical and physical conditions in which the reaction proceeds have been identified and the economic aspects involved in the use of the proposed technology have been evaluated in terms of cost effectiveness and marketability of the main products.

Keywords: Aluminium waste treatment, Post-use aluminium cans, Hydrogen from wastes, Recycling alternatives.

G Di Bella is currently undertaking research into the valorization and re-use of silico-aluminous industrial and municipal wastes.

L Mavilia is researcher of Science and Technology of Materials in the Faculty of Engineering at the University of Reggio Calabria. His research focuses mainly on developing new strategies for the recycle of industrial and municipal solid wastes and on the characterization of materials of historical buildings for conservation and restoration purposes.

C Paradiso is currently undertaking research into the valorization of pumice mine wastes and glass municipal wastes.

I Arrigo is currently undertaking research into the valorization and reuse of silico-aluminous wastes, such as coal fly ash, pumice mine wastes and glass municipal wastes.

F Corigliano is full professor of Chemistry of Commodities at the Faculty of Science, Messina University, Italy. His research focuses mainly on the valorization and reuse the industrial and municipal wastes and on the causes and effects of oil fly ash emissions into the environment.

INTRODUCTION

Aluminium, along with iron, is the most valuable material present in municipal solid waste. The reason for this is that the most expensive stage of the primary aluminium production process, i.e. the electrolysis of alumina, requires at least 14 kwh of electricity per kg of aluminium produced. Recycling aluminium waste makes it possible to bypass this and other stages and significantly reduce production costs and energy consumption. Moreover, all aluminium waste can be melted down to produce new products without impairment of quality.

The recycling of aluminium not only satisfies the need to protect the environment but also constitutes an economic opportunity, as the market value of aluminium waste is usually sufficient to cover the recovery, storage and treatment costs associated with any waste collection and recycling operation. This is not the case where aluminium is widely dispersed, as, for example, with aluminium foil or thin sheets found at non-differentiated municipal solid waste collection points.

Let us consider the differentiated collection and recycling of post-use aluminium cans. They contain three different aluminium-based alloys, which makes it necessary to send them to foundries that are equipped to recycle obsolete aluminium cans. Since collection costs are not much lower than the market value of used aluminium (around 0.50-0.75 €/Kg), when the absence locally of a specialised foundry makes it necessary to transport the waste aluminium to a plant in another location, the burden of this additional cost removes any incentive to collect.

The work involved in setting up an extensive waste collection network and organising transport to a single storage site is a costly business both in terms of educating the public to dispose of domestic waste conscientiously and in terms of the economic resources required. It follows from this that the results of such an undertaking should benefit local industries to compensate for the efforts made. To achieve these aims, in our opinion, it is necessary to overcome the technical and industrial limitations posed by solely depending on recycling and identify an alternative process able to satisfy the demand from local industries. An alternative of this kind would allow maximum exploitation of the economic potential of the product where it has been made available and avoid the waste cans ending up in the rubbish dump even after differentiated collection as, paradoxically, is all too often what happens currently. In the light of the above, we wish to propose a post-use aluminium treatment process as an alternative to traditional recycling in order to take full advantage of the metal recovered even when differentiated refuse collection is practised in some of the more difficult contexts mentioned earlier. Our idea is to convert the metal into products that will be of industrial and economic interest such as hydrogen and aluminium oxide by means of an alkaline attack in caustic soda aqueous solution.

EXPERIMENTAL DETAILS

A feasibility study of the above-outlined idea was carried out in the laboratory by 3-litre capacity jacket reactor inside which whole cans could be placed (crushed – as currently happens during transport for re-use) and identifying the ideal conditions for the reaction:

$$2AC^\circ + 2OH^- + 6H_2O \rightarrow 3H_2 + 2AC(OH)_4^- \uparrow \qquad E^\circ \; Al^\circ/Al(OH)_4^- = -2,35 \; V$$

The reactor jacket was connected to a thermostat to keep the reaction temperature constant. The reactor head was fitted with a gas aspiration and conveyance system.

RESULTS AND DISCUSSION

The operating conditions were optimised as follows [1, 2]:

- a NaOH/Al molar ratio of between 1.70 and 2.25;
- an alkali concentration of $3\pm1M$
- a reaction temperature of between 70 and 90°C;
- a reaction time of between 120 ± 20minutes.

The products, yields and reaction times for the treatment of 100g of aluminium cans with 3M NaOH solutions at two different reaction temperatures are shown in Figure 1 and in Table 1.

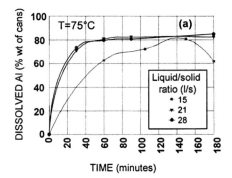 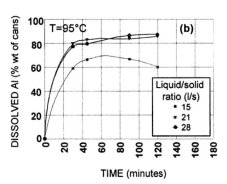

Figure 1   Dissolution kinetics for 100g of cans in various volumes of a 3M NaOH solution at
(a) T= 75°C and (b) T=95°C

Table 1 Solubilised aluminium, expressed as % of original weight of cans

T, °C	l/s	Al	H_2	UNDISSOLVED RESIDUE
	l/kg	%	m^3	g
75	15	74.75	0.1207	30.39
75	21	81.81	0.1220	20.72
75	28	82.09	0.1232	24.22
95	15	60.88	0.1233	42.71
95	21	83.65	0.1225	20.38
95	28	85.53	0.1230	15.82

Since the reaction is exothermal, once triggered it is thermally self-sustaining and complete in the space of 1-2 hours. It must be mentioned that the kinetics of the reaction under the conditions employed are mainly controlled by the small amount of metal surface exposed to the reaction environment, which is limited to the cracks caused by crushing and cutting since the inside of cans is protected by a coating of plastic and the exterior by paint. If this were not the case, the reaction would proceed far more rapidly and tumultuously.

Oxidised aluminium is solubilised as sodium tetrahydroxy aluminate up to super saturation of the liquid phase. If this solution is cooled and concentrated, the aluminium hydrate can be separated as a final product and the most alkali recovered and recycled.

During the reaction extremely pure hydrogen develops, which represents another important and valuable product. The end of the reaction is apparent when hydrogen stops bubbling.

Besides the formation of hydrogen, a solid dark grey residue of fine particles is formed which, when analysed chemically, proved to have the average composition reported in Table. 2. This residue obviously contains the minor metal components of the aluminium alloy as well as the fraction of aluminium oxidised to hydroxide but not dissolved by the alkali in solution. It is accompanied by scales of polymer and paint materials deriving from the surface coatings on the inside and outside of the body of the cans that has not undergone alkali attack.

Table 2 Composition of the Dried Residue

Mn_2O_3	MgO	Al_2O_3	CuO	SiO_2	OTHER OXIDES
%	%	%	%	%	%
33.91	18.75	24.55	1.72	15.54	5.53

The aqueous phase, when filtered from the residue, is yellowish; this colour can be removed by percolation through a column packed with active carbon and is attributable to the leaking of some components of the paint into the solution.

The results appear altogether consistent and confirm the feasibility of the treatment proposed, though many aspects of the process can be improved upon and are worthy of further investigation. There are, for example, studies in progress designed to identify a means of exploiting the minor components of aluminium alloy present in the residue or at least of recovering the aluminium contained therein.

CONCLUSIONS

The main product of the proposed treatment of post-use aluminium waste is very pure gaseous hydrogen; the coproduct is an aqueous solution of sodium tetrahydroxy aluminate, $NaAl(OH)_4$, from which it is possible to derive trihydrate aluminium oxide and recyclable alkali.

For each kg of aluminium that undergoes attack, 1,245 litres of gas (the equivalent of 55.6 moles of hydrogen at ambient temperature and pressure) and 37 moles of sodium aluminate, the equivalent of 2.886 kg of aluminium hydrate, are produced.

The commercial value of aluminium hydroxide is around 0.30-0.40 €/kg and that of high purity industrial hydrogen around 0.30-0.50 €/m^3 (Conservative price estimate, equivalent to that of producer-producer sales). Therefore, considering the average market value of secondary aluminium (0.50-0.75 €/Kg) and an increase of 20-50% to compensate for the greater cost of collecting cans, the treatment would still be economically viable.

Moreover, given the extreme simplicity of the plant required, there would be neither difficulties in circumstances of limited space nor environmental restrictions as to where the plant should be located.

REFERENCES

1.   G. DI BELLA, L. MAVILIA, S. DI PASQUALE, F. CORIGLIANO, XVIII Congr. Merceologia, Verona, Italy, Ott. 1998, vol. II, pp. 183-190 of the Proceedings.

2.   G. DI BELLA, L. MAVILIA, S. DI PASQUALE, F. CORIGLIANO, Conv. Regionale Soc. Chim. Italiana, Sez. Calabria, Rende (CS), Dic. 1998, pp. 8-10 of the Proceedings.

TRACE METAL BIOMAGNIFICATION IN A SOIL-PLANT-ARTHROPOD SYSTEM: A PROBLEM FOR THE AGRICULTURAL RECYCLING OF SEWAGE SLUDGE?

I Green

C Jeffries

Bournemouth University

United Kingdom

ABSTRACT. The effect of crop type on the transfer of Cd and Zn from sewage sludge amended soil in the soil-plant-aphid-system was investigated. The subsequent transfer of the two metals from aphids to green lacewing larvae was also determined. Soil Cd and Zn concentrations were significantly elevated by sludge amendment, but were within the current UK limits. Transfer from the soil did not result in significant accumulation of Cd in wheat and barley shoots or aphids. However, Cd concentrations were significantly higher in both wheat shoots and the aphids feeding on them. In contrast, Zn was significantly accumulated in wheat, barley and aphids. Concentrations in the wheat shoots and the aphids feeding on them were again significantly higher. Transfer of Cd from aphids to the lacewing larvae resulted in the highest concentrations in the system. Zinc concentrations in the larvae were close to the concentrations in the aphids on which they were feeding. For both metals, concentrations in the lacewings were not significantly effected by sewage sludge amendment. Results are discussed in relation to possible risks to predatory arthropods posed by the transfer of Cd and Zn from the re-cycling of sewage sludge to agroecosystems.

Keywords: Biomagnification, Sewage sludge, Agroecosystems, Cadmium, Zinc, Aphids, Lacewings.

I D Green is a research student at Bournemouth University. He is currently completing his PhD on the subject of the transfer and fate of trace metals derived from sewage sludge amended agricultural soils in arthropod food chains.

C Jeffries has recently completed research into the transport of sewage sludge derived trace metals in the soil-plant-arthropod system.

INTRODUCTION

Large quantities of sewage sludge are produced annually in the UK [1]. This poses a disposal problem requiring an economical and environmentally safe solution. In the UK, approximately 50% of sewage sludge production is applied to agricultural land as an organic fertiliser [1]. This beneficially recycles the nitrogen and phosphorous in sludge and can reduce the consumption of limited mineral phosphate sources [2]. However, this practice is restricted by the presence of many potentially toxic elements, including trace metals, in sewage sludges [3]. Concentrations of trace metals in sludges typically exceed those of soils. Consequently, applications of sewage sludge can elevate concentrations of trace metals in the plough layer of soil [4]. To prevent this damaging soil fertility or the health of domestic animals and humans, many countries control the agricultural use of sewage sludge [3].

Of the trace metals present in sewage sludges, Cd and Zn are of greatest concern in agroecosystems [5]. They are the most mobile trace metals in the soil-plant system [6] and consequently they have the potential to be transferred along food chains. Cadmium has no known essential function in metazoan organisms and is considered to be highly toxic [7], whilst Zn is an essential element for plants [8] and animals [9]. However, concentrations of Zn in sewage sludges are typically higher than for any other trace metal [10,1] and it is the accumulation of this metal in soil that most frequently limits the agricultural recycling of sewage sludge [5]. Consequently, although Cd is toxic at lower concentrations, Zn may cause greater toxicity due to higher concentrations in the environment [11].

The biomagnification of trace metals in food chains may potentially endanger predatory arthropods [12]. However, trace metal biomagnification is not a general property of arthropod food chains. The ability of arthropods to regulate the concentration of trace metals in their bodies can differ widely between species [13]. When the component species of a food chain have a poor ability to regulate trace metals, there is the possibility that biomagnification may occur [12]. Accurate prediction of the extent of biomagnification in a food chain requires a detailed knowledge about the regulatory ability of the component species. However, little is known about the transfer of trace metals in arthropod food chains in agroecosystems.

Aphids are common pests on many crops [14] and infestations may result in economic damage [15]. This can be prevented by biological control using the large range of arthropod aphid predators [16]. However, it has been demonstrated that Cd [4] and Zn [4,17] can be biomagnified in the wheat-aphid system. Concentrations of both metals in the grain aphid (*Sitobion avenae*) have been reported at levels over 8 time higher than in the ears on which they were feeding [4]. Consequently, in areas of repeated sewage sludge application, aphid predators may be exposed to elevated concentrations of Cd and Zn. The extent of Cd and Zn transfer from aphids to their predators is poorly understood. The efficacy of the current controls on the agricultural use of sewage sludge has been questioned as a result [4,17,18].

Shoot trace metal concentrations have been reported to vary with crop species [19]. It is not known how this may affect the transfer of Cd and Zn in the soil-crop-aphid system. This study investigated the accumulation of sewage sludge derived Cd and Zn by the grain aphid (*Sitobion avenae*) feeding on two crops (wheat and barley) that are reported to differ in the concentration of trace metals in their shoots [19]. The subsequent transfer of the two metals to a monophagous predator with the potential to control aphids was also investigated.

METHODS AND MATERIALS

A bulk soil sample of a freely draining sandy loam was taken from the Ap horizon of an agricultural field, situated in a rural location in East Lulworth, Dorset (UK). The bulk sample was homogenised divided into three equal parts and each part was then amended with a dried, anaerobically digested municipal sewage sludge at three treatment rates equivalent to 0, 10 or 30 t (dry solids) ha[-1]. The soil and sewage sludge was thoroughly mixed and used to fill 7.5 litre pots, which were subsequently seeded with spring wheat (*Triticum aestivum* cv. Alexander) or spring barley (*Hordeum distichon* cv. Optic) at a rate equivalent to 400 kg ha[-1]. Each sludge amendment-crop combination was replicated six times. Seeded pots were placed in a fully randomised block in a glasshouse. When the plants were between growth stages 37 and 51 [20], cultures were established on the plants by placing 200 grain aphids (*Sitobion avenae*) from laboratory cultures in each pot. Pots were covered with netting to prevent the transfer of aphids between treatments. Aphid cultures were left to establish for 10 days before 10 second instar green lacewing larvae (*Chrysoperla carnea*) were added to each pot containing barley. After a further 11 days all lacewing larvae had pupated. Pupae were then collected and soil, crop shoot and aphid samples were taken from each pot.

Chemical analysis

Five soil samples were taken from the top 15 cm of each pot, bulked, air dried and ground gently before being passed through a 2 mm plastic sieve. All subsequent analysis of the soil was carried out in triplicate using this fraction. Organic matter content was estimated by loss on ignition at 450 oC and pH was determined in the supernatant of 2.5:1 water:soil suspension [21]. Metals were extracted from 0.5 g sub-samples by refluxing in 69 % nitric acid for two days. Refluxed samples were filtered through Whatman No. 42 filter paper and diluted to 50 ml with de-ionised water [17]. Three wheat plants were randomly sampled from each pot, washed once in 0.1 % detergent solution and twice in distilled water before being dried to constant weight at 70 oC. Dried plants were individually digested at 90 oC in 10 ml of 69 % nitric acid. Digests were filtered through Whatman No. 541 filter paper and diluted to 25 ml with de-ionised water [17]. Aphid samples were washed and dried as described for the wheat. Sub-samples of aphids (20 - 40 mg) were digested in 2 ml of nitric acid in glass vessels at a temperature of 80 oC. The clear residue was then diluted to 5 ml with de-ionised water. Lacewing pupae were digested in a similar way, but using 1.25 ml of nitric acid and were diluted to 3 ml. Cadmium and Zn concentration in samples was determined by an ATI Unicam Solaar 939 atomic absorption spectrometer (AAS). Electro-thermal AAS was to determine Cd, whilst flame AAS (both with and without STAT trap) was used to determine Zn. In all instances deuterium background correction was used. Relevant certified reference materials (GBW 07407, BCR 143R, BCR 281 & BCR 60) and reagent blanks were digested and analysed with each batch of soils, wheat plants aphids and lacewings.

Data analysis

Statistical analysis was conducted using SPSS (version 10). Group means were compared using one and two way analysis of variance. The assumptions of homogeneity of variance between treatment groups and normality of distribution were tested. Lacewing Zn concentrations were inversely transformed and aphid Cd log10 transformed to meet these assumptions. One replicate was eliminated as an outlier from the lacewing Cd data. The strength of the relationship between variables was determined by Spearman's rank order correlation (rs).

RESULTS

Soil chemical parameters

Selected chemical parameters of the sewage sludge, unamended soil and amended soil are given in Table 1. Soil pH decreased with sewage sludge amendment, but this was not significant ($F = 0.45$, $P = 0.64$). Loss on ignition values of the soil increased with sewage sludge amendment, reflecting the high organic matter content of the sludge matrix. Differences between treatments were found to be highly significant   ($F = 42.0$, $P < 0.001$). Concentrations of Cd and Zn were increased by 15 % and 11 % respectively in the 10 t ha^{-1} amendment and by 34 % and 50 % respectively in the 30 t ha^{-1} amendment. For both metals differences between treatments were found to be significant ($F = 8.56$, $P = 0.001$ and $F = 14.99$, $P < 0.001$ for Cd and Zn respectively).

Table 1   Total concentration (mg kg^{-1}) of Cd and Zn, loss on ignition (LOI) and pH of sewage sludge, unamended soil and soil amended with the sewage sludge (Mean ± 1 SE).

	SLUDGE	0 t ha^{-1}	10 t ha^{-1}	30 t ha^{-1}
'Total' Cd	2.40	0.18 ± 0.02	0.20 ± 0.01	0.28 ± 0.02
'Total' Zn	725	41.76 ± 1.23	48.22 ± 1.34	55.97 ± 2.61
pH	5.53	4.88 ± 0.26	4.71 ± 0.04	4.69 ± 0.05
LOI	60.0	4.64± 0.08	5.61 ± 0.19	6.17 ±0.06

The effect of crop type on aphid accumulation.

Both the concentrations and pattern of accumulation of Cd in shoots differed between the two crop types (Figure 1). Cadmium concentrations in barley shoots increased in line with sewage sludge amendment, whilst in wheat shoots concentrations peaked in the 10 t ha^{-1} amendment and fell below control concentrations at 30 t ha^{-1}.

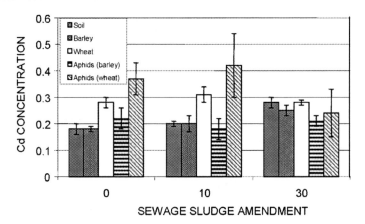

Figure 1   Cd (mg kg^{-1} mean ± 1 SE) Concentration in soil, barley and wheat shoots and the respective aphids feeding on the two crops after the amendment of soil with sewage sludge (t ha^{-1}).

At all amendment rates, Cd concentrations were higher in spring wheat. A two-way analysis of variance indicated that sewage sludge application had no significant effect on the Cd concentration in shoots (F = 0.44, P = 0.65). However, the observed differences in shoot concentration between the two species were significant (F = 4.25, P = 0.049). The interaction between sludge amendment and crop type was not significant   (F = 1.17, P = 0.325), indicating that there was no difference in the way the shoot concentration in the two crops responded to increasing sewage sludge amendment.

Cadmium concentrations in the aphids feeding on barley shoots were marginally lower in the two sludge treatments than in the control (Figure 1). Aphids feeding on wheat shoots had slightly higher Cd concentrations than the control at the 10 t ha^{-1}, but concentrations fell below the control at the highest amendment. Aphids feeding on the wheat plant had higher concentrations than aphids feeding on barley, reflecting the higher shoot concentrations in wheat. A two-way analysis of variance indicated that sewage sludge amendment had no significant effect on the concentration of Cd in aphids (F = 0.36, P = 0.71) and that the concentrations of Cd in aphids feeding on wheat were significantly higher (F = 4.5, P = 0.049). The interaction between sewage sludge amendment and crop type was not significant (F = 1.17, P = 0.33) indicating that aphids feeding on the two crops did not differ in the accumulation of Cd in response to sludge amendment. Shoot Zn concentrations increased in line with sewage amendment in both crops (Figure 2).

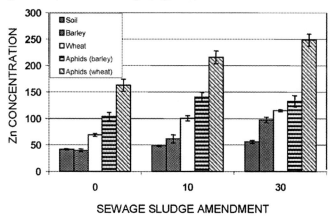

Figure 2   Concentration of Zn (mg kg^{-1} mean ± 1 SE) in soil, barley and wheat shoots and the respective aphids feeding on the two crops after the amendment of soil with sewage sludge (t ha^{-1}).

As with Cd, concentrations were higher in the shoots of wheat in all treatments. A two way analysis of variance indicated that sewage sludge amendment significantly elevated the concentration of Zn in crop shoots (F = 72.9, P < 0.001) and that concentrations were significantly higher in the shoots of wheat plants   (F = 64.3, P < 0.001). The interaction between sewage sludge and crop type was not significant (F = 3.1, P = 0.063), indicating that both crops responded in the same way to increasing sewage sludge amendment.

Zinc concentrations in aphids feeding on both crops were higher in the sewage sludge treatments than in the control (Figure 2). Concentrations of Zn in aphids feeding on wheat shoots increased in line with sewage sludge, whilst in aphids feeding on barley, concentrations peaked in the 10 t ha^{-1} amendment. Differences between crops and treatments were investigated by a two-way analysis of variance. This indicated that sewage sludge amendment had a significant effect on the concentration of Zn in aphids (F = 16.6, $P < 0.001$) and that concentrations in aphids feeding on wheat plants were significantly higher (F = 92.9, $P < 0.001$). The interaction between sewage sludge amendment and crop indicated that the observed difference in accumulation pattern between the aphids feeding on the two crops was significant   (F = 3.9, $P = 0.03$).

The above results indicate that both Cd and Zn concentrations in aphids are related to the concentration in the shoots on which they were feeding. The strength of this relationship was investigated by Spearman's rank order correlation. This indicated that significant relationships existed between the concentration of Cd and Zn in shoots and the concentration in aphids (r_s = 0.49, $P = 0.003$ and r_s = 0.70, $P < 0.001$ for Cd and Zn respectively).

Transfer of Cd and Zn to lacewing larvae.

Cadmium concentrations in the lacewing pupae were 0.34 ± 0.06, 0.42 ± 0.10 and 0.36 ± 0.09 mg kg^{-1} (± 1 SE) for the 0, 10 and 30 t ha^{-1} amendments respectively. These were the highest in the system in all treatments, but observed differences between treatments were not significant (F = 0.33, $P = 0.73$). Zinc concentrations in the lacewing pupae were 105.0 ± 12.4, 162.7 ± 24.0 and 126.6 ± 3.8 mg kg^{-1} (± 1 SE) for the 0, 10 and 30 t ha^{-1} amendments respectively. As with Cd, concentrations in the lacewing pupae from the sewage sludge treatments were higher than the control and were at their maximum in the 10 t ha^{-1} amendment. Differences between treatments were not found to be significant (F = 3.1, $P = 0.10$). In all treatments, Zn concentrations in the pupae were similar to those in the aphids on which they were feeding.

Mean mortality in the lacewing larvae was high in all treatments, ranging between 70 to 80 %. However, there was no significant difference in mortality between the treatments (F =1.54, $P = 0.27$). Pupal dry mass was measured as an indicator of   sub-lethal effects, but again no significant difference between treatments was found   (F = 0.19, $P = 0.83$).

DISCUSSION

The concentration of Zn in the sewage sludge was lower than the fifty percentile value of sewage sludges produced in England and Wales, whilst the concentration of Cd was proportionally much higher, at slightly below the ninety percentile value [1]. The Sewage sludge amendment rates used were equivalent to 0, 2-3 and 6-8 times the typical annual sludge application rate applied agriculturally in the UK [1]. Amendment at these rates resulted in a significant increase in the concentration of Cd and Zn in the soil. However, concentrations of both metals in the soils were well within the current recommended UK limits (3 mg kg^{-1} and 200 mg kg^{-1} for Cd and Zn respectively for soils in the pH range 5.5 – 5.0 [5]).

The transfer of Cd from the soil to the crop plants did not result in significant accumulation in the shoots. It is reported that Cd concentration in plants increase in line with concentration in the soil [22]. However, previous work with this soil found no significant increase in Cd concentration in winter wheat shoots as a result of sewage sludge applications up to 20 t ha^{-1}

[4]. The roots of both wheat and barley freely accumulate Cd [23]. Consequently, the lack of accumulation in the shoots appears to be due to restricted translocation from root to shoot. Concentrations of Cd in barley shoots were lower than in wheat shoots, confirming the results of other workers [19]. Cadmium concentrations in wheat plants were particularly high, exceeding the typical background range for Poaceae of 0.07 - 0.27 [22]. This probably reflects the generally higher metal uptake in pot trials [3] and the low pH of the soil. In contrast to Cd, Zn was significantly accumulated in the shoots of both crops due to sewage sludge amendment. The concentrations in the shoots were greater than in the soil, demonstrating that Zn is biomagnified in the soil-plant system. The higher concentrations in wheat shoots resulted in a greater biomagnification of Zn. Concentrations were at the higher end of reported range, but levels were still within the range considered typical (27-150 mg kg^{-1} [22]).

Concentrations of Cd and Zn in aphids showed a significant correlation with concentrations in the shoots. Consequently, transfer of Cd from the soil to the aphids was effectively restricted by the plant, whilst Zn was readily transfer to and further biomagnified in aphids. The accumulation of Zn in aphids as a result of sewage sludge application is in line with reported studies, which have shown accumulation up to 320 mg kg^{-1} [4]. Concentrations of Cd in the aphids feeding on barley were within the reported range (0.016 – 0.354 mg kg^{-1} [4,17]) for cereal aphids in sewage sludge contaminated environments. Aphids feeding on the wheat were close to or exceeding the upper limit of this range in the sewage sludge treatments. Cadmium accumulation in wheat-aphid systems has been reported [4,24], whilst other work has reported findings inline with the current study [17]. The influence of the plant in the transport of Cd from the soil to higher trophic levels demonstrated in the current study is a possible explanation for this variation.

Concentration of Cd in lacewing pupae was high compared to the rest of the system. In contrast, other work has reported concentrations of Cd in the related lace wing species *Mallada signata* as being below the limit of detection [24]. Cadmium is primary accumulated within the soft proteinaceous tissues in aphids [25]. *Chrysoperla carnea* feeds by injecting digestive enzymes into its prey and sucking out the resultant liquid [26]. Consequently, it consumes the tissues in which Cd is accumulated. This may explain the high concentrations in *C. carnea* and suggests that the total concentration of Cd in aphids may be a poor predictor of the exposure of lacewing larvae to Cd. Despite a significant difference in the concentration of Zn in the aphids on which they were feeding, there was no accumulation of Zn in the lacewing pupae. A similar lack of Zn accumulation in lacewing pupae has been reported in *M. signata* [24]. This indicates that lacewing larvae have an efficient mechanism for excreting Zn. Concentrations of Zn in *C. carnae* were high compared to the range of 37 – 41 mg kg^{-1} found in *M. signata*. However, at the concentrations of Zn and Cd found in this study, there was no observed effect on the lacewing larvae. This would suggest that the concentrations in the lacewings were not excessive.

CONCLUSIONS

The concentrations of Cd and Zn in aphids were dependent on the concentration in the shoots on which they were feeding. Concentrations of both metals were lower in barley shoots than in wheat shoots and consequently concentrations in aphids feeding on the barley were also lower. Crop type therefore has an important influence on the transfer of both Cd and Zn in the soil-plant-aphid system. Cadmium transport throughout the system was effectively regulated by both crop types, limiting transfer to higher trophic levels, whilst Zn was freely mobile in the system up to the level of the aphid. This suggests that the contamination of agroecosystems with Zn poses the greater danger to arthropods. However, neither metal was accumulated by lacewing larvae and no detrimental effects were observed. The results of this study suggest that repeated applications of sewage sludge at rates typically applied agriculturally will not harm aphid predators.

REFERENCES

1. GENENBEIN, A., CARLTON-SMITH, C., IZZO, M., HALL, J., UK Sewage Sludge Survey. Environmental Agency Research and Development Technical Report P165, Bristol, 1999

2. STEEN, I., AGRO, K., Phosphorus availability in the 21st Century: Management of a Non-Renewable Resource. Phosphorus and Potassium Vol. 217, 1998, pp 25-31

3. SMITH, S. R., Agricultural recycling of sewage sludge and the environment, CAB International, Wallingford, 1996

4. MERRINGTON, G., WINDER, L., GREEN, I. D., The bioavailability of Cd and Zn from soils amended with sewage sludge to winter wheat and subsequently to the grain aphid *Sitobion avenae*, Science of the Total Environment, Vol. 205, 1997, pp 245-254

5. MAFF/DoE, Review of the rules for sewage sludge application to agricultural land; soil fertility aspects of potentially toxic elements PB156, HMSO, London, 1993

6. SAUERBECK, D.R., Plant, element and soil properties governing uptake and availability of heavy metals derived from sewage sludge, Water, Air and Soil Pollution, Vol. 57-58, 1991, pp 227-237

7. STOEPPLER, M., Cadmium. In: Merian, E. (Ed) Metals and Their Compounds in the Environment. VCH, Weinheim, 1989, pp 803-849

8. RÖMHELD, V., MARSCHNER, H., Function of micronutrients in plants. In: J.J. Mortvedt, F.R. Cox, L.M. Shuman, R.M. Welch (Eds), Micronutrients in Agriculture 2nd Ed., Soil Science Society of America, Madison, Wisconsin, USA, 1991, pp 297-328

9. MILLER E.R., LEI X., ULLREY D.E., Trace elements in animal nutrition. In: J.J. Mortvedt, F.R. Cox, L.M. Shuman, R.M. Welch (Eds), Micronutrients in Agriculture 2nd Ed., Soil Science Society of America, Madison, Wisconsin, USA, 1991, pp 593-662

10. CIWEM, Sewage Sludge: Utilisation and Disposal, Chartered Institute of Water and Environmental Management, London, UK, 1999

11. HOPKIN S. P., Forecasting the environmental effects of zinc, the metal of benign neglect in soil ecotoxicology. In: P.S. Rainbow, S.P. Hopkin, M. Crane (Eds), Forecasting the environmental fate and effects of chemicals, John Wiley, Chichester, 2001, pp 91-96

12. STRAALEN VAN, N. M., ERNST, W. H. O., Metal biomagnification may endanger species in critical pathways, Okios, Vol. 62(2), 1992, pp 255-256

13. JANSSEN, M. P. M., DE VRIES, T. H., STRAALEN VAN, N. M., Comparison of Cd kinetics in four soil arthropod species, Archives of Environment Contamination and Toxicology, Vol. 20, 1991, pp 305-312

14. SOFFE, R J., The Agricultural Notebook. 19th ed., Blackwell Scientific, Oxford, 1997

15. GEORGE, K.S , GAIR, R., Crop loss assessment on winter wheat attacked by the grain aphid *Sitobion avenae* (F.), Plant Pathology, Vol. 28, 1978, pp 143-149

16. CARTER, N., MCLEAN, I.F.G., WATT, A.D., DIXON, A.F.G., Cereal aphids: a case study and review, Applied Biology , Vol. 5, 1980, pp 271-348

17. MERRINGTON, G., WINDER, L., GREEN, I. D., The uptake of Cd and Zn by the bird cherry oat aphid *Rhopalosiphum padi* (Homoptera: Aphididae) feeding on wheat grown on sewage sludge amended agricultural soil, Environmental Pollution, Vol. 96, 1997, pp 111-114

18. WINDER, L., MERRINGTON, G., GREEN I., Tri-trophic transfer of Zn from the agricultural use of sewage sludge, Science of the Total Environment,   Vol. 229, 1999, pp 73-81

19. JARVIS, S.C., JONES, L.H., HOPPER, L.J., Cadmium uptake from solution and its transport from roots to shoots, Plant and Soil, Vol. 44, 1976, pp 179-191

20. ZADCOKS, J. C., CHANG, T. T., KONZAK, C. F., A decimal code for the growth stages of cereals. Eucarpia Bulletin. Vol. 7, 1974

21. ROWELL, D.L., Soil Science: Methods & Applications, Prentice Hall, Harlow, 1994

22. KABATA-PENDIAS, A., PENDIAS, H., Trace Elements in Soils and Plants, CRC Press, Boca Ranton, 1992

23. GREEN, I.D., Unpublished PhD thesis

24. MERRINGTON, G., MILLER, D., MCLAUGHLIN, M.J., KELLER, M.A., Trophic barriers to Cd bioaccumulation through the food chain: A case study using a plant-insect predator pathway, Archives of Environmental Contamination and Toxicology, Vol. 41, 2001, pp 151-156

25. HUGES, M.K., LEPP, N.W., PHILLIPS, D.W., Aerial heavy metal pollution and terrestrial ecosystems. In: A. MacFadyen (Ed), Advances in Ecological Research, Vol. 2, Academic Press, London, 1980, pp 218-327

26. ROTHERAY, G.E., Aphid predators. Naturalists handbook No. 11, Richmond Publishing, Slough, 1989

INDUSTRIAL ECOLOGY USING AN INTEGRATED CP-EMS MODEL FOR SUSTAINABLE DEVELOPMENT

S M El-Haggar

American University

Egypt

ABSTRACT. Industrial activities have been increasing all over the world. Scientists have been trying to develop new ways to preserve and improve the environment. One of the recent methods is the "industrial ecology", in which the industrial system is managed and operated more or less like a natural ecosystem. Industrial ecology can be achieved by the proper implementation of cleaner production technologies in order to approach zero pollution. The proper implementation of CP can lead to zero pollution (environmental benefits) as well as economical and social benefits, if a good management system exist. Adapting environmental management system (EMS) or ISO 14001 can develop the foundation of a good management system. Therefore the integration of CP with EMS is a must in order to reach zero pollution/waste. In other words, industrial ecology can be implemented within the industrial activates through the implementation of the proposed CP-EMS model. This paper proposes an integrated CP-EMS model or a modified EMS model to integrate cleaner production technologies with environmental management system for sustainable development. The modified EMS model can be implemented not only on industrial processes but also on projects, services and communities in order to reach zero pollution. Two case studies approaching zero pollution in Egypt will be discussed. The first case study is dealing with cement industry and the way to convert it into zero pollution using the newly developed approach for zero pollution. Another case study dealing with Iron & Steel industry will also be discussed.

Keywords: Environmental management system, ISO14001, Cleaner production, Industrial ecology, Sustainable development.

Professor Salah El-Haggar, is professor of Energy and Environment in the Mechanical Engineering Department, The American University in Cairo. His research interest in the area of solid waste management and cleaner production. He developed a new theory for cleaner production to approach zero pollution. He is the author of two series of books, the first series dealing with cleaner production and the second series dealing with basics and mechanisms for sustainable development.

INTRODUCTION

Throughout history, economic activity has taken place through what is called an open system of materials flow. In other words, people have transformed natural resources into products and generating a huge amount of waste which can be disposed in near-by land and forced early peoples to change places as the build-up of wastes rendered existing settlements uninhabitable. Even so, this was reasonable behaviour when the human population was small and uncontaminated lands were fairly easy to find. Now, however, the situation is very different. It is simply no longer possible to avoid the wastes that we create, and disposing of waste is becoming more and more of a burden. This suggests that an open industrial system cannot continue indefinitely, that it will have to be replaced by one that offers more control over the flow of materials. Such a system would involve, among other things, paying more attention to where materials end up, choosing materials and manufacturing processes to generate a more circular flow, and avoiding practices and behaviour that could come back and affect us. Until quite recently, industrial societies have attempted to deal with pollution and other forms of waste largely through laws and regulations such as the Egyptian environmental protection law 4/1994 and the Executive Regulation 338/1995. Although this strategy has been partially successful, it has not really gotten to the root of the problem. To go deeply into the root of the problem and to guarantee a full success a new approach for our industrial system--an "industrial ecology" whose processes resemble those of a natural ecosystem.

In natural ecosystems, materials and energy circulate continuously in a complex web of interactions: Micro - organisms turn animal wastes into food for plants; the plants, in turn, are either eaten by animals or enter the cycle through death and decay. While ecosystems produce some actual wastes (by-products that are not recycled, such as fossil fuels), on the whole they are self-contained and self-sustaining.

In similar fashion, industrial ecology involves focusing less on the impacts of each industrial activity in isolation and more on the overall impact of all such activities. This means recognizing that the industrial system consists of many more than separate stages of extraction, manufacture, and disposal and that the stages are linked across time, distance, and economic sectors. It should also be stressed that industrial ecology incorporates elements of the regulatory approaches. Indeed, it places pollution control, pollution prevention, and recycling in a larger framework that includes additional types of environmentally responsible behaviour.

The objective of this paper is to integrate the main tools necessary for establishing industrial ecology such as Cleaner Production, Zero Pollution, and Environmental Management System in one model. The model was implemented in industrial activities such as cement industry and iron & steel industry.

CLEANER PRODUCTION

The term cleaner production (CP) was launched 1989 by the United Nations Environment Program (UNEP) as a response to the question of how to produce in a sustainable manner. Its core element is prevention vs. clean-up or end-of-pipe solutions to environmental problems. Resources should be used efficiently thus reducing environmental pollution and improving health and safety. Economic profitability together with environmental improvement is the aim. Cleaner Production typically includes measures such as good housekeeping, process modifications, Eco-design of products, and cleaner technologies…etc. The United Nations Environment Program (UNEP) defines Cleaner Production as, *"the continuous application of an integrated, preventative environmental strategy to processes, products and services to increase eco-efficiency and reduce risks to humans and the environment"* [1].

Cleaner Production focuses on conservation of natural resources such as water, energy and raw materials and avoiding the end of pipe treatment. It involves rethinking for products, processes and services to move towards sustainable development.

By considering production processes, cleaner production includes conserving raw materials and energy, eliminating toxic raw materials, and reducing the quantity and toxicity of all emissions and wastes before they leave a process. For products, the strategy focuses on reducing impacts along the entire life cycle of the product, from raw material extraction to the ultimate disposal of the product. Cleaner production is achieved by applying know-how, by improving technology, and by changing attitudes.

The conceptual and procedural approach to production that demands that all phases of the life-cycle of products must be addressed with the objective of the prevention or minimization of short and long-term risks to humans and to the environment.

One factor in defining cleaner production is therefore the reduction in production costs that results from improved process efficiencies. In terms of investment the key difference is that investment in end-of-pipe technologies is nearly always additional investment, whereas investment in cleaner production is usually, at least partly, in replacing existing systems or equipment. This has obvious implications for employment.

A useful definition of cleaner production needs to take account of the distinction between technologies and processes. For example, a process may be made "cleaner" without necessarily replacing process equipment with "cleaner components – by changing the way a process is operated, by implementing improved housekeeping or by replacing a feedstock with a "cleaner" one. Cleaner production may or may not, therefore, entail the use of cleaner technologies. Investment in cleaner production via the implementation of clean technologies is clearly easier to identify than investment in cleaner production by any other means. Whatever the method employed to make production cleaner, the result is to reduce the amount of pollutants and waste generated and reduces the amounts of non-renewable or harmful inputs used.

The Benefits of Cleaner Production

Cleaner Production can reduce operating costs, improve profitability, worker safety and reduce the environmental impact of the business. Companies are frequently surprised at the cost reductions achievable through the adoption of Cleaner Production techniques. Frequently, minimal or no capital expenditure is required to achieve worthwhile gains, with fast payback periods. Waste handling and charges, raw material usage and insurance premiums can often be cut, along with potential risks. On a broader scale, Cleaner Production can help alleviate the serious and increasing problems of air and water pollution, ozone depletion, global warning, landscape degradation, solid and liquid wastes, resource depletion, acidification of the natural and built environment, visual pollution and reduced bio-diversity.

Most countries (developed and developing countries) are working toward zero pollution not only in industrial sectors but also in vehicle emissions to reduce the gaseous emissions to the allowable limits as well as other sectors such as construction sector and agricultural sector. To achieve zero pollution, industry should prevent all pollutants from its effluent. The hierarchy to eliminate all pollutants should start from raw material selection all the way to product modifications in order to avoid end of pipe treatment. It is obvious that zero pollution techniques are good business for industry because:

- It will reduce waste disposal cost.
- It will reduce raw material cost
- It will reduce Health-Safety-Environment (HSE) damage cost
- It will improve public relations / image
- It will help comply with environmental protection regulations.

ZERO POLLUTION

In the past, people's dream was to turn sand into gold. Today though, the gold's secret is not any more in sand but in our waste and pollution. This was a dream until the new hierarchy to approach zero pollution was developed [2] and the 6-R golden rule was initiated [3]. That is, the rule aims at Reducing, Reusing and Recycling of waste. Whereas, the fourth **R** of the 6-R Golden Rule emphasizes the Recovering of raw materials from waste, leading to a partial treatment. The last 2-**R**'s are Rethinking and Renovation where people should rethink about their waste before taking action for treatment and develop renovative techniques to solve the problem. So the 6-R Golden Rule encompasses Reducing, Reusing, Recycling, Recovering, Rethinking and Renovation. The rule provides a theory to manipulate current activities to approach zero pollution and avoid landfill, incineration and/or treatment. Fortunately, with full success, the theory was practically applied in a pilot scale to most of the industrial sectors [4], numerous projects [5] and rural communities [6].

Capital investment, running cost as well as adverse environmental impacts of landfills, incineration and treatment lies at the heart of the implementation of the zero pollution theory. It is very simple, natural and not a newborn theory. Fundamentally, the theory depends on all kinds of recycling (on-site recycling, off-site recycling, partial treatment for possible recycling...etc). This is mainly because recycling is considered a pivotal income generated activity that conserves natural resources, protects the environment and provides job opportunities.

ENVIRONMENTAL MANAGEMENT SYSTEM

An environmental management system (EMS) is a part of the overall management system of an organization, which consists of organizational structure, planning, activities, responsibilities, practices, procedures, process and resources for developing, implementing, achieving, reviewing and maintaining the environmental policy.

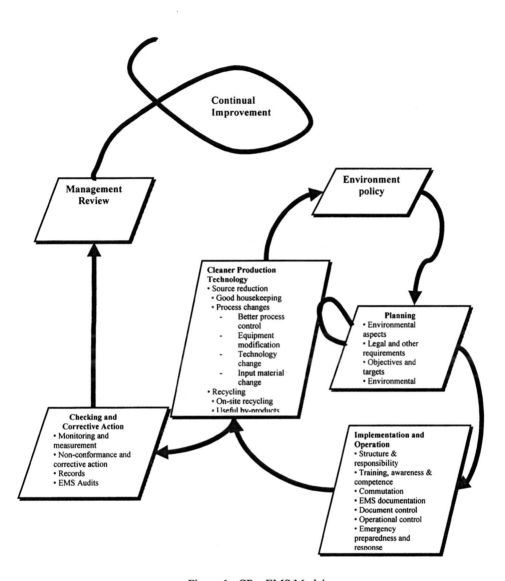

Figure 1   CP – EMS Model

The EMS provides several benefits through continual implementation and development that include:

- Financial benefits: through cost savings as well as increasing local and international market competitiveness
- Improve company's performance and image
- Reducing business risks

In general an EMS should include the following elements as shown in Figure 1:

- Management commitment and environmental policy
- Planning for the environmental Policy
- Implementing the environmental planning
- Evaluation and corrective/preventive actions
- Management review

The EMS can provide a company with a decision-making structure and action plan to bring cleaner production into the company's strategy, management and day-to-day operations. As a result, EMS will provide a tool for cleaner production implementation and pave the road toward it. So, integrating cleaner production technologies with EMS as shown in Figure 1 will help the system to approach zero pollution and maximize the benefits where both CP benefits and EMS benefits will be integrated together.

INDUSTRIAL ECOLOGY

We are talking about sustainable development all the time, but we have lost the direction to our goal because we need to think first how can we develop sustainability. Our real environmental and economical problem in this century is that development of science and technology have increased human capacity to extract resources from nature, then process it, use it, but finally it is not returned back to the environment to regenerate it. Unsustainable human activities are creating an open loop that can not continue and has to reach one day to a dead end. On the contrary, sustainability is the rule that governs natural economics. Natural economics regulates every transaction involving the exchange of any of its resources among the members of the system. Natural economics never pushes nature's inventory of resources beyond critical limits for sustainability. Human economic systems should learn from natural economic systems that there is a need to attain cooperation among individuals to enhance the economic efficiency in the use of resources for each member of the community that will in return maximize economic effectiveness for the community as a whole.

The tool for sustainable human economic systems is industrial ecology, and the means to achieve industrial ecology are cleaner production concepts, environmental management system and zero pollution. Industrial ecology (IE) is the study of industrial systems that operate like natural systems. In a natural system there is a closed materials and energy loop; where animal and plant wastes are decomposed by microorganisms into useful nutrients to the plants, then plants produce food to the animals to consume it, and finally the loop is closed again when animals die and are either converted to fossil fuels or nutrients for the plants. Industries can follow the same system through making one industry's waste another's raw materials, where materials and energy are circulated in a complex web of interactions.

Now, Industrial ecology (IE) can be defined as the study of industrial systems that operate more like natural ecosystems. It can be achieved by the proper implementation of cleaner production technologies in order to approach zero pollution. The proper implementation of CP can lead to environmental benefits (zero pollution) as well as economical and social benefits, if a good management system exists. Adapting environmental management system (EMS) or ISO 14001 can develop the foundation of a good management system. Therefore the integration of CP with EMS is a must in order to reach zero pollution/waste. In other words, industrial ecology can be implemented within the activity through the implementation of the proposed CP-EMS model shown in Figure 1. The proposed CP-EMS model or a modified EMS model to integrate cleaner production technologies with environmental management system as shown in Figure 1 will promote the concept of Cleaner Production technologies throughout the EMS steps starting from the Environmental policy – Planning-Implementation (especially the training component) all the way to checking and corrective action for sustainable development. Top management should be aware of the benefits of cleaner production to their business and the environment.

Industrial Ecology barriers

Even the industrial ecology concept has a lot of advantages from economical, environmental and social points of view, there is still some barriers for implementations. The barriers to industrial ecology fall into six categories [7,8] namely technical, market & information, business & financial, regulatory, legal and regional strategies.

Technical issues require a lot of innovation to convert waste into money or prevent it at the source. The markets for waste materials will ultimately rise or fall based on their economic vitality and can be enhanced through information technology tools. One option for waste markets are dedicated `Waste Exchanges' where brokers trade industrial wastes like other commodities. Developing business plans and providing financial support will help promoting the concept of industrial ecology. The private firm is the basic economic unit and collectively constitutes the mechanism for producing inventions and innovations to practice. The government should regulate planning for industrial ecology and sustainable development. Industries tend to form spatial clusters in specific geographic regions based on factors such as access to raw materials, convenient transportation, technical expertise, and markets. This requires regional strategies provided by local goverorates and federal government.

Industrial ecology can be considered as science of sustainability promises much in improving the efficiency of human use of the ecosystem. Technological improvements are not always better in the full sense of sustainability without taking environment into consideration, where zero pollution is a must for industrial ecology. Cooperation and community are also important parts of the ecological metaphor of sustainability.

CASE STUDIES

Two case studies will be discussed in this paper related to cleaner production to approach zero pollution for sustainable development in two major industrial sectors in Egypt. These case studies are related to cement industry [9] and iron & steel industry [10]. It worth noting that a study was prepared by the World Bank [11] estimated that, the total cost of combating industrial pollution in Egypt is about $1.3 billion. If this amount of money together with the waste generated from different industries will develop a new industrial sector called "waste sector" which is much larger than any industrial sector in the country.

Cement Industry

The cement industry is one of the main industries necessary for sustainable development. It can be considered the backbone for development. The main pollution source generated from cement industry is the solid waste called cement by-pass dust, which is collected from the bottom of the dust filter.

It represents a major pollution problem in Egypt and worldwide where millions of tons per year of cement dust is diffused into the atmosphere causing health problems and environmental damage to the eco-system.

Cement by-pass dust is very fin dust, has no cementing action and naturally alkaline with 11.5 pH value. It represents a major pollution problem not only in Egypt but also throughout the world. The safe disposal of cement dust costs a lot of money and still pollutes the environment because of its nature.

Because of the high alkalinity of the cement by-pass dust, it can be used in the treatment of the municipal sewage sludge, which is considered another environmental problem since it contains parasites and ailing microbes. Although sludge has a very high nutritional value for land reclamation, it might contaminate the land because of parasites and microbes. The safe disposal of sludge costs a lot of money and direct application of sludge for land reclamation has a lot of negative environmental impacts and is very hazardous to health.

Mixing the hazardous waste of cement by-pass dust with the environmentally unsafe sewage sludge will produce a good quality fertilizer. Cement by-pass dust will enhance the quality of the organic waste by killing all microbes and parasites. To enhance the product quality, agricultural wastes must be added to adjust the carbon to nitrogen ratio as well as the pH value for better composting process [9]. The produced fertilizer from composting is safe for land reclamation and free from any parasites or microbes that might exist in raw sludge and provide the land with the required nutrients.

Iron & Steel Industry

The steel industry is another backbone industry for sustainable development. It can be considered as the base for numerous industries that could not have been established without steel industry. Iron dust, slag and sludge are the main solid wastes generated by the iron and steel industry.

The problem of solid waste generated from iron and steel industry is not only hindering the use of millions of square meters of land for more useful purposes but also contaminating it. Many of these waste materials contain heavy metals such as barium, titanium and lead. Also, it is well known that toxic substances tend to concentrate in the slag. The health hazards of heavy metals and toxic substances are well known. Based on the concentration levels, some slags may be classified as hazardous waste materials. Furthermore, groundwater is susceptible to serious pollution problems due to the likely leaching of these waste materials.

Based on the production figures of the major steel plants in Egypt and the generation rates of different waste materials per tonne of steel produced, the volume of the annual waste materials generated in Egypt can be estimated as 1.1 million tonne of slag annually, 40,000 tonnes of dust and 25,000 tonne sludge as well as stock- piles of 20 million tonnes of air-cooled slag laying in the desert [10]. To have an idea about the considerable area needed for slag disposal, it is enough to know that the 20 million tonnes of slags currently available in Egypt are estimated to occupy 2.5 million square meters.

The utilization of different slag types as coarse aggregate replacements in producing building materials such as cement masonry units and paving stone interlock were investigated [10, 12]. Cement masonry specimens were tested for density, water absorption, compression and flexural strengths. While the paving stone interlocks were tested for bulk density, water absorption, compressive strength, and abrasion resistance. They also studied the likely health hazards of all applications. The test results proved in general the technical soundness and suitability of the introduced technology for cement masonry units and paving interlocks. Most of the slag solid brick units showed lower bulk density values than the commercial bricks used for comparison. All slag units exhibited absorption percentages well below the ASTM limit of 13% reached for all masonry groups at 28-day age compared to the control and commercial bricks. All test groups showed higher compressive strength.

CONCLUSIONS

Industrial development and environmental protection cannot be achieved without establishing the concept of industrial ecology. The main tools necessary for establishing Industrial Ecology are Cleaner Production, Environmental Management System and Zero Pollution. The concept of "industrial ecology" will help the industrial system to be managed and operated more or less like a natural ecosystem hence causing as less damage as possible to the surrounding environment. The 6-R Golden Rule encompasses Reducing, Reusing, Recycling, Recovering, Rethinking and Renovation is the basic tool for Industrial Ecology.

The proposed CP-EMS model or a modified EMS model to integrate cleaner production technologies with environmental management system will promote the concept of Cleaner Production technologies throughout the EMS steps and provide a base for industrial ecology.

Establishing industrial ecology within the industrial activities will avoid landfill, incineration and/or treatment. The cost of treatment and safe disposal of waste through incineration or landfills is escalating exponentially. Locating waste disposal sites (landfills) are becoming more difficult and expensive. The environmental and health impacts of landfills and incinerations are becoming more dangerous and disaster for the community and national economy.

REFERENCES

1. UNEP, "Cleaner production AT pulp and Paper Mills: A Guidance Manual Publication in Cooperation with the National Productivity Council", India, 1997.

2. EL-HAGGAR, S.M. and SAKR, D.A., "Globalization for Economical and Ecological Sustainability", AUC research conference, Globalization revisited: Challenges and opportunities, Cairo, Egypt, April 6-7, 2003.

3. EL-HAGGAR, S.M., " New Cleaner Production Hierarchy For Zero Pollution", Enviro 2001, The third International Conference for Environmental Management and Technologies, Cairo, Egypt, 29-31 October, 2001.

4. EL-HAGGAR, S.M., "Zero Pollution for Sustainable development in Egyptian Industries", International Congress on Sugar and Sugar cane by-products, Diversification 2002, Havana 17-22 June, 2002.

5. EL-HAGGAR, S.M., 'Reaching 100% Recycling of Municipal Solid Waste Generated In Egypt", to be presented in the International symposium for Advances in Waste Management and Recycling, Dundee U.K., 9-11 Sept. 2003.

6. EL-HAGGAR, S.M. "Environmentally Balanced Rural Waste Complex For Zero Pollution", Enviro 2001, The third International Conference for Environmental Management and Technologies, Cairo, Egypt, 29-31 October 2001.

7. WERNICK, I.K. and AUSUBEL, J.H., "Industrial Ecology: Some Directions for Research" Pre-Publication Draft, Program for the Human Environment, 1997.

8. EL-HAGGAR, S.M. and G. SALAMA, "Industrial Ecology to Approach Zero Pollution", Submitted to Cairo 8th International Conference on Energy and Environment, Cairo, Egypt, 4-7/1/2003.

9. EL-HAGGAR, S.M., "The Use of Cement By-Pass Dust for Sewage Sludge Treatment", The International Conference For Environmental Hazardous Mitigation, ICEHM2000, Cairo University, Egypt, September 9th –12th 2000.

10. EL-HAGGAR, S. M. and Y. KORANY, "Utilizing Slag Generated From Iron and Steel Industry In Producing Masonry Units And Paving Interlocks", the 28th CSCE annual conference, London, Ontario, Canada, June 7-10, 2000.

11. THE WORLD BANK, "Arab Republic of Egypt: Pollution Abatement Project", Staff Appraisal Report, 15390-EC, April 1996.

12. KOURANY, Y. and EL-HAGGAR, S.M., "Using Slag in Manufacuring Masonary Bricks and Paving Units", TMS Journal, pp 97-106, Sept. 2001.

EXCAVATION OF COVER MATERIAL
FROM OLD LANDFILLS

W Hogland
M Kriipsalu
University of Kalmar
Sweden

ABSTRACT. Landfilling of municipal wastes has great environmental impact. Stringent operational and technical requirements should be implemented in order to minimize any risk to human health, negative effects on surface and ground water, soil and air, and on the global environment. It has been decided that in the European Community only safe and controlled landfill activities should be carried out. However, insecure landfilling will be the dominating waste management method in the world in the foreseeable future. In the Baltic Sea Region alone there exist 75,000-100,000 old landfill sites, and the majority of them are not closed properly. A typical closing procedure consists of some excavation, profiling and capping of the landfill body. In many cases, removal of the whole or a part of the landfill will become more likely. Excavation of a landfill might be interesting in order to retrieve recyclables and other material. One of the most valuable materials secured from landfill excavations is soil, which can be used for daily coverage as well as capping. In this paper excavation of cover materials from old landfills is described. Results of test-excavation from Måsalycke and Gladsax landfills in Sweden are presented and evaluated in accordance to EU Landfill Directive.

Keywords: Landfill excavation, Capping, Remediation, EU Landfill Directive

W Hogland is a Professor at Department of Technology, Division of Environmental Engineering, University of Kalmar. His research has mainly been directed to system analysis of waste management in industries and municipalities, natural systems for leachate and wastewater treatment, and river catchments.

M Kriipsalu is currently undertaking research in bioremediation in oil-contaminated soils and sludges at Department of Technology, University of Kalmar.

INTRODUCTION

The European Council Directive 1999/31/EU [1] (Landfill Directive, LFD) is intended to prevent or reduce the adverse effects of the landfill of waste on the environment, in particular on surface water, groundwater, soil, air, and human health. The Directive states that only waste that has been subject to treatment may be taken to landfill. The amount of biodegradable waste, i.e. any waste that is capable of undergoing anaerobic or aerobic decomposition, going to landfills must be also be reduced. The main objectives are to reduce the global greenhouse effect and to alleviate stability problems of landfills.

When new landfills will be constructed in accordance to stringent rules, old landfills will remain and produce emissions to air, land and water if extensive measures are not taken. Just in the Baltic Sea Region, there exist 75.000-100.000 old landfill sites. As many cities are expanding, landfills that might have been originally located outside of a city are suddenly located within its boundaries, and must be carefully remediated. Landfill excavation and land remediation are potential methods for treatment of waste from old landfill.

LANDFILL EXCAVATION

Landfill excavation (LFE) involves excavation, transfer and processing of buried material (soil, recyclable and hazardous material and residues), and sometimes on-site treatment of contaminated soil or groundwater. Landfill excavation is conducted using techniques similar to those used in open face mining. Excavation is followed by screening and separation of material into various components (Figure 1). Equipment used for the excavation may be a front-end loader, a clamshell, a backhoe, a hydraulic excavator, or a combination of these. The excavated material can be mechanically screened into fractions.

The aims or advantages of LFE may be various and depend on local conditions:

- Energy and material recovery:
- Recovery of soil as cover material for other landfills;
- Removal of contaminated waste and upgrading of the contaminated area;
- Material removal for area or volume reduction for continued operation, or total sanitation (removal of the whole landfill).

Landfills have been excavated and mined throughout the world since the 1950s [2]. The technology was introduced in 1953 in Israel for recovering soil [3]. In the USA, LFE has been carried out mainly for recovering fuel and recyclables [4, 5]. Landfill material (LFM) has been used in Europe and pilot studies have been carried out in England, Italy, Sweden and Germany [6, 7; 8]. Total removal of a landfill has also been reported [9]. In addition to revenues from recyclables, combustibles, and soil, the project may be financed by the increase in the price of land for real estate developments.

The most important variable in LFM is the amount of recovered fine fraction, e.g. soil [10, 5]. This material can be used as filler for road construction, soil amendment, cover material for landfills, or just backfilled in more sustainable way.

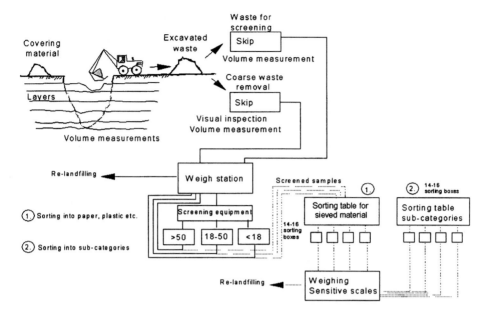

Figure 1 Landfill excavation technology at Måsalycke landfill (Hogland, 2002)

Soil recovery

The level of recovery depends on the chemical and physical conditions in the landfill, and the efficiency of the equipment used. The soil to waste ratio reported at various excavated landfills differs due to the amount of daily and final cover material employed, the size of the openings of the screen, type of landfill and waste, degree of compaction, age of landfill, and local conditions like moisture content in waste, and degree of decomposition. Ratios in the range of 20:80 to 75:25 were found in different LFM projects [11].

The amounts of fines in various US projects have varied between 40% and 60% [5]. Gausebroek [9] has reported the removal of two small landfills with total volume of 85,000 t (54,000 m^3). In total, 80% of the excavated material was extracted for re-use. About 10,000 t of rubble was re-used in road construction; 7,000 t as soil amendment and 50,000 t of contaminated soil were remediated. Only 17,000 t was backfilled as waste and 1,000 t of recyclables was taken to recovery.

Although the research indicates, that large amounts of soil can be extracted, the chemical composition must be carefully investigated. Gausebroek [9] reported contamination of mineral oil and PAH, but Hull et al [5] emphasized the importance of analysing of the material for VOCs, metals, pesticides, and PCB-s. In LFE projects, general aspects Landfill Directive 1999/31/EU [1] on the landfill of waste must be followed. For backfilling, Council Decision 2003/33/EC [21] on establishing criteria and procedures for the acceptance of waste at landfills must be followed. For use as soil amendment, the criteria for compost material to be met. If the target values are not met, bioremediation can be applied to improve its quality.

TRIAL-EXCAVATIONS IN SWEDEN

LFE experiences were gained from trial-excavation from Måsalycke and Gladsax landfills in Sweden. The main objective was to have an inventory of excavated material. The Måsalycke landfill is located in the south-east of Sweden and served the population of 33,000. It was commissioned at the beginning of the 1970s, while the excavated section was in operation during the period of 1975 to 1980. During the excavation the nature and composition of the waste was recorded, as well as the temperature, amount of methane in the air and the conductivity/resistivity of the leachate. Representative samples were taken from different layers of the excavated landfill and analysed.

The Gladsax landfill was closed and sealed in 1975 after more than 30 years of operation. The upper part of the landfill consists of about 4 metres of demolition and construction waste. Underneath this layer, municipal solid waste was found. The mining procedures were same as in case of the Måsalycke landfill.   At the Måsalycke landfill it was found, that the composition of the finest fraction (<18 mm) varied a lot. The percentage of poorly decomposed particles large increased with depth. It was found that the distribution of weights varied with depth of excavation. At the depth 0.5-2 m, the fines made 18.97% of total; at 2-4 m the largest amount of 24.72% was found, at 4-6 m it was reduced to 20.42%, and at 6-8 m it made just 14.81%. The moisture content varied between 22 and 39% by weight. About 70–80% of the material from all levels was in the size range of 1–10 mm (Figure 2). This very well correlates with Hull et al [5], who reported, that at least 40% by weight of the fragments in the soil admixture passed through a 2 mm sieve.

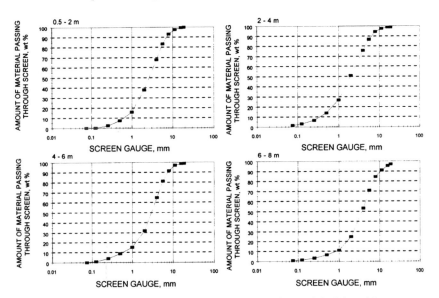

Figure 2   Particle size distribution of the screened material of the <18 mm
fraction at Måsalycke [19]

Returning material to landfills

Excavated waste can not be re-disposed to the landfill of inert waste, as for only selected construction and demolition waste with low contents of other types of materials (like metals, plastic, soil, organics, wood, rubber, etc), which origin is known, is accepted. Also soil and stones from contaminated sites will be excluded from landfills for inert wastes. The waste from Test-excavations revealed that large amounts of organics were present. Soil can be disposed to the class B2 landfill for organic waste, or landfills for mixed non-hazardous waste. The fine fraction can be disposed to separate cells for biomethanisation. This technology is accepted by LFD: the category B2 sites may be subdivided into bioreactor landfills (2003/33/EC) [21].

The humus-like mature material had a relatively high organic content at test-excavations. This can complicate the backfilling of soil in other landfills, as the organic content of wastes is strictly regulated. According to LFD, the content of organic fraction in municipal waste must be reduced by 2002 to 75% by weight in comparison to the 1995 values, by 2005 and 2012 the reduction must be 50% and 35%, respectively. Organic content of backfilled material must be measured, and if unacceptably high, reduced. Biological treatment means composting, anaerobic digestion, mechanical/biological treatment or any other process for sanitising biowaste. Stabilisation means the reduction of the decomposition properties of biowaste to such an extent that offensive odours are minimised and that either the respiration activity after four days (AT4) is below 10 mg O_2/g dm or the dynamic respiration index is below 1,000 mg O_2/kg VS/h [12]. It should be targeted, that the excavated waste is handled as a continuous waste stream, to minimise the need for analyses for hazardous substances.

Material contamination

The most serious problem with soil material is contamination. It was found at Måsalycke that the amount of heavy metals increased with depth; being largest at the depth of 6 m (total height of the landfill was 9 m). The reason might be that the area of the excavated landfill was relatively dry and the percolation of rainwater through the material had been low, which has limited the decomposition.

Table 1   The composition of fine screenings at Måsalycke landfill, compared to the proposed quality classes of compost and stabilised biowaste [10, 12]

		VARIOUS DEPTHS, 0-8 m SCREEN < 18 mm	COMPOST *		STABILISED BIOWASTE
			Class 1	Class 2	
DS	%	71.2-79.5			
Ash	% DS	87.3-90.2			
C	%	6.6-19.2			
N	%	0.3-0.5			
Ptot	mg/kg DS	820-1500			
Hg	mg/kg DS	0.029-0.3	0.5	1	5
Cd	mg/kg DS	0.1-1.2	0.7	1.5	5
Pb	mg/kg DS	47-270	100	150	500
Cr	mg/kg DS	4.7-78	100	150	600
Ni	mg/kg DS	3.1-15	50	75	150
Cu	mg/kg DS	5.5-36	100	150	600
Zn	mg/kg DS	91-230	200	400	1,500
As	mg/kg DS	<0.4			
PCBs	(mg/kg)**	n.a.	-	-	0.4
PAH	(mg/kg)**	n.a.	-	-	3

* normalised to an organic content of 30%
** threshold values to be set in consistence with the Sewage Sludge Directive
n.a. not available

If the material is of high quality, it can be used as soil amendment in agricultural areas. If this is intended, the chemical composition of the soil must be analyzed. In the test-excavations, the concentration of heavy metals was below the recommended limits (Table 1)

If the backfilled waste does not meet the standards, it must be further treated or landfilled in a secure landfill. Member States shall determine which of the test methods and corresponding limit values in the table should be used (Table 2).

Table 2   The limit values for granular non-hazardous waste accepted in landfills for non-hazardous waste (2003/33/EC) [21]

COMPONENTS	MÅSALYCKE LANDFILL, VARIOUS DEPTHS, 0-8 m SCREEN < 18 mm	L/S = 2 l/kg	L/S = 10 l/kg	C_0, PERCOLATION TEST
	mg/kg DS	mg/kg DS	mg/kg DS	mg/l
As	<0.4	0.4	2	0.3
Ba	n.a.	30	100	20
Cd	0.1-1.2	0.6	1	0.3
Cr total	4.7-78	4	10	2.5
Cu	5.5-36	25	50	30
Hg	0.029-0.3	0.05	0.2	0.03
Mo	n.a.	5	10	3.5
Ni	3.1-15	5	10	3
Pb	47-270	5	10	3
Sb	n.a.	0.2	0.7	0.15
Se	n.a.	0.3	0.5	0.2
Zn	91-230	25	50	15
Chloride	n.a.	10,000	15,000	8,500
Fluoride	n.a.	60	150	40
Sulphate	n.a.	10,000	20,000	7,000
DOC	n.a.	380	800	250
TDS	n.a.	40,000	60,000	

n.a. not available

REMEDIATION OF EXCAVATED SOILS

Physical-chemical methods

Soil leaching, also known as soil washing or soil flushing, is a process of gradually cleaning soil by flowing fresh water through the soil particles and extracting hydrocarbons, or other contaminants. Chemical extraction can be used with soil washing. This is based on the mobilization and extraction of contaminants with chemical additives, usually detergents (surfactants), or with solvents, often water or steam, or with CO_2. A treatment facility to treat the effluent water or a recycling system is required or the effluent must be collected. If a surfactant is added to the leaching solution, an oil-water emulsion is formed. Since the emulsions are usually quite stable it is necessary to collect the effluent emulsion and treat it in some manner. Land farming is an attractive option.

Biotreatment

In bioremediation systems, the environmental conditions are modified and the degrading action of microorganisms is enhanced by correcting some of the environmental factors that limit bioactivity [13; 14; 15; 16]. The ultimate goal of bioremediation is to mineralize the contaminants; that is, to transform a hazardous chemical into harmless compounds such as carbon dioxide or some other gas or inorganic substance, water, and cell material of the degrading organisms. Composting has proved to be an extremely versatile biological remediation process that can be adapted to treat a wide variety of organic contaminants. It has the potential to be the most simple and cost-effective treatment method [17, 16]. Bioremediation has been successfully used to treat petroleum-contaminated soils for the last 30 years [18]. Soil treatment processes include landfarming, biopiling of soil, and composting. Composting includes multiple advantages [14; 13; 17]:

- composting systems are relatively easy to design, construct and maintain;
- the contaminants are decomposed aerobically to CO_2, water and biomass;
- remediation can be completed in a relatively short time;
- composting offers a cost-competitive technology compared to secure landfilling;

The technology has limitations [17], which had to be solved. Composting may not be effective for high contaminant concentrations (>50,000 mg/kg TPH). However, during excavation the peak contaminant levels are reduced because highly contaminated soil becomes mixed with surrounding soil that is less contaminated or the sludge can be amended. Five- and six-ring PAHs are considered to be resistant to degradation in compost piles. Also, high heavy metal concentrations (>2,500 mg/kg) may inhibit microbial growth. Excavation may be followed by storage at a permitted landfill or by ex-in situ treatment. The storage area can be modified as a treatment system: soil masses can be placed in windraws, periodically turned, and a sprinkler system used to control moisture content.

Even if a section or a whole landfill will be totally removed, measures must be taken to ensure that the site is decontaminated. The soil underneath the unlined landfill usually contains a mixture of toxic substances. [9] reported that after the removal of the whole landfill, contaminated soil was found in so called 'hot spots'. These 'spots' must be bioremediated in-situ. Biodegradation can take place under aerobic and anaerobic conditions. The goal of in situ bioremediation is to supply oxygen and nutrients to the microorganisms in the soil. Oxygen is frequently limiting for optimal degradation. There exist many processes to deliver chemically bound oxygen to stimulate the activity of naturally occurring microorganisms. The use of oxygen release compounds (ORC) has proved to be successful. Some oil additives and chlorinated hydrocarbons, e.g. perchloroethylene (PCE), need to be degraded anaerobically [20].

Monitored natural attenuation

An open porous soil will offer more options for remediation. Just excavation of the contaminated soil can greatly speed up the biological processes [1]. Unlike heavy metals, petroleum contamination is self-correcting to a degree. Degradation of hydrocarbons by naturally-occurring soil bacterial is called passive biodegradation [20]. When contaminated soil is excavated and hauled to a landfill, this is the process by which restoration occurs. This is the least expensive, and slowest of all the available remedial technologies.

Natural attenuation relies on natural processes to clean up or attenuate pollution in soil and groundwater. At most polluted sites, natural attenuation occurs. The time it takes for monitored natural attenuation (MNA) to clean up a site depends on several factors: the type and amounts of chemicals present; the size and depth of the polluted area, and the type of soil and conditions present. These factors vary from site to site, and the cleanup usually takes years to decades. These conditions must be monitored or tested to make sure that natural attenuation MNA is working. MNA can also be used as a final cleanup step after another method has been used for removing most of the pollution.

CONCLUSIONS

In Baltic Sea Region there exist 75,000-100,000 old landfill sites. Most of them are not closed properly. A typical closing procedure consists of some excavation, profiling and capping of the landfill body. In many cases, removal of the whole or a part of the landfill will become more likely. Excavation of a landfill might be interesting in order to retrieve recyclables and other material, hidden in the landfill. One of the most valuable materials gained from landfill excavations is soil. This can be used for daily coverage as well as capping.

Various soil-to-waste ratios are reported, reaching up to 80%. This, however, depends on screening technology, screen size and moisture content.

From different LFE aspects, backfilling or using of excavated soil-like material, have the potential to confront with EU waste regulations. Problematic areas include:

- Classification of excavated waste (was it waste, soil, cover material, or inert waste)
- Organic content of the waste
- Heavy metal content

The properties of excavated material differ by fractions, and also by layers. Material from screened excavated material and daily covering material is suitable for use as new covering material. A peat-like material with a high organic content was obtained from the Måsalycke landfill. If the concentration of pollutants is low, this kind of material may be used as a soil improver. The analysis of heavy metals at Måsalycke indicated only high concentrations of zinc and there was no significant difference between the fine and the medium-sized fractions.

The EU directive on the landfill of waste requires the organic fraction to be reduced. If its organic matter content is high and the mass receives additional moisture, biogas production starts. In such cases backfill may be problematic and the excavated waste may need to be pre-treated prior to backfilling. Fractions with high content of organics may be treated in anaerobic digesters to generate biogas for energy production.

After the removal of a landfill section, the site itself must be remediated. Monitored natural attenuation seems to be the most cost-effective method. If the site will be used for housing, other in-situ treatment methods should be used.

Excavation of the contaminated soil can greatly speed up the biological processes. When contaminated soil is excavated and hauled to a landfill, this is the process by which restoration occurs. This is the least expensive, and slowest of all the available remedial technologies.

In further landfill excavation projects, sampling should be done in accordance to the directive (2003/33/EC) [21].

ACKNOWLEDGEMENTS

The LFM project in Måsalycke was initiated by Ms Sissi Sturesson at The Foundation for Technology Transfer in Kristianstad, Håkan Bladh, Director at the Simrishamn Municipality and Lars Erik Arvidsson from The Waste Management Company of Österlen. Prof. Emer. Aleksander Maastik, and the PhD candidates Sven Nimmermark and Lars Thörneby are also very much acknowledged as well as the Knowledge Foundation (KK-stiftelsen, Sweden); Kalmar Research and Development Foundation, and The Graninge Foundation, for financial support.

REFERENCES

1. EU LANDFILL DIRECTIVE (1999/31/EU). Council Directive of 26 April 1999 on the landfill of waste. Official Journal L 182, 16/07/1999, p 0001-0019. Luxembourg.

2. HOGLAND, W., SALERNI, E., THÖRNEBY, L., MEIJER, J.E. Landfill mining in Europe and USA – The State of the Art. Proceedings, WASTECON 96, Durban, South Africa, 1996.

3. SAVAGE, G.M., GOLUEKE, C.G., Von STEIN, E.L. Landfill Mining: Past and Present. BioCycle, Vol 34, 1993, Iss. 5, 58-62.

4. SPENCER, R. Mining landfills for recyclables. BioCycle, February 1991, 34-36.

5. HULL, R.M., KROGMANN, U., STROM, P.F. Characterisation of municipal solid waste (MSW) reclaimed from landfill. In proceedings of Eighth International Waste management and Landfill Symposium', CISA publisher, Cagliari, Italy 2001, vol IV, pp.567-578.

6. COSSU, R., MOTZO, G.M., LAUDADIO, M. Preliminary study for a landfill mining project. In Sardinia in T.H. Christensen, R. Cossu, R. Stegman (Eds) Proceedings 'Sardinia Fifth International Landfill Symposium', 1995, vol III   CISA publisher, Cagliari, Italy, pp.841-850.

7. HOGLAND, W., JAGODZINSKI, K., MEIJER, J.E. Landfill Mining Tests in Sweden. Fifth International Landfill Symposium, Cagliari, Sardinia, Italy, 2-6 October, 1995, pp 783-704.

8. RETTENBERGER, G., GÖSCHL, R. Ergebnisse und Betrieberfahrungen aus dem Demonstrationsprojekl "Deponiebrückhan Burghof". Deponieriickbau Seminar 8 November, Stuttgart Germany. 1994.

9. GEUSEBROEK, H.L.J. Complete removal of two landfills in Veenendaal (NL). In: T.H. Christensen, R. Cossu, R. Stegman (Eds) Proceedings of 'Sardinia 2001, Eighth International Waste management and Landfill Symposium', CISA publisher, Cagliari, Italy vol IV, 2001, pp.561-567

10. HOGLAND, W., MARQUES, M., NIMMERMARK, S., LARSSON, L., Landfill Mining and Stage of Waste Degradation in Two Landfills in Sweden. APLAS Fukuoka 2000, Asian-Pacific landfill Symposium, October 11-13, 2000, Fukuoka, Japan.

11. TAMMEMAGI, H. Landfill Mining – Lancaster County's experiment could succeed in Canada. Hazardous Materials Management, Southam Inc, 1999.

12. BIOLOGICAL TREATMENT OF BIOWASTE, 2nd draft, Brussels, 12 February 2001.

13. NORRIS, R.D. In-situ bioremediation of soils and groundwater contaminated with petroleum hydrocarbons. In: Handbook of Bioremediation, by Robert D. Norris (et al), Robert S. Kerr Environmental Research Laboratory. Lewis Publishers, USA. 1994.

14. COLE, G.M. Assessment and Remediation of Petroleum Contaminated Sites, Lewis Publishers, CRC Press, Boca Raton, FL, USA. 1994.

15. PERAMAKI, M.P.P.E., BLOMKER, K.R. Practical design considerations for composting contaminated soil. In: Papers from the Fourth International In Situ and On-Site Bioremediation Symposium. New Orleans, USA. 1997.

16. EWEIS, J.B., ERGAS, S.J., CHANG, D.P.Y., & SCHROEDER, E.D. Bioremediation Principles. McGraw-Hill International, New York, USA. 1998.

17. Von FAHNESTOCK, F.M., WICKRAMANAYAKE, G.B., KRATZKE, R.J., & MAJOR, W.R. Biopile design, operation, and maintenance handbook for treating hydrocarbon-contaminated soils. Battelle Press Ohio, USA. 1998.

18. RYAN, J.R., LOEHR, R.C., RUCKER, E. Bioremediation of organic contaminated soils. Journal of hazardous materials, vol. 28, 1991, pp 159-169.

19. HOGLAND, W. Remediation of an Old Landfill Site: Soil Analysis, Leachate Quality and Gas Production. In: Selected papers of the First Baltic Symposium on Environmental Chemistry, Tartu, Estonia, 26-29 September 2001. Special Issue: Environmental Science and Pollution Research. 2002, 1, pp 49-55.

20. KOENIGSBERG, S.S., NORRIS, R D. Accelerated bioremediation using slow release compounds. Selected Battelle Conference papers: 1993-1999. Battelle, USA, 1999.

21. COUNCIL DECISION 2003/33/EC on establishing criteria and procedures for the acceptance of waste at landfills. Official Journal L 011, 16/01/2003 P. 0027 – 0049.

POTENTIAL CONTRIBUTION OF COMBUSTIBLE DOMESTIC WASTE TO THE ENERGY SUPPLY OF LITHUANIA

G Denafas	H Seeger
I Rimaitytė	A Urban
Kaunas University of Technology	Kassel University
Lithuania	Germany

ABSTRACT. The Republic of Lithuania is divided into 10 districts which differ widely in terms of population density of 3.5 million. The average annual amount per head of domestic waste produced in Lithuania is 300 kg in large cities, 220 kg in small cities and 70 kg in the rural areas. The energy potential of domestic waste in Lithuania is 1411G Wh annually. In the case of the introduction of extensive material recycling of domestic waste, this amount would be reduced to 727G Wh per year. Two variants of thermal waste treatment processes were taken into consideration: incineration by grate furnaces, and gasification followed by the incineration in gas power plants. The calculation of the necessary capacities for thermal treatment of the domestic waste of every district is based on an annual availability of the plants of 75%. Finally 4 scenarios are considered: incineration on grate furnaces and gasification in combination with the current energy potential of domestic waste, and the potential after the introduction of extensive material recycling.

Keywords: Domestic waste production, Waste content, Recycling quotas, Calorific value, Waste energetic potential, Incineration, Fire grate, Gasification, Thermal waste treatment plants.

Dr Gintaras Denafas is a lecturer in the Department for Environmental Engineering, Kaunas University of Technology, Lithuania.

Ms Ingrida Rimaitytė is a masters student at the Department for Environmental Engineering, Kaunas University of Technology, Lithuania.

Mr Hendrik Seeger is a PhD student at the Department for Waste Treatment, Kassel University, Germany.

Dr Arndt Urban is a Professor and Head of Department for Waste Treatment, Kassel University, Germany

INTRODUCTION

With a territory of 65.301 km^2 and a population of 3.5 million, Lithuania is the largest of the three Baltic nations. Lithuania is divided into 10 districts (Figure 1) which differ widely in their population density (Table 1). Together with their direct neighboring nation Latvia, Lithuania belongs to the EU candidate-nations. The main topic concerning European integration of Lithuania is its economical development. However environmental protection including waste management, which is related to economical development, is also of importance.

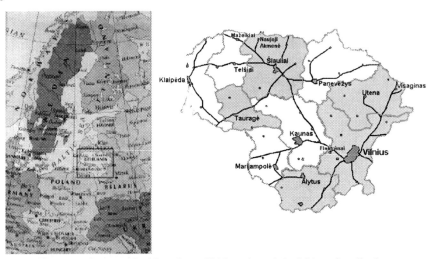

Figure 1   Geographical location of Lithuania and the Lithuanian districts

Table 1   Number of inhabitants (thousands) in the Lithuanian districts [1]

DISTRICT	IN LARGE CITIES	IN SMALL CITIES	IN VILLAGES	TOTAL
Vilnius	542	123	184	850
Kaunas	379	132	190	702
Klaipeda	193	90	103	386
Šiauliai	134	93	143	370
Panevezys	120	59	121	300
Telšiai	-	106	73	180
Utena	-	101	85	186
Marijampole	49	45	95	189
Alytus	71	39	77	188
Taurage	-	55	80	134
In Lithuania total	1488	844	1152	3484

Since the recovering of independence, the first steps to modern waste management policy have been made. Waste management legislation has existed since 1998 and has been developed according to EU legislations [2, 3]. Waste collection has been privatized. The recovering of materials, mostly paper, glass and plastic is performed too. However waste is mostly dumped into unprotected landfills, although some landfills have already been modernized, including installation of drainage and landfill gas procuring systems. Due to the rapidly economical development of the EU candidate nations in Central and EasternEurope, it is likely that they will be confronted with similar problems of waste management as WesternEurope. So the adoption of modern waste treatment technologies in Lithuania including thermal waste treatment is possible.

Along with the waste disposal, thermal waste treatment usually also recovers and uses the energy content of the incinerated waste. Due to the future closing of the Ignalina Nuclear Power Plant, which now produces approximately 80% of the electricity in Lithuania, a restructuring of the energy economy is necessary. Since Lithuania has very small resources of fossil fuel, energy recovering from waste could be of interest.

WASTE IN LITHUANIA

The amount of domestic waste, manufacturing waste and dangerous waste produced in Lithuania has been registered from 1992. The development of the waste production is presented in Figure 2. Of more concern are data about the waste composition. Only some research has been done in the past.

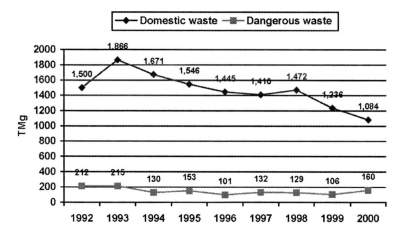

Figure 2 Development of the production of domestic waste and dangerous waste in Lithuania from 1992 to 2000 [5]

The two charts in Figure 3 present compositions of domestic and similar waste were used as base data for the following considerations: the waste composition of Kaunas as a model for large Lithuanian cities (>50,000 inhabitants) and the waste composition of Visaginas as a model for small Lithuanian cities [5].

However, establishing the composition and amount of waste to be disposed is problematic, because practically no data exists regarding collection quotas of recovering materials from domestic waste. For calculation of waste energy potential two initial scenarios have been considered. For the first scenario, we assume that at the moment the recovering of material from waste has not reached a level that it has a significant impact on the composition of waste.

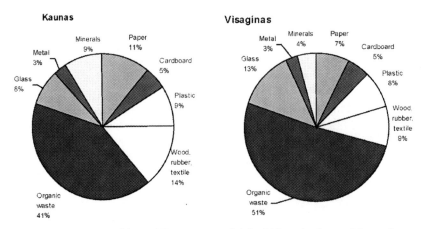

Figure 3   Waste composition of Kaunas as model for Lithuanian large cities and waste composition of Visaginas as model for Lithuanian small cities

This scenario will be referred to as "current situation". For the second scenario we assume that in the future recycling of materials will reach a similar level similar to that in Germany, with the exception of the adoption of a system for the collection and composting of organic waste. The proposed recycling quotas for this scenario are presented in the Table 2 [6]

Table 2   Proposed future recycling quotas in Lithuania

WASTE FRACTION	Glass	Paper	Organic waste	Metal	Plastic	Light waste	Other
RECYCLING QUOTA, %	64	64	75	50	40	0	20

ENERGETIC POTENTIAL OF DOMESTIC WASTE

The annual energetic potential of waste is calculated from the annual amount and the calorific value of the waste. The data of the Lithuanian Environmental Ministry show that annual amounts of domestic waste per capita are: 300 kg in large cities, 220 kg in small cities, 70 kg in rural areas [7]. The high difference between the waste amounts in cities and in rural areas is explained by the use of organic waste as feedstuff for animals and of the burnable waste (paper, wood, plastics etc) as fuel for households in the rural areas. Therefore, it can be concluded that waste from rural areas contains practically no burnable waste and because of that it is not considered in the subsequent calculations.

Though the amount of waste is known, establishing the calorific value due to heterogeneous composition of the waste is not without problems and arouses scientific discussion [8,9]. From a known waste composition, the calorific value of mixed waste can be calculated from the calorific values of the separate material fractions. These calorific values have been experimentally established [10].

The calorific values of separate waste fractions that were used for the calculation of the calorific value of the Lithuanian waste are presented in the Table 3. The origins of these data are partly from Lithuanian investigations [11, 12, 13], and since calorific values of Lithuanian waste fractions are not significantly different from German waste fractions and only the whole composition is different, calorific values established in Germany both experimentally and from the literature were also used for calculation of the energetic potential of the waste.

Table 3   Calorific values of different waste fractions

FRACTION	CALORIFIC VALUE kJ/kg	WATER CONTENT %	ASH CONTENT %	BURNABLE, %
Paper, cardboard	11,600	10	5	84
Plastic	24,900	2	10	88
Other burnable (wood, rubber, textile)	18,000	5	8	87
Organic waste	4,300	72	4	24
Glass	0	0	100	0
Metal	0	0	100	0
Minerals	0	0	100	0

From these presented calorific values, the waste composition data and the amount of waste produced, the energetic potential of the domestic waste was established. The calculated values for every Lithuanian district regarding the current situation and regarding possible recycling quotas are presented in Table 4.

Table 4   Energetic potential of domestic waste in Lithuanian districts, GWh

DISTRICT	LARGE CITIES		SMALL CITIES		TOTAL	
	Current situation	After recycling	Current situation	After recycling	Current situation	After recycling
Vilnius district	379.00	202.14	54.35	26.74	433.35	228.88
Kaunas district	264.84	141.25	58.22	18.16	323.06	159.41
Klaipeda district	134.85	71.92	39.48	19.42	174.33	91.35
Šiauliai district	93.57	49.91	41.11	20.22	134.68	70.13
Panevezys district	83.69	44.64	26.11	12.85	109.80	57.48
Telšiai district	0.00	0.00	46.87	23.06	46.87	23.06
Utena district	0.00	0.00	44.40	21.84	44.40	21.84
Marijampole district	34.02	18.14	19.59	9.64	53.61	27.78
Alytus district	49.96	26.65	17.35	8.54	67.32	35.18
Taurage district	0.00	0.00	24.08	11.85	24.08	11.85
Lithuania total:	1039.94	554.66	371.55	172.30	1411.49	726.96

The data of Lithuanian investigations predicts a heat demand of 14600 GWh for 2020 [14]. If the efficiency of the waste-to-heat conversion would be 80%, around 10% of the annual heat demand could in theory be covered by the waste-to-heat conversion of domestic waste. In practice, the heat demand is seasonal. Since the waste production fluctuates only marginally and domestic waste cannot be stored for a long time, in reality only a smaller part of the heat demand can be satisfied by waste-to-heat conversion. Moreover, electric power generation from waste compared with electric power production from fossil fuels is less efficient and thus less economic. If the efficiency during waste-to-power conversion is 30%, only 4-6% of the annual electric power demand of 13,700 GWh can be satisfied.

SCENARIOS FOR THE ENERGETIC USE OF
DOMESTIC WASTE IN LITHUANIA

Based on the energetic potential of domestic waste from Lithuanian districts, 4 scenarios for the installation of thermal treatment plants were developed.

Two main criteria influenced the selection of potential sites for the plants. Because the production of heat is a byproduct of the production of electricity, sites where a potential heat demand could be expected were considered. The second criteria was close proximity to appropriate infrastructure. In particular, sites connected to railway routes were favoured.

Two technologies for waste-to-energy conversion have been proposed. At first incineration on the fire-grate is an internationally prevalent technology. However municipal and industrial power plants in Lithuania have mostly gas/oil boilers and this will mostly remain the case in the future. Therefore, the alternative possibility would be to use a waste gasification system connected to a gas/oil boiler where the gas can be co-combusted together with natural gas/oil. For this second technology, an additional criteria for the selection of a plant site was the existence of a power plant.

The main technical data for Lithuanian power plants are presented in Table 5.

Table 5 Installed and exploited capacities of Lithuanian power plants [14]

POWER PLANTS	INSTALLED CAPACITIES	EXPLOITED CAPACITIES
Ignalina NPP	3000	2600
Lithuanian PP*:	1800	1800
150 MW units	600	600
300 MW units	1200	1200
Vilnius CHP*:	384	364
CHP-2	24	24
CHP-3	360	360
Kaunas CHP*:	178	178
CHP-2	170	170
Petrasiunai CHP	8	8
Mazeikiai CHP*	194	99
Klaipeda CHP*	11	11
Kaunas HPP	101	101
Other HPP	8	8
Kruonis HAPP	800	760
Industrial CHP*	61	61

*thermal power plants

It is necessary to point out that the Lithuanian network of power plants is not evenly distributed. The largest power plant producing 80% of all electricity in Lithuania is located in Northeast Lithuania, Utena district, near the crossing of the Lithuanian, Latvian and Byelorussian borders. This power plant will be gradually closed from 2004 to 2009. Over this period the exploitation of thermal power plants will be increased significantly. The output of the Ignalina Nuclear power plant will be fully replaced by those of the other Lithuanian power plants. This power plant is located in Elektrenai between the Lithuanian capital Vilnius and the second largest city Kaunas. Both of these cities also have thermal power plants. Thus, the largest thermal power plant capacities are situated in the Middle-South part of Lithuania. Only Mazeikiai CHP is situated in North-West Lithuania beside an oil processing enterprise. Its capacities are the highest in North West Lithuania. Small industrial CHP are mostly situated mostly near Vilnius and Kaunas. All Lithuanian thermal power plants have been adapted for gas/oil burning, but not for burning of solid fuel on fire grates or fluidized-bed furnaces.

The sizes of required incinerators for every district have been established by recalculation of waste energetic potential to incineration capacities by assuming that all non-recovered waste will receive thermal treatment and that plant operation availability would be 75%. If the necessary treatment capacities were too small for a single plant in some districts, these districts' waste would be transported to treatment plants in neighbouring districts. In some districts the necessary capacity would be distributed between several sites.

Finally we received 4 scenarios:

1. Incineration on the fire grate based on the current situation of waste production (Figure 4).
2. Incineration on the fire grate based on the future waste recycling quotas (Figure 5).
3. Application of gasification technology based on the current situation of waste production (Figure 6)
4. Application of gasification technology based on future waste recycling quotas (Figure 7).

In practice implementation of thermal waste treatment in Lithuania would be a mixture of these 4 scenarios. Also, it is not necessarily the case that all domestic waste would be thermally treated.

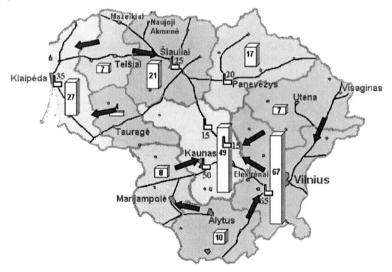

Figure 4   Scenario 1 – potential domestic waste thermal treatment capacities for all districts and proposed fire grate capacities (MW) based on current domestic waste production

During the selection of plant sites, both for fire grate and for gasification, we considered that fire grate should only be used in modernized or new-built thermal power plants. To these belong the current Vilnius and Kaunas CHP, and the future Klaipeda, Siauliai and Panevezys

CHP and industrial CHP in Jonava and Kedainiai cities near Kaunas. Unfortunately the Lithuanian PP in Elektrenai between Vilnius and Kaunas is to be modernized in the near future, so here no fire grate incineration will be implemented because current incineration capacities are already sufficient.

The installation of waste gasification technologies is a less difficult case because gasification products can be co-incinerated together with natural gas or heavy fuel oil. Thus Mazeikiai CHP, which will ultimately be modernized, can also be equipped with waste gasification.

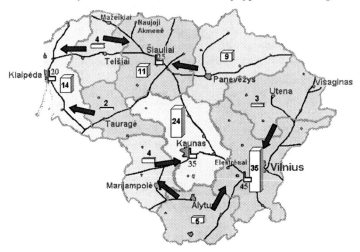

Figure 5   Scenario 2 – potential domestic waste thermal treatment capacities for all districts and proposed fire grate capacities (MW) based on possible domestic waste recycling quotas

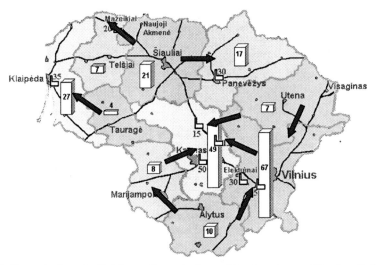

Figure 6   Scenario 3– potential domestic waste thermal treatment capacities for all districts and proposed gasification capacities (MW) based on current domestic waste production

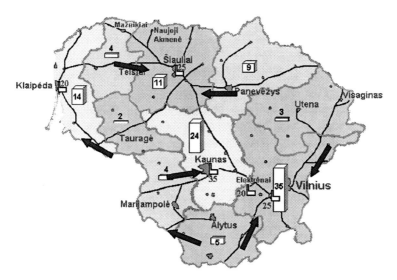

Figure 7   Scenario 4– potential domestic waste thermal treatment capacities for all districts and proposed gasification capacities (MW) based on possible domestic waste recycling quotas

SUMMARY AND CONCLUSIONS

With 3.5 million inhabitants, Lithuania is the largest of the three Baltic states. Lithuania is divided into 10 districts which differ widely in their population density. The average annual amount per head of domestic waste produced in Lithuania is 300 kg in large cities, 220 kg in small cities and 70 kg in the rural areas.

Since currently a large part of the energy of Lithuania (around 80% of electricity demand) is produced by the Ignalina Nuclear Power Plant, which has to be closed in 2 years, a restructuring of the national energy sector is imminent. Because Lithuania possesses no fossil fuels, it is worth considering the extent to which the energetic use of waste by thermal treatment can contribute to the energy supply.

The energy potential of domestic waste in Lithuania is 1411 GWh annually. In the case of the introduction of extensive material recycling of domestic waste, this amount would be reduced to 727 GWh per annum.

For the selection of potential locations for thermal waste treatment plants, 2 criteria have been used. Because heat is a byproduct of the generation of electricity by thermal waste treatment, the plants should be located near to areas of high demand for heat. Secondly, close proximity to infrastructure was desirable. In particular, a good connection by railway was seen as important.

Two variants of thermal waste treatment processes were taken into consideration: incineration by grate furnaces and gasification followed by the incineration in gas power plants. The calculation of the necessary capacities for the thermal treatment of domestic waste of every district was based on an annual availability of plants of 75%. Finally, 4 scenarios were considered; both incineration on grate furnaces and gasification in combination with current energy potential of domestic waste, and the potential after the possible introduction of extensive material recycling in the future.

REFERENCES

1. http://www.statistika.lt

2. LITHUANIAN LAW FOR WASTE DISPOSAL. – Vilnius, 16.06.1998, Nr. VIII-787)

3. MAIN DEMANDS FOR WASTE INCINERATION. LAND 19-99. – Lithuanian Normative Document for Environmental Protection. Official Edition. – Vilnius, 1999.

4. PROGRAM FOR WASTE COLLECTION AND DISPOSAL IN VISAGINAS CITY. – JSC "Baltijos ekostudija", Vilnius, 1999.

5. MERL, The Ministry of Environment of the Republic of Lithuania: Environment 2000 part II, state of environment, main change trends and protection measures, Vilnius, 2001

6. FRIEDEL, M.; URBAN, A. Ökologische Aspekte beim Einsatz aufbereiteter Abfälle in thermischen Anlagen in Faulstich, M.; Bilitewski, B.; Urban, A. [Hrsg.]: 3.Fachtagung Thermische Abfallbehandlung, Berichte aus Wassergüte- und Abfallwirtschaft Nr. 137, Technische Universität München, München 1998

7. DATA from Lithuanian Environmental Ministry.

8. ROTTER, S. Schwermetalle im Haushaltsabfall- Potenzial, Verteilung und Steuerungsmöglichkeiten durch Aufbereitung. Promotion an der Technischen Universität Dresden; Schriftenreihe des Institutes für Abfallwirtschaft und Altlasten Band 27; Dresden, 2002

9. THOMANETZ, E.: Das Märchen von der repräsentativen Abfallprobe. In Müll und Abfall Nr. 3 (2002); S. 136-142

10. KOST, T. Brennstofftechnische Charakterisierung von Abfällen. Promotion an der Technischen Universität Dresden; Schriftenreihe des Institutes für Abfallwirtschaft und Altlasten Band 16; Dresden, 2001

11. RESEARCH SUPPORT SCHEME. Report of co-operative project RSS 1631/2000" The Environmental Consequences of Use of Biomass and incinerable Wastes in the Baltic Region – Baltic States and Kaliningrad" – Kaunas, Riga, Tartu, Kaliningrad – 2000 – 2001.

12. DENAFAS, G.; ZALIAUSKIENE, A.; REVOLDAS, V. The Use of Ecological Compatibility of Waste and Biomass for Energetically and Transport Requirements (Lith.) // Environmental Research, Engineering and Management Nr. 4(18), 2001. P.30-39.

13. DENAFAS, G.; REVOLDAS, V.; ZALIAUSKIENE, A., BENDERE, R.; KUDRENICKIS, I.; MANDER, U.; OJA, T.; SERGEEVA, L.' ESIPENKO, A. Environmental Consequenses of the Use of Biomass and Combustible Waste in the Baltic region. // Latvian Journal of Physics and Technical Sciences, 2002, Nr. 2, Riga, P.24-45 Nordic Grant Scheme. Interim report for international Research Project ""Regional and Local Environmental Impact caused by Closing of Ignalina Nuclear Power Plant: Incorporating Environmental Considerations in the Decision Making Process", - Kaunas, December, 2001

LCIs FOR NEWSPAPER WITH DIFFERENT WASTE MANAGEMENT OPTIONS – CASE HELSINKI METROPOLITAN AREA

H Dahlbo   S Koskela

T Myllymaa   T Jouttijarvi

J Laukka   M Melanen

Finnish Environment Institute (SYKE)

Finland

ABSTRACT. This paper introduces preliminary results of life cycle inventories (LCIs) carried out to determine the environmental load of different newspaper waste management options. An LCI was performed for five cases, in which newspaper waste was divided into two streams: separately collected paper, and newspaper in mixed waste. A common element in each product system was newspaper manufacturing, while the other subprocesses were modified case-specifically. The chosen waste management options included various combinations of material and energy recovery and landfilling. Energy consumption and main emissions to air - methane, fossil carbon dioxide, nitrogen oxides and sulphur dioxide - are analysed in this paper. The avoided emissions were calculated by using the average Finnish electricity and heat production as the substitute. The ecological sustainability of different waste management options cannot be assessed with the inventory results, though. This will be carried out later together with the impact assessment results.

Keywords: Life cycle inventory, Recycling of paper, Material recovery, Energy recovery, Landfilling, Municipal solid waste management, Case study.

H Dahlbo is a senior researcher in the Research Department of the Finnish Environment Institute (SYKE). She is preparing her doctoral thesis on the ecological sustainability of different waste management solutions.

T Myllymaa is a research engineer currently undertaking research into the waste management systems in the Research Department of SYKE.

J Laukka is a graduate student working in the Research Department of SYKE.

S Koskela is a senior researcher in the Research Department of SYKE. She is specialised in life cycle assessment.

T Jouttijärvi is a research engineer in the Research Department of SYKE. His special field is forest industry.

M Melanen works as a Research Professor in the Research Department of SYKE. His current research activities cover a wide spectrum of integrated resources management and integrated pollution prevention and control.

INTRODUCTION

Increased recycling and energy recovery of waste materials have raised the profile of waste management as one of the critical sectors contributing to sustainable management of natural resources. The environmental dimension of sustainability in waste management addresses on one hand pollution and on the other hand resource conservation. Concerns over conservation of resources have led to calls for a) general reductions in the amount of waste generated, i.e. waste minimisation or waste prevention, and b) ways to recover the material and/or energy in the waste, so that they can be used again. Recovery of resources from waste should slow down the depletion of non-renewable resources, and help to lower the use of renewable resources to the rate of their replenishment. [1].

In the waste policy of the European Union, waste prevention has been set as the first priority of waste management. Prevention is hierarchically followed by material recycling, recovery as energy and, as the last option, safe final disposal. These options all have different effects on the life cycle of products and materials entering the waste streams.

Life cycle assessment (LCA) is an environmental management tool that aims to predict the overall environmental interventions of a product, service or function. In a research project described in this paper, namely "Life cycle approach to the sustainability of waste management - a case study on newspaper, the LCA-WASTE project", LCA methodology is applied to analyse the ecological sustainability of different waste management solutions of newspaper in the Helsinki Metropolitan Area, Finland. The paper summarizes part of the inventory results of the project. Economic aspects are also studied within the LCA-WASTE project, but they are not included in this paper.

GOAL AND SCOPE

The project is carried out as a case study using the Helsinki Metropolitan Area (HMA) as the region and newspaper as the product. The primary objective of the project is to develop a methodology for assessing the effects of different waste management systems on the ecological impacts caused by a product or a material during its life cycle. The methodology should be generally applicable to different regions and waste fractions. Another objective of the case study is to generate information on the impacts of different waste management strategies needed in waste policy concerning paper waste. In total, five newspaper waste management solutions and their environmental interventions are compared.

Newspaper was selected for the study for two reasons: 1) paper is one of the largest fractions of municipal solid waste (MSW), and 2) its manufacture is based on the use of forests, the most important renewable natural resource in Finland. The HMA was selected on the basis of data availability and quality. It is the most densely populated area in Finland: the area is only 0.2% of the whole country, but approximately 20% of Finland's population live there (997 000 in the year 2001). Helsinki Metropolitan Area Council has the overall responsibility for waste management in the HMA, where 1.1 million tonnes of waste are produced annually [2]. The current waste management system is based on separation at source. The separately collected fractions are biowaste, paper, cardboard, liquid board packaging, paper/board based packaging, glass, metals, electrical goods, electronic scrap, clothes and hazardous wastes.

APPLIED METHODOLOGIES

For the purposes of the LCA-WASTE project, the procedures of internationally standardised LCA for products (ISO 14040 - 14043) were combined with integrated solid waste modelling. KCL-ECO [3], a software designed for life cycle inventories of particularly forest industry products, was the basic tool used for the inventory analysis. The calculations of environmental interventions caused by collection and transportation of wastes (both separately collected paper and mixed waste) from residential properties and commercial establishments were, however, carried out using a methodology developed by the Finnish Environment Institute for analysing MSW management based on source separation [4, 5]. The results of these calculations were then used as input data for the waste collection and transportation module included in the product system constructed with the KCL-ECO software.

SYSTEM BOUNDARIES

Figure 1 shows the product system studied in the LCA-WASTE project. The system consists of all the sub-processes within the life cycle of newspaper, i.e. the manufacturing of newsprint, the printing of newspaper and the collection, transportation and waste management of newspaper waste. The functional unit of the product system, against which the interventions were calculated, was defined as one tonne of newspaper delivered to consumers.

In contrast to several other LCA studies focusing on waste management, e.g., [6, 7], the manufacturing of newspaper was included in the studied product system. In this way, the significance of the different life cycle phases, namely forestry, newsprint manufacturing, newspaper printing and newspaper waste collection and treatment, could be assessed.

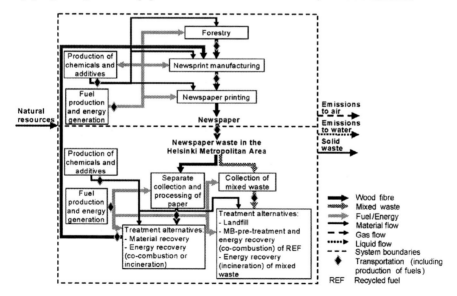

Figure 1 The product systems boundaries in the LCA-WASTE project.

WASTE MANAGEMENT OPTIONS

The five alternative waste management options (cases) studied were formulated by combining various treatment methods applicable to newspaper in the separately collected paper fraction (newspapers and magazines from residential properties and commercial establishments) and newspaper in mixed waste. The options were as follows (Figure 2):

Case 1 describes the current situation in the HMA:
* material recovery of newspaper in separately collected paper and landfilling of mixed waste;

Case 2 is based on the plans in the HMA to set up new treatment facilities by 2010:
* **a)** material recovery of newspaper in separately collected paper combined with mechanical-biological (MB) pre-treatment of mixed waste and energy recovery of the combustible fraction (REF, recycled fuel);
* **b)** material or energy recovery of newspaper in separately collected paper combined with mixed waste treatment identical to Case 2a;

Case 3 is a theoretical alternative:
* **a)** material recovery of newspaper in separately collected paper combined with incineration (with energy recovery) of mixed waste;
* **b)** material or energy recovery of newspaper in separately collected paper combined with incineration (with energy recovery) of mixed waste.

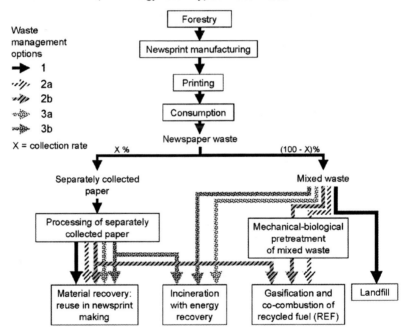

Figure 2   The waste management cases studied in the LCA-WASTE project.

In newsprint manufacturing, data from a process using both thermo-mechanical pulp (TMP, processed from spruce pulpwood and chips) and deinked pulp (DIP, processed from separately collected newspapers and magazines) was used. The processing of separately collected paper in the recovered paper treatment plant included baling and other handling of paper.

REF (recycled fuel) is a product of the mechanical-biological pre-treatment of mixed waste. Other fractions, such as metals, biowaste and inorganic reject were assumed to include no paper. Co-combustion of REF was assumed to take place in a pulverised coal combustion (PCC) plant equipped with a REF gasifier.

Energy recovery in mixed waste incineration plant was allocated to newspaper based on its elemental composition. Emissions from incineration was modelled by using partitioning coefficients of elements [8].

RESULTS AND DISCUSSION

In the inventory, data on the inputs (use of natural resources and materials) and outputs (products, emissions into the atmosphere and watercourses, wastes) of the processes forming the product system were compiled and quantified. The results presented in this paper are thus based on emissions and resource use exclusively. After finishing the inventory phase in August 2003, a life cycle impact assessment (LCIA) will be performed. In the LCIA the potential environmental impacts caused by emissions will be assessed. In addition to emissions, other aspects such as saved forest area with its implications, can be considered in the LCIA.

The inventory results were calculated per the functional unit of the product system, which was one tonne of air-dry (10% humidity) newspapers delivered to consumers. For presenting the results, the product system was divided into the following five life cycle phases (Figure 3):

- Forestry, including
  - silviculture, sawmill, use and manufacturing of lubricating oils, fertilisers and fuels needed, and generation of electricity.
- Paper mill (newsprint manufacture), including
  - paper mill (debarking, chip processing and screening, de-inking plant, thermo-mechanical pulp (TMP) process, the paper machine producing newsprint, water purification and demineralisation plant, wastewater treatment plant, power plant),
  - use and manufacturing of chemicals and fuels needed, generation of electricity and heat.
- Printing, including
  - printing the newspaper, use and manufacturing of chemicals, aluminium and fuels needed, generation of electricity and heat.
- Waste treatment, including
  - treatment of newspaper in mixed waste with the following methods in different cases: 1) landfilling, flaring the landfill gas, use and manufacturing of fuels needed, generation of electricity (Case 1); 2) MB pre-treatment, gasification and

co-combustion of REF at the power plant, use and manufacturing of chemicals and fuels needed, generation of electricity (Cases 2a and 2b); and 3) incineration of mixed waste, use and manufacturing of chemicals and fuels needed, generation of electricity (Cases 3a and 3b).
   o treatment of newspaper in the separately collected paper fraction with the following methods in different cases: 1) processing of separately collected newspaper (baling and loading), use and manufacturing of fuels needed, generation of electricity (Cases 1, 2a, 2b, 3a, 3b); and 2) gasification and co-combustion (Case 2b) or incineration (Case 3b) of separately collected processed newspaper, use and manufacturing of fuels needed, generation of electricity.

- Transportations, including
   o transportations of products, chemicals, raw materials, fuels and waste needed in the product system, use and manufacturing of fuels needed.

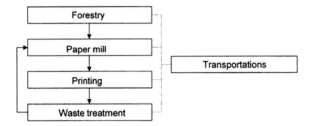

Figure 3   Life cycle phases of the product systems studied in the LCA-WASTE-project.

The most important assumptions made during the inventory analysis were as follows:
- Material recovery of newspaper was assumed to take place in a closed-loop. In other words, all newspaper collected separately for material recovery in the HMA returned to the paper mill for de-inking and recovery in the newsprint manufacturing.
- 76.2% of the newspapers delivered to consumers were assumed to enter separate collection, 3.2% ended up in small-scale recycling (e.g., as kindlings, biowaste bags or wrappings) and the remaining 20.6% entered mixed waste.
- The efficiency of the MB pre-treatment plant in separating newspaper from mixed waste into the REF fraction was assumed to be 100%.
- In cases 2b and 3b 50% of the separately collected and processed newspaper was combusted (Case 2b) or incinerated (Case 3b). This was assumed to decrease the use of de-inked pulp and increase the use of virgin fibre based thermo mechanical pulp in newsprint manufacturing.
- In cases 2a and 2b we assessed the gasification, co-combustion and resulting emissions of the REF fraction equal to the amount of newspaper entering the REF fraction in the MB pre-treatment plant. In cases 3a and 3b, on the contrary, emissions from incineration were allocated to newspaper (based on the composition of newspaper). This has to be taken into account when comparing the results of the cases.
- In modelling the generation of electricity and its emissions, the fuel mix of Finland's average electricity generation and imported electricity during 2000-2002 were used [9].
- For heat production, the fuel mix of the average district heat production in Finland in 2000 was used [10].

- Gasification and co-combustion of REF and incineration of mixed waste produce energy that substitutes other energy production. The fuel used for the substituted energy determines the avoided emissions derived by energy recovery of waste. In Figures 4 and 5 the energy recovered from wastes substituted average Finnish electricity and heat production.

Fuel Consumption

Figure 4(a) shows the total fuel consumption of the product systems and 4(b) the division of the consumption into renewable and non-renewable fuels. Biofuels, wasteliquids, hydropower and wind power were considered renewable fuels and coal, oil, natural gas, peat, nuclear power and gas from blast furnace and coking plant were considered non-renewable. In the whole product system the fuel use was dominated by newsprint manufacturing and newspaper printing, which together used well over 90% of all fuels. Thus only small differences could be detected between the cases. Cases 2b and 3b used fuels most, because increased use of virgin wood material resulted in higher demand of purchased electricity at the paper mill. The use of non-renewable fuels in the purchased electricity generation increased their total consumption in b-cases. On the other hand, in Cases 2b and 3b also the fuel consumption avoided through the energy recovery of wastes was the highest. As a result, the total fuel consumption decreased.

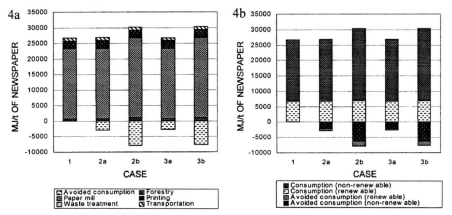

Figures 4   The fuel consumption of the product systems. In Figure a, the total fuel consumption is divided into life cycle phases. In Figure b the consumption is divided into renewable and non-renewable fuels. The avoided fuel consumption was calculated assuming that the energy recovered from wastes substitutes average Finnish electricity and heat production.

Atmospheric Emissions from the Whole Product Systems

Figures 5(a-d) show the emissions to air of methane (CH_4), fossil carbon dioxide (CO_2), nitrogen oxides (NO_x) and sulphur dioxide (SO_2). The methane emissions were dominated by landfilling in Case 1 (Figure 5(a)). No energy recovery for landfill gas was assumed, but the gas was flared. However, energy recovery will be studied in a later phase of the project.

Regarding the emissions of carbon dioxide and nitrogen oxides the studied cases differed only slightly (Figures 5(b) and (c)). In Cases 2b and 3b the energy recovery of separately collected paper increased the emissions originating in forestry, since recovered paper was substituted by virgin wood materials in the newsprint manufacturing. These were, however, almost totally compensated by the decrease of emissions originating in paper mill. The decrease was due to the increase of demand for purchased electricity and its lower CO_2 and NO_x emissions compared to the power generation of the paper mill. Hence the minor differences between cases, seen most clearly in Figure 5(c), were due to emissions originating in waste energy recovery. In Cases 2b and 3b the amount of material combusted was higher due to combustion of separately collected paper. Therefore the emissions of these cases were higher in comparison to Cases 2a and 3a. Emissions of sulphur dioxide were highest in Case 2b (Figure 5(d)). As previously in Cases 2b and 3b, also emissions of sulphur dioxide increased in forestry and decreased in paper mill. In Cases 2a and 2b, however, the emissions from waste treatment were not totally comparable with the other cases, because the emissions from energy recovery were not allocated to newspaper. In all Figures 5(a-d), the emissions avoided by energy recovery of wastes were the greatest in Cases 2b and 3b, in which also the amount of material combusted was greatest. Incineration produced 3.5% electricity and 96.5% heat whereas the co-combustion produced 30% and 70% respectively. Furthermore, the average production of heat caused higher emissions than the average production of electricity, because of the use of nuclear and hydropower for the latter. As a result, the emissions avoided by incineration (Cases 3a and 3b) were greater than by gasification and co-combustion of REF. At this project phase, however, the ecological implications of the increased use of forest and wood in Cases 2b and 3b were not considered. These aspects will be assessed in the LCIA phase.

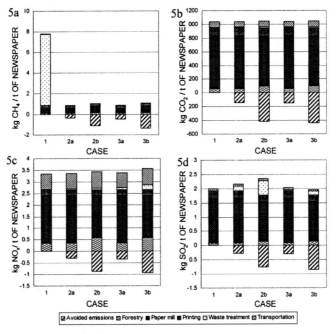

Figures 5(a-d) Emissions to air from the product systems: methane (CH_4), fossil carbon dioxide (CO_2), nitrogen oxides (NO_x) and sulphur dioxide (SO_2).

Atmospheric Emissions from Transportations

Transportations were further divided into different transportation phases (Figures 6(a-d)). The total emissions generated by the different transportation phases were influenced by the transported distance and load and the quality (type, age) of vehicles used for the transportations [11]. The transportations constant in all cases were: delivery of newspapers to consumers, separate collection and transportations of newspapers and collection and transportations of newspaper in mixed waste. Hence, the remaining transportations caused the differences between cases.

The methane emissions (Figure 6(a)) were dominated by the delivery of newspapers to consumers due to extensive use of small petrol-driven cars. The differences in carbon dioxide and nitrogen oxides emissions (Figures 6(b) and 6(c)) were due to the increase of wood transportations in Cases 2b and 3b. Although the transportations of separately collected processed newspaper simultaneously decreased, they did not compensate for the increase of wood transportations. Consequently, the total emissions from transportations increased. Sulphur dioxide emissions (Figure 6(d)), on the other hand, decreased in Cases 2b and 3b, due to the decrease of sea transportations of bentonite needed in the de-inking process.

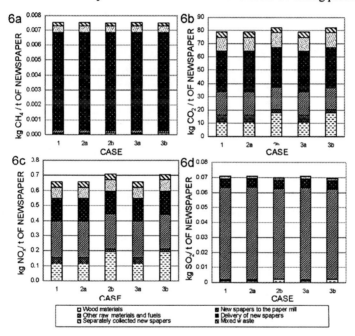

Figures 6(a-d)   Emissions to air from the transportations: methane (CH_4), fossil carbon dioxide (CO_2), nitrogen oxides (NO_x) and sulphur dioxide (SO_2). The transportation phases are: transportations of wood materials to the paper mill, transportations of the separately collected, processed newspapers to the paper mill, transportations of other raw materials, fuels and chemicals, delivery of newspapers to consumers, separate collection and transportations of newspapers and collection and transportations of mixed waste.

The ecological sustainability of different waste management options cannot be assessed with the inventory results until together with the impact assessment results.

ACKNOWLEDGEMENTS

The financial contribution of the National Technology Agency (Tekes) to the LCA-WASTE project is gratefully acknowledged.

REFERENCES

1. MCDOUGALL, F., WHITE, P., FRANKE, M. & HINDLE, P., Integrated solid waste management: a life cycle inventory. 2nd ed. Cornwall, Blackwell Science, 2001.

2. YTV, Helsinki Metropolitan Area Council. http://www.ytv.fi/english/waste/index.html, 2003.

3. KCL, Oy Keskuslaboratorio - Centrallaboratorium Ab, the Finnish Pulp and Paper Research Institute. www.kcl.fi/eco/softw.html, 2003.

4. TANSKANEN, J.-H. An approach for evaluating the effects of source separation on municipal solid waste management. Helsinki, Finnish Environment Institute. Monographs of the Boreal Environmental Research No. 17. Doctoral dissertation, 2000.

5. TANSKANEN, J.-H. & MELANEN, M., Modelling separation strategies of municipal solid waste in Finland, Waste Management & Research, Vol. 17(2), 1999, pp 80-92.

6. FINNVEDEN, G., JOHANSSON, J., LIND, P., MOBERG, Å., Life Cycle Assessments of Energy from Solid Waste, Ursvik, 2000.

7. EDWARDS, D. W. & SCHELLING, J., Municipal waste life cycle assessment. Part 2: Transport analysis and glass case study. TransIChemE, PartB 77(5), 1999, pp 259-274.

8. BJÖRKLUND, A., Environmental systems analysis of waste management with emphasis on substance flows and environmental impact, Licenciate thesis, Royal Institute of Technology, Stockholm, 1998.

9. HEIKKINEN, A., Enprima Engineering Oy, a notification on 17.2.2003.

10. TATTARI, K., VTT Building and transport, a notification on 31.3.2003.

11. VTT Technical Research Centre of Finland, LIPASTO Database of traffic emissions, http://lipasto.vtt.fi/indexe.htm, 2003.

A RESPONSIBLE APPROACH TO THE RECYCLING OF BY-PRODUCT WASTES INTO COMMERCIAL CLAYWARE

F Moedinger

Ziegel Gasser GmbH

Italy

M Anderson

Staffordshire University

United Kingdom

ABSTRACT. As a consequence of rapidly declining landfill space throughout the EU, increasing interest is being shown in developing more environmentally acceptable disposal procedures for a broad range of industrial and urban by-product wastes. Opportunities for the inclusion of many such materials exists in the ceramic industry which uses large quantities of non-renewable clays, marls and shales to produce a wide variety of fired structural ceramic products. As pressure on landfill space continues to grow it is anticipated that increasing interest in adopting this alternative approach will become more widespread within the EU. Consequently, it is of paramount importance that the adoption of this technology is carried out in a responsible way and within internationally recognized guidelines. Ziegel Gasser GmbH-Sri is a brick manufacturer in Southern Tyrol that is establishing itself as a market-leader in pioneering this alternative approach by developing a systematic and audit-traceable procedure linked to its manufacture of bricks containing a range of by-product wastes. A wide variety of such materials are theoretically suitable to be incorporated into the normal feedstock of a typical brick factory, but success is dependent on careful evaluation. In the approach reported, the methods of selection, based on a company specific standard of reference of potentially suitable wastes is described. The contractual requirements with the particular suppliers of such materials are also examined.

Keywords: Waster recycling, Wastes in brick manufacture, Methods of utilizing wastes, Contractual requirements for using wastes, Methods of analysis, LCA Life Cycle Assessment.

F Mödinger, is Plant Manager and also in charge of R&D at the Ziegel Gasser GmbH. brickworks in South Tyrol, Italy.

M Anderson, is Manager of the Ceramic Technology Group within the School of Engineering and Advanced Technology at Staffordshire University.

INTRODUCTION

The use of waste materials to partly substitute for traditional ceramic raw materials within the feedstock of a brick factory can be very worthwhile, both from a technical as well as a financial point of view. It is nevertheless essential to first establish a proper framework within the company operating structure to evaluate which wastes can be beneficially used and which must be rejected. As part of this process it is also of importance to thoroughly investigate the contractual, financial, quality-control and supply implications of such undertakings at an early stage. The essential control and quality assurance procedures required to be put in place to achieve this objective are described in this paper.

ASSESSING THE USE OF WASTES

Analysis requirement

In most cases in developed countries waste disposal laws and regulations usually require that the wastes to be disposed of are accompanied by some form of analysis data (in many instances some form of leaching information relating to the wastes is also required).

Table 1   Required analytical data for evaluation of wastes for use in brick products

Chemical properties		Chemical properties (continued)	
Aluminium oxide	Al_2O_3	Cadmium	Dd
Sulphur trioxide	SO_3	Chrome	Cr
Antimony oxide	Sb_2O_3	Mercury	Hg
Barium oxides	BaO	Nickel	Ni
Boron trioxide	B_2O_3	Copper	Cr
Calcium Oxide	CaO	Selen	Se
Iron oxide	Fe_2O_3	Lead	Pb
Phosphorus Pentoxide	P_2O_5	Zinc	Zn
Magnesium oxide	MgO	Ancillary properties	
Manganese oxide	MnO	Moisture content	H_2O
Lead oxide	PbO	Acidity/alkalinity	pH
Potassium oxide	K2O	Dry solids	%
Silica	SiO_2	Ash content	%
Sodium oxide	Na_2O	Fineness	-
Titanium dioxide	TiO_2	Plasticity	-
Zinc oxide	ZnO	Odour (on drying)	-
Zirconium oxide	ZrO_2	Firing colour	-
Chlorides	Cl	Shrinkage (drying – firing)	-
Arsenic	As	Evaluation of possible emissions	-

However, such analytical data is frequently still insufficient to determine whether the waste may be used in brick products as it usually furnishes very little detail about the vital nature of the waste itself. For a typical brick factory, the broad information required is listed in Table 1. In some instances accompanying standard documentary information may be acceptable. However, other more critical data must be ascertained by actual analysis.

Quality Control

It is essential that the quality and composition of the waste even when it comes into use at a brickworks, should continue to be regularly monitored to avoid the possibility of any sporadic fluctuations in composition on a day-to-day basis, which could otherwise detrimentally affect the manufactured product. In this context, the following guideline assessment or 'check-list' is recommended.

- The regular determination of the oxides and their individual levels of presence is considered essential to establish if the intended use of the waste will detrimentally affect the firing behaviour of the product in which it will be contained. In particular, the concentration of chrome and chlorates which might be present, are of great importance with respect to their possible volatilization on firing and their subsequent concentration in the flue gases. This would bring both the risk of the rapid corrosion of the exposed refractories and metallic ductwork they encounter, as well as the even more important emissions consideration.

- Particle-size distribution, water absorption, plasticity and carbonates content must be established because of their recognized influence on any brick product involved.

- 'Product-wise' it is also important to know if the soluble salts content of the waste displays variations on a 'load-by-load' basis - in order to assess the risk of subsequent undesirable fluctuating efflorescence that can lead to consumer complaints relating to the final marketed product.

Some of the potentially detrimental results on brick products caused by various wastes can be offset by the use of appropriate 'corrective' additives. But such additives have a cost implication that must be offset by the additional revenues generated from incorporating the wastes. If the additional revenue does not cover this cost, then it is not sensible to consider the use of the waste. Taking risks with the factory's regular production output is not recommended!

Useable Wastes

From the experience at the Ziegel Gasser brick factory, a large number of waste materials can potentially be incorporated within clay bricks. The most successful wastes so far investigated include paper sludge (both from recycled product[1] and virgin cellulose sources); ashes from various industrial/municipal combustion processes; container glass wastes (bottles); ceramic process sludges; water and waste-water treatment sludges; fruit-processing incineration by-products and anaerobic digester residues. All offer potential inclusion, but in

---

[1] A problem with sludge from a paper recycling operation is the use of bleaching agents in the process operation that can lead to damage of the metallic parts in the brick factory dryer and also to detrimentally affect the refractories in the pre-heating zone of the kiln.

each case, experience has shown that it is essential to match a particular waste to a particular brick product – particularly at the firing stage. Apart from revenue generation, the other benefits of using waste materials in brick manufacture are numerous. For example, it can have the benefit of minimizing Local Authority environmental impact assessment findings relating to the operation of the factory - due to reduced quarry activity through the wastes replacing a proportion of the normally quarried raw material. The use of wastes possessing significant calorific value can also noticeably reduce the quantity of fuel needed during the firing process. Moreover, the take-up of wastes which would have otherwise been destined for landfill disposal extends the useful life of the landfill sites themselves.

Life Cycle Assessment

The Life Cycle Assessment (LCA) approach to evaluating wastes can have an extremely beneficial impact on their future long-term utilization. For example, for each citizens of the EU Member States, an average of 70 tonnes of raw materials are annually generated. Consequently, satisfactory recycling of the resulting end-of-life products will present a growing pressure on future economies. Thus, every waste must be individually evaluated and tested for potential reuse in some manner. However, sometimes wastes arising from the same process origin (but from different locations) have been found to possess significantly different properties - which may only become apparent once added to the brick feedstock. Thus, some standard of reliable referencing must be considered to address this problem.

THE ZIEGAL GASSER STANDARD OF REFERENCE

No 'universal' standard of reference can exist for the above purpose due to the wide diversity of clays, shales and marls used in brickmaking throughout the EU Member States and the very different range of products being manufactured at such factories. Thus, it is necessary to establish an individual 'tailor-made' standard to meet the needs of each particular brick factory. The first step towards meeting this goal is to obtain a complete evaluation of the original raw materials in use there and their individual properties and characteristics which affect the brick products. For example, at the Ziegel Gasser brickworks it is know that the chrome value in the newly manufactured bricks exceeds the proscribed limits set for such a product. However, it is also known from long-term testing carried out (yet to be published) that over time the soluble hexavalent chromes originally present are oxidized to non-soluble trivalent chrome, thus there will be no problem associated with recycling these bricks in the future.

Table 2 shows a 'maximum permissible' leaching norm built-up from empirical data collected over a three year period, which Ziegel Gasser now uses to quality-control its bricks. Over this extended period of time the information gathering has embraced not only the chemical analysis of the quarried feedstock materials, but also the monitoring of the firing conditions in the kiln and regular checking of the composition of the finished products. Currently within the EC no laws or regulations exists requiring information on the chemical composition or specific leachate limits for any marketed building products - although regulations of this kind do exist for recycled materials. The assumption can therefore be made that the leaching values of conventional building products must at the end of their life-cycle be compliant with those limits already set for recycled building materials. Thus, the values shown in Table 3 are considered by Ziegel Gasser to be environmentally acceptable maximum values for their products and provided that these values are not exceed, no problem is expected to be encountered with excessive leachate levels from the bricks, or from excessive exhaust emissions from the kiln flue-gases. Hence they are taken by the Company as a Standard for evaluating whether a particular waste material can be accepted for evaluation.

Table 2   Leaching Standard established for Ziegel Gasser bricks

Species		Maximum limits
Fluorides		< 2,00
Chlorides	mg/l	< 200,00
Nitrates		< 50,00
Sulphates		< 250,00
Arsenic		< 25,00
Barium		< 15,00
Berillium		< 10,00
Cadmium		< 2,50
Cobalt		< 2,50
Total chrome		< 300,00
Mercury	µg/l	< 0,50
Nichel		< 25,00
Lead		< 10,00
Copper		<10,00
Selenium		< 10,00
Vanadium		< 250,00
Zinc		< 100,00

Table 3   Maximum permissible values for Ziegel Gasser raw materials

	Species		Limits		Species		Limits
H₂O	Humidity		-	SiO₂	Silica		<65,00
%	Dry substance		-	Na₂O	Sodium oxide		<1,50
%	Not combusted		-	TiO₂	Titanium dioxide		<15.000,00
Al₂O₃	Aluminium oxide		<25,00	ZnO	Zinc oxide		<1.000,00
SO₃	Sulfur trioxide		<5,00	ZrO₂	Zirconium oxide		<450,00
Sb₂O₃	Antimon oxide		<5,0	Cl	Chlorates		<100,00
BaO	Barium oxides	% dry	<1.000,00	As	Arsenic	% dry	<50,00
B₂O₃	Boron trioxide	substance	<500,00	Dd	Cadmium	substance	<1.50
CaO	Calcium Oxide	mg/kg d.s.	<10,00	Cr	Chrome	mg/kg d.s.	<200,00
Fe₂O₃	Iron oxide		<10,00	Hg	Mercury		<2,00
P₂O₅	Phosphorus Pentoxide		<0,5	Ni	Nickel		<100,00
MgO	Magnesium oxide		<5,00	Cr	Copper		<500,00
MnO	Manganous oxide		<0.25	Se	Selen		<2,00
PbO	Lead oxide		<150,00	Pb	Lead		<20,00
K2O	Potassium oxide		<25,00	Zn	Zinc		<1,00

An Example of Applying the Ziegel Gasser Standard

Among a variety of by-product wastes offered to the Company has been the fluidized-bed ash produced from the incineration of fruit residues which are produced in large quantities in the surrounding area and represent a significant disposal problem. The typical analysis results of this ash are shown in Table 4 compared with standard bricks and an experimental product containing 20% volume replacement of fruit ash. Applying the Ziegel Gasser Standard, a replacement of 20% in volume of this ash for the normal feedstock was successfully carried out in the confidence that this level of introduction would not significantly alter the leachate value of the fired bricks as confirmed in Table 5. Under previous circumstances (without the benefit of this background information) a starting point of 3-5% volume replacement would have been considered the prudent approach. Based on applying this Standard, if all the relevant criteria are met, the waste under consideration can from a technical point of view, be accepted for use according to the procedure route shown schematically in Figure 1.

Table 4   Comparative analysis of fruit ash and fruits ash bricks with normal bricks

Species		Average 3 years (>50 samples)	"Zipperle" fruit combustion ash	"Zipperle" fruit combustion ash 20% in feedstock
Humidity	H_2O			
Dry substance	%			
Not combusted	%			
Aluminium oxide	Al_2O_3	< 25,00	3,90	12,80
Sulfur trioxide	SO_3	< 5,00	2,80	0.7
Antimon oxide	Sb_2O_3	< 5,00	2,90	< 1,20
Barium oxides	BaO	< 1.000,00	800	650,00
Boron trioxide	B_2O_3	< 500,00	260	82,00
Calcium Oxide	CaO	< 10,00	9,2	10,80
Iron oxide	Fe_2O_3	< 10,00	1,4	6,30
Phosphorus Pentoxide	P_2O_5	< 0,50	14,34	1,21
Magnesium oxide	MgO	< 5,00	4,30	2,40
Manganous oxide	MnO	< 0,25	0,05	0,09
Lead oxide	PbO	< 150,00	41,00	73,00
Potassium oxide	K2O	< 25,00	15,50	3,30
Silicae	SiO_2	< 65,00	32,60	39,50
Sodium oxide	Na_2O	< 1,50	1,40	0,70
Titanium dioxide	TiO_2	< 15.000,00	1.800,00	10.000,00
Zinc oxide	ZnO	< 1.000,00	600,00	230,00
Zirconium oxide	ZrO_2	< 450,00	340,00	290,00
Chlorates	Cl	< 100,00	< 50,00	< 50,00
Arsenic	As	< 50,00	< 1,00	13,00
Cadmium	Dd	< 1,50	1,00	< 1,00
Chrome	Cr	< 200,00	190,00	140,00
Mercury	Hg	< 2,00	< 1,00	< 1,00
Nickel	Ni	< 100,00	77,00	79,00
Copper	Cr	< 500,00	350,00	69,00
Selen	Se	< 2,00	< 2,00	< 2
Lead	Pb	< 20,00	85,00	73,00
Zinc	Zn	< 1,00	260,00	230,00

(% dry substance mg/kg d.s.)

Table 5   Leaching results of bricks containing 20% volume replacement of fruit ash compared with the Ziegel Gasser Standard

Species		Maximum limits	Fruit combustion ash (20% vol. in feedstock)
Fluorides		< 2,00	1,50
Chlorides	mg/l	< 200,00	38,00
Nitrates		< 50,00	19,00
Sulphates		< 250,00	170,00
Arsenic		< 25,00	5,00
Barium		< 15,00	6,00
Berillium		< 10,00	< 0,50
Cadmium		< 2,50	< 0,50
Cobalt		< 2,50	< 1,00
Total chrome		< 300,00	125,00
Mercury	μg/l	< 0,50	< 0,10
Nichel		< 25,00	1,00
Lead		< 10,00	< 1,00
Copper		<10,00	< 1,00
Selenium		< 10,00	< 3,00
Vanadium		< 250,00	55,00
Zinc		< 100,00	6,00

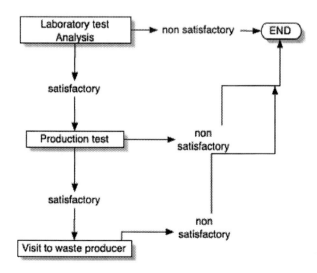

Figure 1   Flow diagram summarizing the overall Ziegel Gasser waste acceptance

THE PROCEDURAL STAGES LEADING TO FINAL WASTE ACCEPTANCE

A summary of the full evaluation approach used by Ziegle Gasser from initial contact to the final acceptance of a specific waste material into their brick factory is described below.

Supply Contract

The enterprise generating the waste desires to arrange disposal of it in the cheapest way possible. The brick factory on the other hand consider such wastes as potential new raw materials. Not always are these two different points of view compatible.

Waste Producer Visit

No waste should be accepted 'on-trust' without a visit to the plant it originates from. It is important to get to know the process involved and the people with which the Company will be potentially dealing. With any such a waste producer it is also necessary to agree at an early stage the gate-fee that the producer will have to pay, plus the wastes transport arrangements, conditions and responsibilities in case the waste does not meet the agreed specifications etc.

Once a provisional agreement has been reached, it is then necessary to establish whether the use of the offered waste is possible under existing authorizations or if a new or special authorizations needs to be requested. A second question concerns whether the waste offered is a once-only 'spot' supply, or will be provided on a continuous basis. Non-continuous supplies are generally not to be recommended as the period between shipments carries the risk of undesirable variability detrimentally affecting the host brick product. This can be true even for wastes of ostensibly the same origin but from different supply operations – for example paper sludges from different paper mills. A third question requiring checked is whether the waste is conveyed to the brick factory directly from the waste producer - or if some intermediate storage or unloading/loading operation is involved. In the latter cases due to the potential risk of contamination of the waste if stored with other similar by-products, then it should be refused. This stage of decision-making and available options are shown in Figure 2.

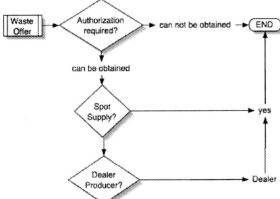

Figure 2   Sequence of steps to establish the legal requirements and supply pattern of the waste under evaluation.

If the above points have been satisfactorily resolved and the waste has been established as being supplied directly by the producer, then the vetting assessment can be progressed by following the quantitative data collection procedure shown in Figure 3.

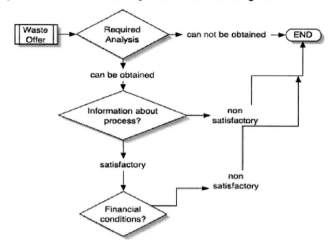

Figure 3   Steps to establish quantitative information on the waste under assessment

At this stage it is necessary to conduct a comparative assessment between the waste and the standard feedstock that has already been comprehensively characterized. Variations in results in the range of +/- 20% of the established norm are usually acceptable. It is also essential for a quantity of all samples examined during this phase of the study to be retained (preferably frozen) as a reference source so that they can be re-examined in the future if need be. The surrounding financial aspects also need careful assessment to confirm commercial viability.

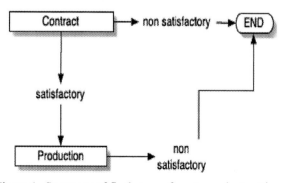

Figure 4   Summary of final stage of waste product vetting

Should all the vetting procedures carried out up to this point of the investigation prove positive, the final situation shown in Figure 4 can be concluded. At this point, a final visit to the waste producer needs to be conducted to formalize the proposed business relationship and to gain a final mutual appreciation that what the producer considers waste, is about to become a new raw material for the brick manufacturer. Should such an understanding prove elusive, it is better to decline the waste - as it is essential that a collaborative understanding is established which embraces a willingness of the partnership to jointly solve any unexpected problems in quality or supply that might arise in the future. Such a contract between the waste producer and the brick maker should also cover the financial conditions and a back-up guarantee for the user in case the delivered waste is not found to be comparable with the samples supplied during testing. In this case the brick maker must have the right to return the waste.

On-Going Factory Procedures

During the period of the contract it is important to continue to maintain a high level of control over the waste and its regular use in the brick feedstock as shown schematically in Figure 5.

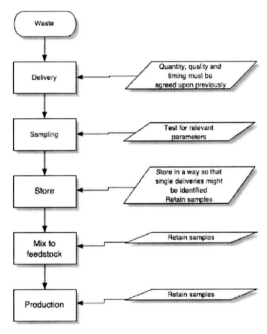

Figure 5   Schematic diagram of the sampling and monitoring procedure conducted during the period of contract while the waste is being supplied to the brick factory

Maintaining a Regular Check and Control System

Once the waste has been satisfactorily vetted and introduced into the factory process-line a rigorous sampling and testing regime (shown in Table 6) needs to be put in place to ensure its quality is maintained during the life of the contract.

Table 6   Sampling and testing scheme in-place at Ziegel Gasser brick factory to ensure quality control and audit tractability for brick products containing wastes

Frequence of tests	Raw materials	Wastes	Feedstock	Dried	Fired
Basic chemical analysis	Yearly or when other values that are tested more frequently change by more than 15%	Once every three month	Once every three month	Monthly	Monthly
Seger calculation	-	-	Every analysis or when feedstock composition varies	-	-
Firing tests	Yearly or when other values that are tested more frequently change by more than 15%	-	When feedstock composition varies	-	-
Leaching test	-	-	-	-	Monthly
Plasticity	Monthly	-	Daily	-	-
Water content	Several times a day	Several times a day	Severals times a day	-	-
Carbonates	Yearly or when other values that are tested more frequently change by more than 15%	Yearly or when other values that are tested more frequently change by more than 15%	Monthly	-	Monthly
Fineness	Yearly or when other values that are tested more frequently change by more than 15%	-	Yearly or when other values that are tested more frequently change	-	-
Weight loss on drying	Weekly	Weekly	Daily	Daily	-
Shrinkage	Weekly	Weekly	Daily	Daily	-
Shrinkage	-	-	-	-	Daily

Table 6 (continued)

Frequence of tests	Raw materials	Wastes	Feedstock	Dried	Fired
Compressive strength *	Yearly or when other values that are tested more frequently change by more than 15%	–	–	–	Monthly
Water absorption *	Yearly or when other values that are tested more frequently change by more than 15%	–	–	–	Daily
Freeze / thaw *	Yearly or when other values that are tested more frequently change by more than 15%	–	–	–	Quarterly
Porosity *	Yearly or when other values that are tested more frequently change by more than 15%	–	–	–	Monthly
Efflorescence *	Yearly or when other values that are tested more frequently change by more than 15%	–	–	–	Monthly
* To be tested according to EN 771.1 'Clay Masonary Units'					

CONCLUSIONS

Increasing opportunities are now being presented to brick manufacturers within the EU Member States to incorporate a wide variety of industrial and urban by-product wastes into their products. The adoption of this practice will not only contribute towards helping the sustainability of many traditional brickmaking raw materials, but also offers to generate a new income-stream capable of allowing such brick manufacturers to reduce their operating costs by receiving revenue from collaborating waste producers. However, by necessity, a high priority must be placed on the quality of all raw materials used in brickmaking processes, any significant variations in specifications from one day to the next carries the potential risk of causing serious production problems and loss of revenue. Confronted with this issue Ziegel Gasser GmbH has developed an innovative approach involving the selection and vetting of a variety of wastes using a standard of reference that they have developed over the last three years. Now in place, this control system has allowed the factory to produce new brick products containing a number of suitable wastes and successfully market these into the commercial building products sector.

CHARACTERISATION AND LEACHABILITY OF SEWAGE SLUDGE ASH

C A Johnson

Swiss Federal Institute of Environmental Science and Technology (EAWAG)

Switzerland

ABSTRACT. This study focuses on the characterisation of sewage sludge ashes from 2 sites in Switzerland. The elemental composition and the acid neutralising capacity (ANC) were determined and mineral components identified. The leaching behaviour of major and minor inorganic components was assessed as a function of liquid:solid (L:S) ratio and leachate pH. The major elements are Ca, Al, Si, and P, all having concentrations around and above 5% w/w. In the SSA samples, the Fe concentrations are high (around 15%) and this is associated with the treatment of sewage sludge. Organic carbon concentrations are around 0.5% as HCHO. Carbonate concentrations lie between 0.1 and 0.7% for all samples. The heavy metal and metalloid components in the highest concentrations are Cu, Zn, Mn, Pb, Cr and Ni, which are all in the g kg^{-1} range. In the mg kg^{-1} range are V, Co, Sb, As, Mo, Cd, W and Se. The solubility of the heavy metal and metalloid ions is low due to the precipitation of solid phases and adsorption to Fe oxides.

Keywords: Sewage sludge ash, SSA, Characterisation, Heavy metals, Acid neutralising capacity, Leachate.

C A Johnson works in the Water Rock Interaction Group at the Swiss Federal Institute of Environmental Science and Technology.

INTRODUCTION

The treatment of sewage sludge has come increasingly into focus as wastewater treatment plants increase in number in Europe and the disposal of increasing volumes of sewage sludge becomes a problem. The use of sewage sludge as a fertiliser for agricultural purposes is limited because of its heavy metal content. Alternatively sewage sludge may be landfilled. The move away from landfilling organic-rich materials means that sewage sludge treatment involves some kind of mineralisation process. The incineration of sewage sludge is becoming increasingly important. The partially dried sludge may be incinerated alone in especially designed plants, together with other wastes in municipal solid waste incinerators or added to raw meal in cement production. The first option yields a relatively homogenous granular material that may be used as a secondary product or landfilled. However, little is known about this material and an environmental assessment is necessary.

The characterisation of wastes is an essential step in the assessment of either their potential use as secondary construction materials or of landfilling strategies and procedures [1,2,3,4]. The aim is to assess the contaminant potential (or concentrations) and the likelihood of long-term release of contaminants to the environment. The latter depends of course on the environment under which the material is stored or utilised. In the case of essentially inorganic waste materials the assessment usually centres on the heavy metal content and on the factors that control their mobility.

Heavy metal mobility can only be fully understood in context of the mineralogical, geochemical and hydrological processes, the most important factor being the composition of the matrix components. These components determine the leachate pH and other parameters upon which the heavy metal solubility largely depends. The acid neutralising capacity (ANC), which is largely determined by the Ca mineralogy, is the prime factor. Ageing processes of matrix components can affect the long-term mobility of the heavy metals. These are subject to different time scales. Whereas oxidation processes and initial clay formation may occur within 10 to 20 years, weathering processes, in analogy to soil systems, proceed over thousands of years. Within this context, field and laboratory studies encompass the initial leaching and ageing stages.

The geochemical properties of the heavy metals can be very different and must be individually assessed. A combination of dissolution/precipitation, sorption and complexation reactions control heavy metal solubility over time. Heavy metal cations such as Cu, Pb, Cd and Zn sorb strongly in alkaline conditions and also precipitate as mixed oxides and carbonates. Heavy metal anions, on the other hand, sorb in acidic conditions and can be quite mobile in alkaline solutions.

An assessment of the contamination potential of a waste material entails the characterisation of the matrix components, the assessment of the contamination potential of heavy metals and an investigation of their present and potential leachability. Such a study entails the following steps:

i)   the determination of elemental composition
ii)   the determination of matrix minerals
iii)   the determination of ANC
iv)   an investigation of the potential leachability of heavy metals
v)   model calculations or estimates of the long term leachability

In this paper sewage sludge ash (SSA) is assessed according to the steps listed above. The results presented here are discussed in greater detail together with further examples in Johnson and Canonica [5].

SAMPLING AND ANALYTICAL PROCEDURES

Sampling Procedure

The samples KA1 and KA2 are sewage sludge ash (SSA) samples from the fluidised-bed furnace, ARA-Hard, Winterthur, Switzerland. At this plant sewage sludge is used to produce biogas in digester for about 20 days, after which time excess water is removed by centrifugation (a flocculation agent is employed prior to centrifugation) and the solid residue incinerated. The sample KA1 is a 1:1:1 mixture of single samples taken on one day over the period of one hour. Sample KA2 is a 1:1:1 mixture of 3 samples, each a composite of ash over a one-hour period [6].

The samples KA3 and KA4 are SSA samples taken from a multi-platform furnace in Dietikon, Switzerland. The same procedure is used to prepare the digested sewage sludge for incineration as described above. Samples KA3 and KA4 were sampled as described for KA2 [6].

Due to the granular nature of the ashes, no sample preparation was required.

Elemental Composition

Samples were digested in duplicate using two methods; aqua regia and hydrofluoric acid. In the aqua regia digestion samples (100 g) were boiled under reflux for 4 hours in aqua regia (25 mL, 1:3 mixture of concentrated HNO_3 and HCl). The mixture was then filtered through paper filters to remove residual particles and diluted to 100 mL with deionised water. The hydrofluoric acid digestion was carried out in a microwave digester (MLS 1200, Mikrowellen-Labor-System). The sample (1000 mg) was placed in a teflon vessel together with the acid mixture (3 mL 65% HNO_3, 3 mL 32% HCl, 3 mL 40% HF) and digested (700 W) for 5 minutes. The sample was stabilised with 10 mL of 5% H_2BO_3. Duplicate blank samples underwent both digestion procedures.

Sodium, K, Fe, Pb, Ni, Cr and Cd were analysed by Atomic Absorption Spectrometry (AAS, Perkin Elmer 5100 PC), the last element being measured by graphite furnace AAS. Magnesium, Ca, Al, Cu, Zn and Mn concentrations were determined by Inductively Coupled Plasma Optical Emission Spectrometry (ICP-OES, Spectroflame M, Spectra). Molybdenum and W were analysed by Inductively Coupled Plasma Mass Spectrometry (ICP-MS, PE Sciex Elan 5000, Perkin Elmer). Ortho-phosphate was determined in the aqua regia digestions by flow injection analysis (ASIA-system, Ismatec) at 440 nm as a phospho-vanado molybdenum complex [7]. Arsenic was determined in the aqua regia digestions by Atomic Fluorescence Spectrometry (AFS, PSA 10.055 Millenium Excalibur).

Total C and N were determined with a CNS-analyser (Carbo Erba ANA 1500). Carbonate was determined coulometrically (Coulometric Inc. 5011). For Si, samples (100 mg) were digested for 3 h at 600 °C with 2 g NaOH. Water (10 mL) was added to dissolve the components at room temperature overnight and the samples were then diluted to 250 mL without filtering. Dissolved Si was determined by flow injection analysis as a molybdenum blue complex [8]. For the determination of total S samples (100 mg) were placed in an autoclave together with 30 drops of paraffin oil, 100 mL H_2O_2 and 10 mL 0.5 M NaOH. The mixture was placed in an O_2 atmosphere (O_2-Station C7048, IKA-Analysentechnik) and ignited, keeping the pressure at 30 bar with. The resulting sulphate was analysed using ion chromatgraphy (IC, Sykam, System 2).

Concentrations of Cl, Co, Hg, Sb, Se and V were determined or verified by X-ray fluorescence (XRF, Spectro, X-lab2000). Calibration curves were routinely compared to standard addition curves in order to avoid matrix problems.

Determination of Acid Neutralising Capacity (ANC)

Automated titrations were carried out on all samples using a pH-Meter (Metrohm 713) and Dosimat titrator (Metrohm 665) controlled by an IBM personal computer. The electrode was calibrated as described above. The sample (1 g) was placed in deionised water (100 mL) and immediately flushed with argon to prevent CO_2 entering the system. Titrant (0.1 mL of 1 M HNO_3, Titrisol) was added every 6 hours. Up to 5 mL of titrant were added to the system in 0.04 mL steps. The duration of a titration was approximately one week.

X-ray Diffraction

An X-ray diffraction spectrum was taken of each sample (Scintag XDS 2000 diffractometer, Cu-Kα, 1.5406 nm, 2000 W, 45 kV, 40 mA). The analyses were qualitative. The slides were prepared by dropping a small portion of each sample ground in ethanol onto a glass plate.

Leaching Experiments

Suspensions (160 mL) of KA1 were prepared in duplicate with a L:S ratio of 20:1, 10:1 and 5:1 and equilibrated for 3 days. After this they were filtered (0.45 mm, Sartorious), one half portion being acidified (1 mL concentrated HNO_3 per 100 mL sample, Suprapur, Merck) for the analysis of metals and metalloid components (Table 3.5). The other portion of solution was not acidified but used for the measurement of pH, dissolved organic carbon (DOC), Si, NH_4 and major anionic components.

Suspensions (160 mL) of KA1were prepared in duplicate at a L:S ratio of 100:1 with the addition of acid. A total of 8 samples with acid addition of 0.4, 0.8, 1.2, 1.6, 2.0, 2.4, 2.8 and 3.2 meq g^{-1} sample was prepared. All samples were equilibrated for 3 days and treated as described above.

Leachate pH values were determined with a Ross combination electrode (Orion Model 81-72BN) calibrated using a titration using a pH-Meter (Metrohm 713) and Dosimat titrator (Metrohm 665) controlled by an IBM personal computer. Standard potentials and slopes were determined by linear regression from the calibration data over a pH range of 2 to 11, and the log [H^+] values of aqueous solutions were determined from the calibration constants. Such a calibration results in the direct determination of hydrogen ion concentration, rather than hydrogen ion activity.

Major cationic components were measured with AAS, graphite furnace AAS and ICP-OES. Heavy metals were analysed by ICP-MS. Both DOC and carbonate were determined were analysed as soon as possible using a Total Organic Carbon Analyser (Shimadzu, TOC-5000A, Shimadsu Europe). Chloride, SO_4, NO_3 and acetate were determined by ion chromatography (Sykam, System 2; LCA AO4 column from Stagroma; Stagroma; eluent, 1:1 mixture of 12 mM Na_2CO_3 and 15 mM $NaHCO_3$). Ammonium was determined colourimetrically using the indophenol blue method [9].

RESULTS AND DISCUSSION

Solid-Phase Characterisation

The elemental composition of the SSA samples is surprisingly consistent, particularly considering that the samples come from different plants (Table 1). The major elements are Ca, Al, Si, and P, all having concentrations around and above 5% w/w. The alkalis (Na and K) and Cl are concentrated by incineration. The Fe concentrations are high (around 15%) as the Swiss process of SS digestion involves addition of Fe in order to immobilise sulphide. Organic carbon is around 0.2% w/w as C and 0.5% as HCHO. Carbonate concentrations lie between 0.1 and 0.7% for all samples.

Table 1   Elemental Composition of the SSA samples

	Na g kg^{-1}	K g kg^{-1}	Mg g kg^{-1}	Ca g kg^{-1}	Al g kg^{-1}	Fe g kg^{-1}	Si g kg^{-1}	C g kg^{-1}	CO$_3$ g kg^{-1}
KA1	3.32	4.42	12.35	112.5	37.2	144.3	116.2	2.46	3.5
KA2	3.17	2.97	8.20	86.0	26.5	132.7	118.0	4.34	1.4
KA3	3.45	4.21	9.10	99.0	40.6	171.1	82.9	4.46	2.6
KA4	2.39	3.68	8.65	84.0	37.4	153.3	85.1	2.72	4.0

	N g kg^{-1}	S g kg^{-1}	P g kg^{-1}	Cl mg kg^{-1}	Zn g kg^{-1}	Mn g kg^{-1}	Cu g kg^{-1}	Cr g kg^{-1}	Pb g kg^{-1}
KA1	0.20	9.42	61.1	0.16	2.87	2.00	0.95	1.84	0.26
KA2	0.16	12.63	64.3	0.28	2.48	1.47	1.18	0.54	0.18
KA3	0.39	10.32	76.1	0.38	2.66	0.48	0.85	0.49	0.20
KA4	0.19	5.46	67.3	0.32	2.08	0.39	0.73	0.69	0.22

	Ni g kg^{-1}	Cd mg kg^{-1}	Co mg kg^{-1}	Mo mg kg^{-1}	W mg kg^{-1}	Sb mg kg^{-1}	As mg kg^{-1}	Se mg kg^{-1}	V mg kg^{-1}
KA1	0.10	18.25	32.4	14.8	10.2	15.6	9.2	4.9	55.1
KA2	0.12	3.07	56.0	15.1	8.3	16.1	12.1	4.0	62.7
KA3	0.11	0.80	72.0	11.7	7.2	12.9	12.3	3.2	30.5
KA4	0.12	1.49	61.0	14.1	5.0	13.6	9.8	2.6	32

The heavy metal and metalloid components in the highest concentrations are Cu, Zn, Mn, Pb, Cr and Ni, which are all in the g kg^{-1} range. In the mg kg^{-1} range are V, Co, Sb, As, Mo, Cd, W, Se and Hg (Table 1). With the possible exception of Cr, the metals in the first group are cationic and can be expected to be relatively immobile in the neutral pH-range, unless they are strongly complexed. The metals and metalloid elements in the second group are predominantly anionic in nature and though these elements could be relatively mobile, they are present in low concentrations.

The acid neutralising capacity (ANC) of the samples varies quite widely. The ANC_{pH7} (the amount of acid that is required to lower the pH of the suspension to pH 7) of samples are listed in Table 2. It should be noted that these values are operational and dependent on equilibration time. Components that contribute to the ANC are primarily soluble hydroxides and carbonates. For predictive purposes, the carbonate content is the most reliable measure of buffering capacity.

Table 2   Acid buffering properties of the SSA samples.

SAMPLE	INITIAL pH (W:S = 100)	ANC_{pH7} meq/g	CARBONATE CONTENT mmol CO_3/g
KA1	9.87	1.0	0.29
KA2	9.47	0.1	0.12
KA3	9.41	0.6	0.22
KA4	8.42	0.3	0.33

The minerals found by X-ray diffraction in the Fe-rich SSA products are Ca (Fe) phosphates, haematite (α-Fe_2O_3) and maghemite (γ-Fe_2O_3), quartz, anhydrite. It should be pointed out that other minerals may exist either in quantities lower than approximately 5% or in non-crystalline forms.

Table 3   Possible average mineral composition (%) of the SSA products. The estimates are based on elemental composition (Table 1) and are approximate values.

COMPONENT	SSA
SiO_2	22
org C (as HCHO)	8
$Ca_3(PO_4)_2$	34
Al hydroxide	8
$CaCO_3$	3
$CaSO_4$	6
Fe hydroxide	20

The possible mineral composition can be estimated from the elemental composition and the information available from XRD analysis. The resulting estimates (Table 3) show that the solid phase is rich in quartz, Fe hydroxides and phosphates.

Leaching Behaviour

The L:S ratio dependence of the measured component concentrations provides an indication of the whether or not a component concentration is solubility controlled or not. A inverse relationship between the L:S ratio and a component concentration is an indication that a particular component is not solubility controlled by reactions such as sorption, precipitation or ion exchange. The composition of a solution equilibrated with the sample at a L:S ratio of 10 has a pH value of 9.11, indicating that bases, most probably $Ca(OH)_2$ or related basic soluble phases, are present in the solid phase in small quantities. The major components in solution are Ca(II) and SO_4^{2-} and the solution is buffered by calcite ($CaCO_3$), which makes up approximately 3 wt. % of the solid phase. Phosphates, Quartz and Fe hydroxides are relatively insoluble in the neutral to alkaline pH range. Concentrations of Zn, Ni and Mn and of the anionic species of As, Sb and V are between $10 - 100$ μg L^{-1}. Concentrations of Cu, Mo, W and Co are between $1-10$ μg L^{-1} and Pb and Se are not detectable. The mechanisms controlling the solubility of the above-mentioned solution components are explored in the following sections.

Components such as Na, Cl, K, NO_3 and NH_4 are known to be very soluble and a direct relationship between L:S ratio and dissolved concentration is seen. Major components such as Ca, Fe, Al, Si, CO_3, SO_4 or PO_4 are solubility controlled (for example as the solid listed in Table 3). Aluminium and Fe concentrations are most probably controlled by the precipitation of $Al(OH)_3$ or $Fe(OH)_3$. The solutions are undersaturated with respect to $CaSO_4$ and saturated with respect to $CaCO_3$ and β-$Ca_3(PO_4)_2$.

The relationship between the dissolved concentrations of heavy metals and metalloid species and L:S is illustrated in Figure 1. A direct relationship between W:S and dissolved concentration is only seen :for Ni, Mn and Se. The solubility of other species appears to be independent of W:S, indicating solubility control. Solubility may be controlled by precipitation of solid phases containing the heavy metal species or by adsorption to the surface of solids such as Fe (hydr)oxides. Solubility control is to be expected for most species with the exception of possibly the soluble species of Cr (as Cr(VI)) and Se (as Se(VI)). The W:S dependency of Ni may be explained by the formation of strong solution complexes.

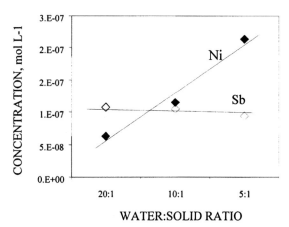

Figure 1   Dissolved concentrations of Ni and Sb as a function of water:solid ratio

Dissolved heavy metal and metalloid species concentrations are often determined as a function of pH and then compared to model calculations in order to gain insights as to which solubility-controlling phases may be important. Good agreement between theoretical and measured concentrations can be an indication of solubility control. Concentrations that are lower than predicted indicate that either an unknown phase, or sorption to the solid phase or limited availability is responsible. A flattening of dissolved concentrations at lower pH values is an indication of the limited availability. The relationship between metal/metalloid concentrations and pH can be divided into roughly 2 groups, the cations that tend to increase in solubility in acidic solutions and the anions that are not so pH sensitive but may sorb under acidic conditions. The pH-dependency of the solubility of heavy metal cations is shown in Figure 2. It should be noted that with such high phosphate concentrations, the precipitation of phosphates is likely. In most cases the phosphates are more insoluble than solids common for heavy metals, such as hydroxides or carbonates. However, with the possible exception of Co, the solubility of most species lies well below that of the corresponding phosphate and it is highly likely that the cations are adsorbed to Fe (hydr)oxide surfaces. The sorption of ions to Fe (hydr)oxides is difficult to model because we do not have an idea of the concentration of surface sites available for adsorption.

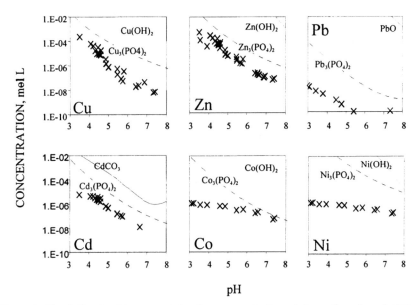

Figure 2   Total dissolved heavy metal cation concentrations (×) as a function of pH together with the calculated maximum concentrations of possible solubility controlling phases.

The metalloid components As and Se are undersaturated with respect to $Ca_3(AsO_4)_2$ and $CaSeO_4$ and $CaSeO_3$ respectively. It is possible that an unknown phase is controlling the solubility of As und neutral pH conditions, but at lower pH values the concentrations reach the limiting values set by the availability of these species. For W and Mo, solubility may be controlled by Ca metallate precipitation (Figure 3). It should be noted that dissolved concentrations of W, Mo, Sb and V all decrease with pH as these species sorb to the solid phases in acidic solutions. There is no reliable thermodynamic data available for modelling V and Sb solubility so it is not possible to determine solubility-controlling phases. Chromium is very soluble as chromate and as Cr(III) sorbs strongly to solid phases. Since less than 0.1% is found in solution, Cr is most likely to be present in the samples as Cr(III). The observed pH dependence of the dissolved concentrations of the KA1 leachate samples may be explained as follows; at neutral pH values, the small fraction of Cr that is present as Cr(VI) is found in solution, though The solubility could be explained by $Cr(OH)_3$ precipitation. At lower pH values Cr(VI) is sorbed and as the acidity of the solutions increases Cr(III) is seen to come into solution, albeit at very low pH values.

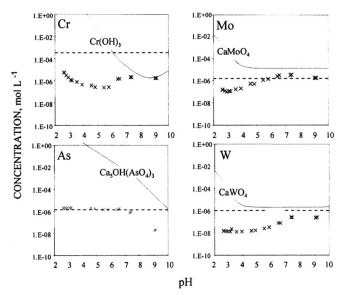

Figure 3   Total dissolved heavy metal anion concentrations (×) as a function of pH together with the calculated maximum concentrations of possible solubility controlling phases. The dashed line represents the total available concentration.

CONCLUSIONS

The composition of sewage sludge ash is greatly affected by the production and treatment of the sludge. The samples investigated here are of surprisingly similar composition. The ash is not only rich in Ca, Al and Si, but also in oxidised Fe, in sulphates and phosphates. Any secondary use of such a material would have to consider whether or not the latter components pose a problem in combination with other materials. The low organic carbon content is an asset of this material. The heavy metal components, Cu, Zn, Mn, Pb, Cr and Ni are are all in the g kg^{-1} range, whilst the concentrations of V, Co, Sb, As, Mo, Cd, W and Se are in the mg kg^{-1} range. The ashes have a relatively low ANC. Were the samples to encounter acidic conditions, heavy metal cations would be released. However, the solubility of anions and cations is quite low as a result of their sorption to Fe (hydr)oxides surfaces. The precipitation of phosphates may also be responsible for the low solubility of heavy metal cations. It is possible that mixed phosphates are produced, which have a lower solubility than the pure phases modelled here.

ACKNOWLEDGEMENTS

Edi Medilanski and Hasan Belevi are thanked for providing the samples. Vivendi is acknowledged for financial support.

REFERENCES

1. BELEVI H. and BACCINI P.. Long-term behaviour of municipal solid waste landfills. Waste Manag. Res., 1989, 7, 43-56.

2. CHRISTENSEN T.H., KJELDSEN P., ALBRECHTSEN H.J., HERON G., NIELSEN P.H., BJERG P.L. and HOLM P.E.. Attenuation of landfill leachate pollutants in aquifers. Crit. Rev. Environ. Sci. Technol., 1994, 24(2), 110-202.

3. MEIER W., and SCHURTER M., Deponieverhalten der Feinfraktion aus Bausperrgut-Sortieranlagen. Abfall-Spektrum, 1993, 2/93, 16-18.

4. HJELMAR O. Characterisation of leachate from landfilled MSWI Ash, Proceedings Int. Conf. Municipal Waste, Hollywood, Florida 1989. Minister of Supply and Services Canada, 1989.

5. JOHNSON C.A. and CANONICA L.. Characterisation of sewage sludge incineration products. 2003. In preparation for J. Environ. Qual.

6. MEDIANSKI E., Einfluss der Verbrennungstechnik in Wirbelschicht-, Etagen- und Drehrohröfen. Diploma study, Swiss Federal Institute of Science and Technology, (ETH), Zürich, Switzerland, 1998, pp. 53.

7. APHA (American Public Health Association, American Water and Wastewater Association and Water Environmental Federation). Standard methods for the examination of water and wastewater. A.D. Eaton, L.S. Clesceri, A.E. Greenberg (Eds.), 19th Edition, APHA, 1995, 4-111 - 4-114.

8. APHA (American Public Health Association, American Water and Wastewater Association and Water Environmental Federation). Standard methods for the examination of water and wastewater. A.D. Eaton, L.S. Clesceri, A.E. Greenberg (Eds.), 19th Edition, APHA, 1995, 4-118 - 4-119.

9. APHA (American Public Health Association, American Water and Wastewater Association and Water Environmental Federation). Standard methods for the examination of water and wastewater. A.D. Eaton, L.S. Clesceri, A.E. Greenberg (Eds.), 19th Edition, APHA, 1995, 4-81 - 4-82.

INCREASING SUSTAINABILITY OF THE FRANCHISE SOLID WASTE MANAGEMENT SYSTEM IN A LOW-INCOME NEIGHBOURHOOD IN LAHORE, PAKISTAN

I Ahmad

Norwegian University of Science and Technology

Norway

ABSTRACT. The municipal government of Lahore, Pakistan strongly felt the need in June 2001 to adopt a door-to-door collection method considering it as a best method of municipal waste collection. They envisaged psychological barriers, public education, manpower, finances and supervision as the major problems in initiating a door-to-door method in the entire municipal limits of Lahore. To cope with these problems, they resorted to formal private sector participation at the primary collection stage in the solid waste management system as the only solution to bring manpower, resources and new vision. Therefore, they decided to launch a pilot project in nine main areas in the city of Lahore under the franchise system for a trial period of one-year. The main objective of the study described in this paper was to scrutinize the franchise system in a low-income neighbourhood in the light of inter-relationships existent among different stakeholders. The stakeholders identified during the course of this study include City District Government (public sector), Youth Commission for Human Rights organization (formal private sector), waste collectors (informal private sector) and community (service users, non-service users, area/elected representatives). This study encompasses the participation of these stakeholders defining their roles and associated strengths, weaknesses, opportunities and constraints. It also portrays an interesting blend of cause and effect relationships of the issues that emerged in the franchise system as a result of interaction among different stakeholders. The major issues highlighted in this research study are weak co-ordination between the public and formal private sectors and weak co-operation between the community and formal private sector. These issues need to be tackled in order to increase the sustainability of the franchise system in a low-income neighbourhood as well as other neighbourhoods.

Keywords: Franchise system, Low-income neighbourhood, Stakeholder, Inter-relationship, Sustainability

Iftekhar Ahmad completed his Masters Degree in Urban Ecological Planning at the Norwegian University of Science and Technology, Trondheim, Norway in June 2003. He did his bachelor degree in City and Regional Planning from the University of Engineering and Technology, Lahore, Pakistan. His research interest is in solid waste management in developing countries context. He has three years working experience in the field of solid waste management in Pakistan.

INTRODUCTION

The management of municipal solid waste is becoming one of the major concerns particularly in the cities of low-income countries, which contribute about 40% of the total municipal solid waste produced the world over. The quantity is increasing sharply by virtue of unprecedented population growth and economic development. The municipalities spend about 25 - 50% of their total municipal expenditure on solid waste management. Despite this only one third of the total waste is collected and even less than 5% is properly disposed of. The effects of such poor performance on public health, environment and economic growth are manifolds and accelerating [1]. The low-income countries have lagged far behind in devising innovative solutions to face this dilemma. Sustainable solid waste management with the overriding goal to minimize the impact on the environment in socially and economically acceptable ways is the challenge for decades to come [2].

Solid waste management can not only be viewed in terms of technological solutions to solve physical problems as practised in developed countries. This technocentric approach works in that context where it is entirely consistent with the particular economic and social circumstances, cultural practices and regulatory capabilities of the country [3]. Solid waste management systems in developing countries are commonly plagued by excessive staff, obsolescent equipment, cumbersome procurement procedures for spare parts, inflexible work schedules, limitations on management changes, inadequate supervision, and strong worker unions [4]. It is difficult for the public sector to implement the changes necessary to match efficiency of the private sector. Private sector participation, therefore, brings significant resources in solid waste management [5]. It is not meant to solve all government's solid waste management problems.

PRIVATE SECTOR PARTICIPATION IN SOLID WASTE MANAGEMENT

Solid waste management is a public service aimed at delivering its services to all citizens. It is not appropriate to leave those people unserviced who do not pay. Reason being that public cleanliness, collection and disposal of wastes are essential services to safeguard public health and the environment. Needless to say, solid waste management is a public good and municipal governments are responsible for it. The question arises where the role of private sector comes in? Involving private sector in the delivery of solid waste management services can reduce the government activity. Where the existing service delivery is either costly or ineffective, the involvement of private sector should be viewed as a means of enhancing efficiency, lowering the cost, mobilizing private investment thus eventually adding more resources for existing infrastructure and equipment [4].

The private sector can take part in solid waste management activities through four different methods such as contracting, concession, franchise and open competition [4]. The franchise system will be discussed in detail in this paper. According to this system, the government awards a finite term zonal monopoly contract to private firm for solid waste collection service. The private firm qualifies for this award after a competitive process. The private firm deposits performance bond with the government and pays a license fee in order to meet the expenses of the government for monitoring. The private firm bears its cost and earns profit through direct charge to the households and other establishments that are serviced. The government controls over the tariff charged to the consumers through development of adequate competition and control of price collusion as well as price regulation. In this method, economy of scale is attained only if the waste is collected along contiguous routes or exclusive zones. In developing countries, the franchise system is applicable in those areas where people can readily be educated about public cleanliness. This is the only way to get people's full co-operation as well as ensure full cost recovery. The franchise options are waste collection service to residents; and waste collection service to commercial and industrial establishments [1].

NEED FOR THE FRANCHISE SYSTEM IN LAHORE

Lahore is a large sprawling city surrounding an area of 1770 square kilometers and hosting an estimated population of 7 million. The total waste being generated from different sources in Lahore is 3850 tonnes per day while 2306 tonnes of waste is being collected and the rest, 1544 tonnes is left to accumulate at different places to pollute the environment as well as constituting a nuisance to public health. The average generation rate of solid waste in Lahore is 0.55 kilogram per capita per day [6].

Needless to say, solid waste management comprises of different activities starting from generation, storage, collection, transfer and transport right through to disposal. When we talk about collection, it means both primary and secondary waste collection. In developed countries, door-to-door waste collection method is being adopted for primary waste collection. This method is, so far, considered to be the best method for removing waste from the source but it is an expensive and non-feasible option at government level in developing countries. In Pakistan, this method is not widely used. Municipal sweepers are supposed to sweep the streets every day. It boosts general public to throw the garbage out thinking that municipal sweepers would collect the waste again next day. This cycle never ends.

Understanding the importance of a door-to-door collection method, the municipal government in Lahore felt the need to incorporate this method in the prevalent solid waste management system with the aim of providing a state-of-the-art waste management system in the Metropolis. The government's belief was that lack of management and not resources was holding back implementation. The problems associated with this method were psychological barriers, public education, manpower, finances, and supervision. The psychological barrier was that this method was not possible in the city of Lahore as well as other cities of Pakistan. Almost all the municipal officers had this firm belief. Public education involved wide public co-operation, which was difficult to achieve without intensive efforts. This method needed trained and efficient manpower while the existing sanitary workers were used to working in a more traditional manner for many decades. Therefore, it was difficult to train them. It was also a myth that shortage of finances would hamper the use of this method. The last problem was supervision. As this method involves direct interaction with the community so there was a need to have more efficient and vigorous supervisory mechanism compared to the traditional system.

In view of the above problems, the local municipal government resorted to formal private sector participation in the hope that it would bring manpower, resources and new vision. They thought that they needed only management because the private sector already existed in the market. After going through a number of case studies on privatization the world over as well as in Pakistan mainly the cities of Lahore, Islamabad (Capital of Pakistan), Karachi, and Peshawar, the government devised franchise systems as the most appropriate method of privatization for primary solid waste collection in Lahore considering local traditions and requirements as well. Street sweeping was also included as an additional service under the franchise system. This was the first system of its kind introduced in Lahore. The most distinguishing feature of this system was to service all low, middle and high-income neighbourhoods at a city-wide scale. The low-income neighbourhoods are not usually touched upon by the formal private sector as well as by the government for waste management facilities. They are considered to be less productive and lucrative areas because they are inhabited by low-income people, who cannot afford a door to door collection service and produce insufficient recyclable materials. As a matter of fact, they turn into blights producing large quantity of solid wastes (mainly organic component). There is a dearth of knowledge on the application of the franchise system in low-income neighbourhoods not only in Pakistan but also in other developing countries because it is a relatively new idea/approach, which has not been widely adopted. This idea was brainstormed in the last part of the year 2000 and the beginning of 2001 but the system was formally launched in June 2001 for a trial period of one year. It was a landmark in the history of Pakistan.

Before introducing the franchise system at city-wide scale, a prototype was first tried in some areas namely Wahdat Colony, Government Quarters, Gazetted Officers Residences (GOR-III), Shadman Colony and Block A of Johar town. It proved to be successful after a trial period of four to six months. Then a pilot project was formally launched in nine main areas of Lahore namely Garden Town, Muslim Town, Faisal Town, Allama Iqbal Town, Gulshan-e-Ravi, Township, Shadman, Shah Jamal and Model Town Extension. These areas comprised of predominantly high-income neighbourhoods. They were considered to be the feasible areas for the application of this system. These were the areas where the municipal government of Lahore was delivering services with its limited staff, which was withdrawn after the implementation of the franchise system.

WORKING OF THE FRANCHISE SYSTEM IN A LOW-INCOME NEIGHBOURHOOD

By the end of the contract, there was only one private contractor named Youth Commission for Human Rights (formal private sector) still running this system in the area of Township in a hope to get more units to be serviced after the renewal of the contract. The private contractor was servicing a number of low, middle and high-income neighbourhoods in Township though not covering it as a whole. Other contractors withdrew from this system one after another as the time elapsed and could not sustain it by the end of the contract due to unknown reasons. Considering the importance for delivering waste collection service in low-income neighbourhoods, the scope of my study was confined to only one low-income neighbourhood namely Christian Colony in Township (Lahore). While studying the system in a low-income neighbourhood, I have focussed on the inter-relationships among different stakeholders rather than analyzing the system purely in economic terms. These relationships form the basis to make the system economically sustainable. The concept of sustainability in the context of integrated solid waste management can better be understood as:

"A system that best suits the society, economy and environment in a given location, a city in most cases. The concept of integrated sustainable solid waste management not only takes technical, financial/economic sustainability into account as is conventionally done, but it also includes socio-cultural, environmental, institutional and political aspects that influence overall sustainability of waste management. It also stands for strategic and long-term approach." [7]

Operational Mechanism

The City District Government Lahore (public sector) awarded a contract to the Youth Commission for Human Rights (formal private sector) to carry out solid waste collection task in Township area. The formal private sector deposited earnest fee i.e. Pakistani Rupees (PKR) 10,000 (US$ 1 = 57.8 PKR) and performance bond/security in the form of cash costing the very amount (refundable) to get license for the delivery of the services. After the award of the contract, the task of the formal private sector was to mobilize the community to take part in the franchise system in first three months of the launch of the project. They also motivated the commercial shopkeepers to pay for the franchise services. Consequently, they were successful in getting some membership to start service. When I visited that community, 250 houses out of 412 were availing themselves of the franchise services. The shopkeepers were left unserviced because they were not willing to pay for these services.

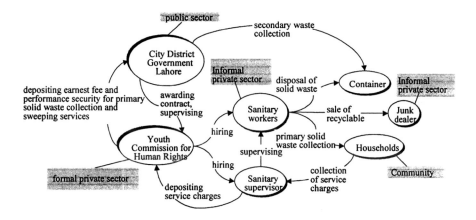

Figure 1   Operational diagram of the franchise system in Christian Colony defining the roles of the actors and their inter-relationships.

To implement the system in this area, the formal private sector hired one sanitary worker (scavenger) and one sweeper (female Christian) to serve 250 houses charging PKR 25 per month directly from the service users. The equipment used for waste collection was three-wheeler handcart. The sanitary worker was supposed to collect solid waste from households on daily basis except Sundays according to terms and conditions stipulated in the contract. He would sort out the waste in the handcart soon after its collection from each house. The recyclable thus sorted out were sold to Junk Dealer nearby the area by the sanitary worker and other wastes (organic and street refuse) were disposed of to the municipal government's designated containers. The salary of the sanitary worker was negotiated depending upon the quantity of recyclable in this area. The sweeper was supposed to sweep the streets at least twice a week. An area supervisor was exclusively hired to collect service charges from the users of the service and also ensure good quality service in the area. He was supposed to deposit collection charges with the formal private sector. A complaint cell was established to listen to complaints and redress them promptly. (Figure 1)

Identification of the Stakeholders

The terms 'actors', 'urban actors' and 'stakeholders' are used quite interchangeably but the term 'stakeholders' is used more formally. Above all, these three terms refer to people and organizations like households, enterprises, and all others, who have an interest in good solid waste management and participate in the activities that make it possible [8]. The stakeholders are key persons, groups or institutions with an interest in a project or program [9].

During the course of the study, a number of stakeholders were identified and grouped into three types i.e. primary, secondary and external (Figure 2). Primary stakeholders are those, who are directly affected, either positively or negatively, by the implementation of a solid waste management program [9] such as households (including service users, non-service users, community activists in my research study). Secondary stakeholders play some intermediary role and may have an important effect on the program's outcome [9]. In my research study, they appear as local municipal government (public sector), private contractor (formal private sector), and waste collectors (informal private sector). External stakeholders are not directly involved into the program but may nevertheless be affected [9] such as junk dealer, area representatives in my research study. This classification may vary depending upon the nature of the program.

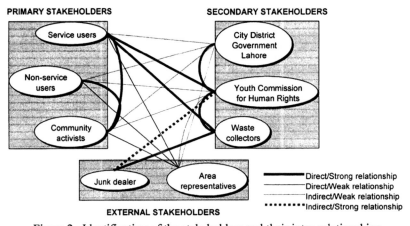

Figure 2   Identification of the stakeholders and their inter-relationships

This figure illustrates that there exists direct and strong relationship between City District Government Lahore and Youth Commission for Human Rights, the eason being that the public sector awarded the contract to the private contractor (formal private sector) to accomplish the primary solid waste collection task. The secondary waste collection and supervision were held the responsibilities of the government itself. The strength of the relationship implies the intensity of the interaction with each other. Similarly, there is a direct and strong link between the formal private sector and the service users. This sector was supposed to ensure good quality service to increase the participation of the community members in the ongoing service and also charge them directly. A direct and strong connection is also vivid between the formal private sector and the community activists. These activists had strong relationships with service users and non-service users. They mobilized the community to take part in the system depending upon the quality of the services.

Conversely, there can be seen direct/indirect but weak (almost non-existent) connections of the public and formal private sectors with area representatives, the formal private sector with non-service users, area representatives with service users, non-service users, and community activists. These representatives had no connection to this system because they were not empowered by the public sector. According to the franchise contract, the public sector was obliged to penalize the waste littering defaulters identified by the formal private sector with the help of community but it was not practically implemented.

There can be found direct and strong relationship within informal private sector between waste pickers and junk dealer. Both of them transact businesses of selling and buying of recyclable. It shows indirect relationship between formal private sector and informal private sector too because the former hires waste pickers to enable them to earn their living selling recyclable to the junk dealer.

Roles of the Stakeholders and Strengths, Weaknesses, Opportunities, Threats (SWOT)

Analysis was carried out to highlight the roles, strengths, weaknesses, opportunities and threats of different stakeholders in four sectors such as public, formal, informal and community. The information gathered through different primary sources involving questionnaires (to solicit information from the service users and non-service users), in-depth interviews with key informants, and direct observations helped carry out SWOT analysis.

Public Sector

The roles of the public sector identified in the franchise system are to award contract to the formal private sector, supervise and co-ordinate (carrying out secondary waste collection) with them. The major strength of this sector carries statutory powers. The weaknesses of the system are imparted in the form of lack of enforcement of by-laws, lack of monitoring, lack of interaction with the community. Further opportunities lie in the form of distributing powers to the area representatives to keep an eye on formal private sector's operations, collaborate with the community to legitimize the services and increase their participation. The major threats can be posed as biased attitude of the area representatives, exploitation of powers with them in the pursuance of these opportunities.

Formal Private Sector

The major role of this sector is to deliver the franchise services. The major strengths lying with this sector include management flexibility, contestability, utilization of private resources. The weaknesses are highlighted in the form of inability to provide adequate franchise services, insufficient capacity building. The opportunities can be explored by subletting the contract to the informal private sector, working with local community leaders. The threats to these opportunities can be credibility barrier, legal bindings in the contract.

Informal Private Sector

The major roles of this sector are to collect waste from the households, sort and sell it to the junk dealer while working under the formal private sector. The major strengths of this sector encompass know-how of waste types, its sorting and handling, willingness to work in solid waste. This sector is badly stricken by poor economic status, low job security, low job wages, poor working conditions.

The possible future opportunities can be traced in the form of working with the formal private sector on contractual basis, training opportunities. These opportunities can be stalled by unwillingness of the formal private sector due to contractual obligations with the public sector, credibility barrier.

Community

The major roles associated with the community are referred to as service users, non-service users holding particular strength like freedom of choice. The community activists (service users/non-service users) and area representatives are also considered to be the part of community. The strengths of these actors can be seen as community's support and legal backing. The weaknesses highlighted on the part of the community as a whole include lack of interaction among community members, non-existence of neighbourhood committee, unaffordability to pay, lack of power to ensure good quality services, inability to stop waste polluters in the area, cultural differences of community members, unauthorization of the area representatives to intervene in the system. A number of opportunities can be sought in response to these weaknesses such as formation of the neighbourhood committee (active roles of the community activists), highly subsidized services, empowerment to monitor the system and take necessary action (area representatives). The major impediments to these opportunities can be occurred in the form of conflicting interests of the community members, loosing trust of the community due to inefficiency and incapability of community activists, unwillingness of the public sector to bring about changes in the system.

Cause and Effect Relationships of the Issues in the Franchise System

While identifying the inter-relationships and carrying out stakeholder analysis through SWOT, two major issues popped up such as weak co-ordination between the public and formal private sectors and weak co-operation between the community and formal private sector. In order to analyze these issues, I adopted Logical Framework Approach (Figure 3). It is an analytical planning tool, which is used to comprehend the issues within rational framework.

When we talk about the first issue, it implies that poor monitoring and non-compliance of contractual obligations are major causes behind failure in delivering the franchise services efficiently. These problems persist owing to lack of accountability of the public sector, formal private sector, and community. The unaccountability factor pops up in the wake of lack of interest of higher officials, which reveals low political priority of the solid waste management by the public sector. The effects of these issues are depicted in the form of indiscriminate dumping of the solid waste within service areas leading to environmental pollution eventually giving rise to nuisance and distraction to service users.

While dealing with second issue, we find that some members are unwilling to avail for the franchise services owing to poor quality services and unaffordability. The quality of service is badly affected due to inefficient operational management. In addition to it, there are other reasons such as feeble interaction with the community activists, and unawareness about the system. The element of affordability also contributes toward ultimate low participation of the community in the system.

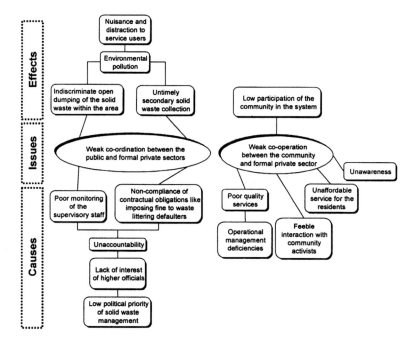

Figure 3   Cause and effect relationships of the issues in the franchise system

CONCLUSIONS

The working of the franchise system in a low-income neighbourhood presents a complex web of inter-relationships among different stakeholders whose strengths and weaknesses vary depending upon the intensity of interaction. The highly interactive nature among the three major sectors (public, formal private, community) brings to light two major issues such as weak co-ordination between the public and formal priate sector and weak co-operation between the community and formal private sector. These two major issues significantly contribute towards the decreasing sustainability of the franchise system in a low-income neighbourhood.

In order to increase sustainability of the franchise system in a low-income neighbourhood as well as other neighbourhoods, certain sets of measures need to be taken; for example, distribution of responsibilities to the area representatives, encouragement to form neighbourhood committees, creation of an enabling environment for the formal private sector, transparency and accountability, mass awareness, education and information campaigns. It is worthwhile to mention here that the franchise system cannot work in low-income neighbourhoods in isolation. It has to be strongly linked to servicing middle and high income neighbourhoods not only for the sake of subsidization but also to meet the costs likely to incur as a result of the improvements in the system. Hence, the improvements should be made in an integrated form.

ACKNOWLEDGEMENTS

I am gravely indebted to State Educational Loan Fund, Norway whose financial support under Quota Program enabled me to undertake my Master's research work in Lahore, Pakistan. I extend my deepest gratitude to Professor Hans Christie Bjønness (my supervisor and course co-ordinator) for his invaluable guidance and incessant support in the successful completion of this work.

REFERENCES

1. SWISS CENTER FOR DEVELOPMENT CO-OPERATION IN TECHNOLOGY AND MANAGEMENT (SKAT), Promotion of Public/Private Partnerships in Municipal Solid Waste Management in Low-Income Countries, Workshop Report, 22-23 February 1996.

2. LUDWIG, C. et al., Municipal Solid Waste Management: strategies and technologies for sustainable solutions, Springer-Verlag, Berlin, 2003, p 1.

3. THOMAS-HOPE, E., Solid Waste Management: critical issues for developing countries, Canoe Press, Jamaica, 1998, p 8.

4. COINTREAU-LEVINE, S. J., Private Sector Participation in Municipal Solid Waste Services in Developing Countries, Urban Management Program and the Environment, Vol. I., The Formal Sector, Washington D.C., The World Bank, April 30, 1994. (http://www.worldbank.org/html/fpd/urban/cds/mf/private_sector_part.html).

5. UNITED NATIONS ENVIRONMENT PROGRAM (UNEP), Newsletter and Technical Publications: municipal solid waste management, Division of Technology, Industry, and Economics, no-date, p 1 (http://www.unep.or.jp/Ietc/ESTdir/Pub/MSW/SP/SP3/SP3_6.asp).

6. CITY DISTRICT GOVERNMENT LAHORE (CDGL), Report on Solid Waste Management in Lahore, Feb 15, 2002, pp 5-6.

7. KLUNDERT, A.V.D. AND ANSCHUTZ, J., The Sustainability of Alliances between Stakeholders in Waste Management: using the concept of integrated sustainable waste management, Working paper for UWEP/CWG – Draft, 30 May 2000, p 3 (http://www.wwresourcecentre.net/pdf/ISWM.pdf)

8. MULLER, M. AND HOFFMAN, L., Community Partnerships in Integrated Sustainable Waste Management: tools for decision-makers, Experiences from the Urban Waste Expertise Program (1995-2001), 2001, p 12 (http://www.wase.nl/docpdf/tools_compart.pdf).

9. ALI, S.M. AND SNEL, M., Stakeholder Analysis in Local Solid Waste Management Schemes, Task No: 69, Water and Environmental Health at London and Loughborough (WELL), March 1999, p 4 (http://www.lboro.ac.uk/well/resources/well-studies/summaries-htm/task0069.htm).

USE OF RECYCLED CONCRETE AGGREGATE IN CONCRETE PAVEMENT CONSTRUCTION: A CASE STUDY

R K Dhir

K A Paine

S O'Leary

University of Dundee

United Kingdom

ABSTRACT. This paper describes a recently completed project to construct a new access road on the premises of a former Jute Mill, which was used to demonstrate the feasibility of using recycled concrete aggregates in new concrete construction from a technical standpoint and in terms of effecting cost savings (the large volume of concrete would normally have been crushed before going to landfill, incurring tipping charges and landfill tax,) and meeting policies of sustainable development. In addition, to the requirement to use recycled concrete aggregates (RCA) within the construction, the client also required that the finished concrete resemble the existing road from an earlier phase of development, where red coloured block paving had been used. Pigmentation and patterning of the RCA concrete was, therefore, required. Prior to construction, extensive laboratory tests were carried out to assess the suitability of the RCA for use in concrete pavement construction, and to develop suitable concrete mix proportions that would ensure adequate fresh, engineering and durability properties. The main pavement construction consisted of a 95m access road consisting of 19 panels (5m x 6.1m), and a turning circle (radius of 13m) consisting of 6 larger panels. The pavement consisted of 150mm RCA concrete with an imprinted surface on top of a 300mm deep recycled aggregate (RA) sub-base.

Keywords: Recycled concrete aggregate, Recycled aggregate, Pavements, Polypropylene fibres, Air-entrainment, Colouring pigment, Demonstration project

Professor Ravindra K Dhir is Director of the Concrete Technology Unit, University of Dundee. He specialises in binder technology, permeation, durability and protection of concrete. His interests also include the use of construction and industrial wastes in concrete to meet the challenges of sustainable construction.

Dr Kevin A Paine is a Lecturer in the Concrete Technology Unit, University of Dundee. His research interests are concerned with the use of waste and recycled materials as cement and aggregate components in concrete to facilitate sustainable urban development.

Mr Sean O'Leary is a Research Student at the University of Dundee. He is currently carrying out research into the potential of using construction and demolition waste in new forms of concrete construction.

INTRODUCTION

Ninety-three million tonnes of construction and demolition wastes are produced annually in the UK [1], a rise of 30% since 1999 and indeed the amount of waste is likely to increase further. At present, approximately 10% of these wastes are recycled for use as aggregate, mainly as fill, drainage or sub-base material [2]. The remaining potentially useful material goes to landfill. The environmental and economic implications of this are no longer considered sustainable and, as a result, the construction industry is under increasing pressure to reduce this landfill.

Recycling and reuse of these wastes as aggregate (*i.e coarse recycled concrete aggregate (RCA) and coarse recycled aggregate (RA)*) in new concrete construction offers an environmentally responsible route. In this regard, the UK government has introduced various measures to promote their wider use, including landfill and possible extraction taxes to improve economic viability, and support of appropriate research and development works. However, doubts about the performance and difficulties of producing concrete with a material such as RCA and RA, deters many potential producers and users. It is widely acknowledged that demonstration projects will give the necessary confidence to give it "a go".

This paper describes a recently completed demonstration project carried out to show the use of RCA in new pavement construction. The demonstration was part of Dundee City Council's (DCC) proposals for a second phase of development of a Biotechnology Park, comprising of a new access road, treatment of boundary walls and ground preparation. This required require removal of the existing factory floors and the underground foundations to what was formerly the largest Jute Mill in Dundee. The existing floor (Figure 1) contained a large volume of concrete, and this would normally have been crushed before going to landfill, incurring tipping charges and landfill tax. The alternative was to seek ways of re-using the crushed material on site - effecting cost savings and meeting DCC's stated policies of sustainable development. Four main options for recycling were originally identified:

1. General upfill to form hardstanding for building and associated car parks
2. Low grade concrete in composite masonry gravity retaining structures
3. Medium grade concrete in general concrete works
4. High quality concrete in the new access road and footways

However, following excavation and crushing of the material it was clear that there was insufficient material for all four options. RA (consisting mainly of brick) was therefore used as general up-fill throughout the site (Option 1) and unbound sub-base, whilst RCA was used as aggregate in pavement quality concrete (Option 4).

CONCRETE MIX

Specification

The main work comprised construction of the 95m length access road and a decision was made to specify the concrete for a XF4 environment with a minimum air content of 3.5% (20mm aggregates), relating to a condition of "horizontal concrete surfaces, such as roads and pavements exposed to freezing and deicing salts either directly or as spray or run off".

In addition, the client required that the finished concrete resembled the existing road from the first phase of development, where red coloured block paving had been used. Pigmentation and patterning were, therefore, required.

Figure 1 Site of the demonstration project prior to excavation
and crushing of the concrete floor

Materials

The properties and characteristics of the RCA used in this demonstration, obtained from crushing of the factory floor and foundations, are given in Table 1. Demolition was carried out in a controlled manner to avoid contamination of the RCA with masonry from brick piers and other potential sources of foreign material. Screening was also carried out.

The acid-soluble sulfate content of 0.6% was lower than the limit in BS 8500: Part 2, however the acid-soluble chloride content of 0.12% was higher than that recommended (BS 882) for use in reinforced concrete. For this reason a decision was made to construct an unreinforced pavement. The only steel in the pavement, therefore, were 400mm long stainless steel dowel bars at the expansion and contraction joints.

Because of the high water absorption of the RCA, it was suggested following laboratory tests that to ensure the required concrete performance, it would be necessary to use the RCA in a saturated condition. Therefore, during the pavement construction phase, the stockpiles of RCA were maintained in a wet condition by frequent spraying.

Table 1 Characteristics of RCA used in the project

	20-10 mm		10-5 mm	
	Mean	Std Dev	Mean	Std Dev
Constituents				
Masonry	-	-	5.0	4.0
Concrete/stone	100	0	95.0	4.0
Lightweight	-	-	-	-
Foreign materials	-	-	-	-
Particle Size-Distribution				
37.5	100	-	100	-
20.0	98	1.3	100	-
14.0	23	1.3	100	-
10.0	1	0.4	90	1.7
5.0	-	-	3	0.6
0.3	-	-	1	-
Density, kg/m^3				
Relative	2445	5.3	2440	4.2
Apparent	2670	9.6	2670	3.7
Loose bulk	1200	12.8	1260	7.6
Water absorption, %	5.9	0.3	5.5	0.2
Acid-soluble sulfates, %	0.60	0.26	0.62	0.24
Acid-soluble chlorides, %	0.12	0.01	0.12	0.02
Alkalis, Na$_2$Oeq	-	-	-	-
LOI, %	-	-	-	-
TFV, soaked [†], kN	151	2.4	-	-
AIV, soaked [†], %	22	2.4	-	-
Setting time*, min				
Initial	340			
Final	400			

[†] 14-10 mm only   * 20-5 mm (50% NA)

For laboratory studies, natural gravel and sand aggregate were used to provide a benchmark from which to compare the performance of RCA concrete. However, in the main pavement sections, two natural aggregates were used as an alternative to RCA in some sections, these were (i) natural gravel aggregate and (ii) whinstone aggregate. In all sections, the fine aggregate consisted of natural Fife sand.

Red-brick pigmentation was achieved by use of a coloured water-reducing set-controlling admixture supplied in the form of a powder.

Polypropylene fibres were used to (i) provide enhanced freeze/thaw resistance and (ii) assist in preventing cracking due to drying shrinkage.

LABORATORY MIXES

Prior to construction of the main pavement, a number of laboratory mixes were cast and investigated at the University of Dundee to ascertain the best proportions of constituents to be used. Mix proportions for these laboratory mixes are given in Table 2. General comments on the findings of the laboratory study are given below.

Table 2 Mix proportions for laboratory tests

COARSE RCA CONTENT %	MIX CONSTITUENTS, kg/m³								w/c
	PC	Filler	Water	Aggregates					
				Sand	NA 10mm	RA 10mm	NA 20mm	RA 20mm	
(A) Air-entrained (AE)									
0	415	0	165	545	385	0	770	0	0.40
25	415	0	165	545	290	90	580	180	0.40
50	445	0	165	520	190	180	385	360	0.37
100	445	0	165	520	0	360	0	720	0.37
(B) Non air-entrained (NAE)									
100	380	0	180	580	0	390	0	780	0.47
100	380	50	180	530	0	390	0	780	0.47
100	380	100	180	480	0	390	0	780	0.47
(C) Pigment[†] and NAE									
100	380	0	180	580	0	390	0	780	0.47
(D) Pigment[†], NAE, and fibres[§]									
0	380	0	180	580	840	0	420	0	0.47
25	380	0	180	580	630	100	315	200	0.47
50	380	0	180	580	420	200	210	395	0.47
100	380	0	180	580	0	395	0	790	0.47
(E) Pigment[†] and AE									
100	390	0	155	575	0	390	0	780	0.40

[†] Pigment added at proportions of 8 kg per 250 kg of cement
[§] Fibres added at dosage of 0.9 kg/m³

Fresh Properties

The slump for all concretes was measured to be below the 70 mm target. However, all results were within ± 25 mm and over half of the mixes were within ±15 mm of the target. It was noted that use of RCA tended to give slightly less workable mixes. Despite this, all concretes were classified as cohesive using the visual assessment technique described in BS 1881: Part 102, and smooth finishes were achieved in all concretes when assessed using a visual method consisting of running passing a float ten times over the surface of slumped concrete.

It was noted that the pigment admixture had some air-entraining properties, and concretes containing pigment had approximately 2-3% air content. In combination with air-entraining agent this led to very high air contents (i.e. > 8%).

Concretes containing pigment admixture achieved a good colour match with the block pavig used in the earlier stage of development.

Engineering Properties

Cube strength and flexural strength results for each mix are given in Table 3. For all concretes, the 28-day cube strength was greater than 35 N/mm^2. It was noted that the use of fibres and pigment (despite the entrained air)had little effect on strength. Because of the high entrained air content, Mix e (using pigment and air-entrainer) had the lowest strength.

Flexural strengths were generally found to decrease or increase in line with cube strength.

Table 3 Engineering Properties of laboratory mixes

COARSE RCA CONTENT, %	CUBE STRENGTH, N/mm^2			FLEX. STRENGTH, N/mm^2
	3 days	7 days	28 days	28 days
(A) Air-entrained				
0	24.5	31.0	36.0	4.15
25	24.5	30.0	36.0	4.35
50	23.5	30.0	35.0	4.00
100	23.0	29.0	35.0	3.45
(B) Non air-entrained				
100	25.5	31.0	37.0	4.35
100	26.0	30.0	43.0	5.20
100	26.5	33.0	48.0	5.40
(C) Pigment[†] and NAE				
100	25.0	27.5	39.0	4.35
(D) Pigment[†], NAE, and fibres				
0	32.5	36.5	45.0	5.20
25	29.0	40.0	42.5	5.20
50	28.0	36.0	42.0	4.70
100	26.0	35.0	39.0	4.50
(E) Pigment[†] and AE				
100	16.0	20.5	30.0	3.45

Durability Properties

Tests were carried out for freeze/thaw scaling and initial surface absorption. The effect of mix type on day freeze/thaw scaling resistance, for concrete containing 100% RCA, is shown in Figure 2. Only air-entrained concretes passed the Swedish criteria [3] of less than 1 kg/m² of scaling after 56 cycles. Concrete containing no air-entrainment (Mix B) had very poor scaling resistance, and resistance was not improved when the pigmentation admixture (that entrains about 2-3% air) was used (Mixes C-D). Although polypropylene fibres gave significant improvement in resistance to scaling, the amount of scaling after 56 days was still unsatisfactory according to the Swedish criteria. Use of air-entrainment and pigmentation admixtures in combination (Mix E) gave excellent scaling resistance (< 0.1 kg/m² afte 56 days).

One negative aspect of using pigmentation, however, was that it increased the initial surface absorption (Figure 3). This suggests that it may make the concrete less durable.

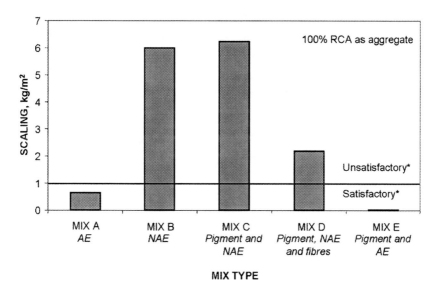

Figure 2 Effect of mix type on freeze/thaw scaling resistance and comparison with Swedish criteria, denoted by *

MAIN PAVEMENT SECTION DETAILS

The main pavement section was approximately 95m in length, and consisted of 19 panels (5m x 6.1m), whilst the turning circle at the end of the pavement (radius of 13m) consisted of 6 larger panels. For sustainability reasons the aim was to construct the majority of the pavement using RCA, however, due to a lack of material it was necessary to use natural aggregate (either (i) gravel aggregate or (ii) whinstone aggregate) for a small number of panels.

However, this did provide the opportunity to compare how the natural aggregate and RCA concrete sections perform. In total eleven RCA panels were constructed, eleven gravel aggregate panels and two whinstone aggregate panels.

The pavement consisted of a 150mm deep RCA concrete with an imprinted surface, laid on top of a 300mm deep recycled aggregate (RA) sub-base (Figure 4). Construction joints were formed between each 5m panel and expansion joints at every third panel (i.e. every 15m). Joints were formed by use of 400mm long stainless steel bars (in an attempt to minimise corrosion problems from the relatively high chloride content of the RCA). No other reinforcement was used in the pavement.

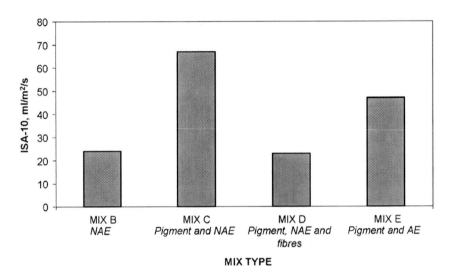

Figure 3 Effect of mix type on initial surface absorption after 10 minutes (ISA-10)

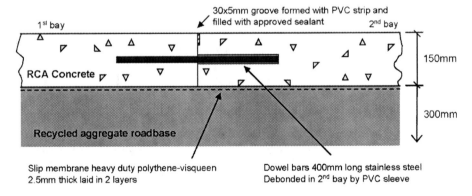

Figure 4 Cross-section of typical pavement panels and detailing of formed contraction joints

PAVEMENT PERFORMANCE

The pavement was competed in December 2002, and the finished pavement is shown in Figure 5. Results of tests on cubes and other specimens taken from selected mixes to check for cube strength, flexural strength, drying shrinkage and accelerated freeze/thaw resistance showed good performance. The mean cube strength of $35N/mm^2$, and mean flexural strength of $3.9 \ N/mm^2$ were consistent with performance of the laboratory mixes. Virtually no freeze/thaw scaling was detected after 56 cycles, thus demonstrating the excellent performance of air-entraining agent and polypropylene fibres working in tandem.

Figure 5 Completed RCA concrete pavement

At time of writing the pavement has been subject to natural conditions for six months and from visual inspection all sections appear to be performing equally well. Note, however, that to date (May 2003) the pavement has not been open to traffic. The client was happy with the surface finish achieved, particularly on the RCA sections and the colour of the pavement has weathered to match the initial phase of development constructed using block paving. Indeed, due to the success of this demonstration the client has used RCA and patterned concrete in a number of other recent projects in the city of Dundee.

SUMMARY

RCA has been successfully used as aggregate in a pavement section designed for a XF4 environment. Following extensive laboratory preparation work, the concrete mix used incorporated (i) pigmentation to produce the colour of red-brick pavers, (ii) air-entrainment and (iii) polypropylene fibres. In addition, the concrete was successfully patterned to give the appearance of an old red brick pavement. Test panels have been returned from the site and long-term and accelerated durability tests are on-going.

ACKNOWLEDGEMENTS

This demonstration was carried out as a partnership between Dundee City Council, Tayside Contracts and the University of Dundee, and part contributed towards a Partners in Innovation collaborative research project. The authors are grateful to the Department of Trade and Industry, Aggregate Industries, AMEC Capital Projects, BAA plc, Dundee City Council, Quarry Products Association, RMC Aggregates (UK) Ltd. and Scott Wilson Kirkpatrick and Co. Ltd. for funding the research project. Particular thanks are given to George Gray and Pat Hamilton of Dundee City Council for their support in carrying out this demonstration.

REFERENCES

1.   NEW CIVIL ENGINEER. Debate: Cutting down on wastes. The facts. 17-24 April 2003, p13.

2.   DEPARTMENT OF THE ENVIRONMENT. Sustainable Development: The UK Strategy. DOE Command Paper CM2426, HMSO London, 1994.

3.   SWEDISH STANDARDS. SS 13 72 44. Betonproving – Hrdnad betong – Avflagning vid frysning

TOWARDS ZERO:
SMART THINKING FOR WASTE ELIMINATION

I Gulland

Alloa Community Enterprises Ltd

United Kingdom

ABSTRACT. In nature the production of waste that is of no use to anything else spells ecological disaster. Humankind has not only ignored this golden rule but we have also neglected the need for a cyclical use of resources to maintain a balance between demand and supply. The Zero Waste philosophy has therefore been given prominence as a touchstone for a better, more sustainable new order.

Keywords: Zero waste, Elimination, Assets, Responsibility

Iain Gulland is the Development Manager for Alloa Community Enterprises Ltd and is currently chairman for the Community Recycling Network for Scotland.

INTRODUCTION AND SUMMARY

"Man is the only species capable of generating waste – things that no other life on earth wants to have" Gunter Pauli, Industrial Ecologist.

In nature the production of waste that is of no use to anything else spells ecological disaster. Humankind has not only ignored this golden rule but we have also neglected the need for a cyclical use of resources to maintain a balance between demand and supply. The Zero Waste philosophy has therefore been given prominence as a touchstone for a better, more sustainable new order.

The underlying principle of the Zero Waste philosophy is to eliminate waste altogether. It focuses on the redesigning of our one-way industrial system into a circular system, similar to the successful strategies found in nature where waste becomes an asset rather than a liability.

But, as well as addressing the issue of waste and its deleterious effect on our environment, Zero Waste goes beyond the rubbish bin and seeks to enhance, community and corporate responsibility, local economic development, business efficiency, social inclusion, public health, and environmental stewardship.

The Zero Waste message is now being translated into local government policy on waste around the world. Almost half of the local governments in New Zealand have set a target of zero waste to landfill by 2015, Canberra in Australia have signed up to no waste by 2010, Toronto, Canada and several US towns, cities and countries including the populous state, California, have adopted Zero Waste goals.

The key principle of these Zero Waste Policies is to maximize the recycling, reuse and composting of waste and to act to eliminate waste which can't.

The policy is driven by the defining of responsibilities between the key players. As this is a whole system approach not just a waste management tool for avoiding landfill, there is a need to engage with partners at all levels.

Community Responsibility – the public need to make use of the recycling and reuse services provided as well as begin to make changes to their purchasing behavior to focus in on materials which can be recycled and reused and do not have a negative effect on the environment.

Industrial Responsibility – Business needs to redesign products for longer use and with recyclability in mind. Eliminating waste during production will also increase productivity and so spell greater economic growth.

Government Responsibility – national and local government have a role to encourage the use of secondary materials rather than virgin resources through the use of legislation and economic instruments.

Are these principles realistic? Can we get the public to recycle all of their waste all of the time? Can we get Industry to design waste out of the system? And can we influence government enough? Can Zero Waste be achieved?

The question is not can it be achieved but how can it be achieved? Zero Waste is both about creating a new set of rules but also is about changing mindsets. The real challenge about Zero Waste is not so much with Industry nor is it with the general public but with the waste management profession. Industry will not take long to understand and adjust to the ramifications of a Zero Waste Policy. Regulation and new economic instruments might cause some resistance within corporate boardrooms but business will invariably spy the opportunities and develop accordingly. Increasing penalties for using toxic materials, which cannot be safely recycled, will also increase the drive for a cleaner production industry. Zero Waste is about a signal of intent to alter the status quo and define a new order of resource conservation and smart systems. Industry will respond with research and innovation focused on achieving this goal.

As for the general public the message of Zero Waste is very much about common sense. Waste generation is an accepted part of life only because there is no alternative. Providing accessible and reliable services for recycling will be accepted by the majority. Allowing communities to take advantage of their own waste stream in terms of job creation and economic gain will also contribute to higher recycling and greater consumer responsibility.

Zero Waste is also not anti-consumerism. Current waste awareness promotion tends to suggest that consumerism is bad and that over consumption by individuals is at the heart of our waste problem. Although the principles behind this message are genuine it often alienates the public by attacking their lifestyle and their inability to reduce consumption in a fast and consumer orientated world. Zero Waste acknowledges that there should be a reduction in global consumption of certain products but asks only that the public recognizes the need to recycle, reuse and compost their waste and to make consumer choices to ensure that what waste they do produce can be re-circulated within the economy with no or little environmental impact. This might appear only a subtle change in focus but ultimately this will be significant. After all we should still expect to be reading newspapers and drinking wine from bottles in 20 years time.

But changing the mindset of our waste industry will indeed be a challenge. Imposing a new set of rules is never easy. Waste management is an old and established profession and the thought that it might one day cease to exist is perhaps abhorrent to many. Waste management is changing and an element of recycling is now commonplace within most waste management firms as it is within local authorities up and down the country. But Zero Waste thinking is no gradual movement away from landfill or incineration nor is it a "be seen to be green" statement. It calls for a radical overhaul of our systems to enable rapid and massive waste reduction outcomes rather than incremental change. The adoption of a Zero Waste policy is the maturing of the waste profession. It is the coming to terms with the problem and the acceptance that the management of waste once it has been produced is un-sustainable. By stating the case for Zero Waste and its cornerstone principles, local government and national government will reap the benefit.

In addition, a Zero Waste Policy cannot sit in isolation within the waste management department of an individual council or the waste policy team at SEPA or Victoria Quay. It needs to be shared right across departments and networked with neighbouring councils and nations. It needs to be disseminated to the local communities and businesses and promoted in schools and at the doorstep. Arguing for more money and tougher legislation to help change direction is one thing – knowing where we are going to go next is another. Zero Waste has to be that new destination.

ARE YOU SITTING COMFORTABLY?
IF SO IT'S TIME TO WAKE UP!

S Aston

Environment and Heritage Service Northern Ireland
United Kingdom

ABSTRACT: If you are sitting comfortably, you shouldn't be! The world of waste is changing at a phenomenal pace and we are all struggling to keep up! New legislation from Europe is shifting the focus of controls form tightening standards of regulation to deepening involvement and responsibility along the supply chain. This summary paper details the key areas in which legislation is affecting us all.

Keywords: Waste, Legislation, Europe, Wake Up to Waste

Stephen Aston is the head of the Waste Management and Contaminated Land Unit, Environment and Heritage Service, Northern Ireland and currently President of the Chartered Institution of Wastes Management.

INTRODUCTION AND SUMMARY

If you are sitting comfortably, you shouldn't be! The world of waste is changing at a phenomenal pace and we are all struggling to keep up! New legislation from Europe is shifting the focus of controls form tightening standards of regulation to deepening involvement and responsibility along the supply chain. This summary paper details the key areas in which legislation is affecting us all.

Low cost, low technology methods of disposal have been the dominant face of waste management for many years, and although the systems and techniques have progressed considerably, technological solutions are not the only consideration .The real issues are about perceptions and behaviour and the influence of true cost economics on materials management.

Technological advances have resulted in a paradox whereby waste products can be managed so effectively there is no longer a perceived need to change behaviour and reduce waste. This sustains the mental segregation of what is waste and what constitutes a resource, rather than a culture of material separation and re-use where necessary to sustain the environment.

The skill and expertise in collection and disposal techniques has enhanced the 'out of sight out of mind philosophy which does nothing to engender a sense of responsibility in the waste producers. Both in business and at the household level there is a weak connection between goods purchased and items for disposal. Materials are discarded because they are perceived to be of no value.

With the exception of a minority of environmentally minded individuals interest in home waste only occurs when collection systems fail or a proposal is made for a waste disposal facility in the locality. At all other times we are almost completely disconnected from the environmental problems associated with the provision of raw materials and residual waste disposal. The impact on both current and future generations is therefore too remote for most individuals to relate to. As a consequence persuading people to change their lifestyles to enable better resource use is almost impossible.

At household level few mechanisms exist in the UK to encourage occupiers to take greater responsibility for the waste they produce. The most successful schemes to date in Europe and the USA apply direct or variable charging. Recent research has demonstrated that such schemes can be successful in reducing residual waste, however in the UK charges for waste disposal and collection are encompassed in the rates or Council Tax and bear little relation to direct costs or reflect the use of the service by the householder.

In addition to an absence of direct incentives the lack of a common message nationally underpinned by often disparate local level campaigns, has done little to change an already clouded public perception. This is not to belittle the success stories of numerous regional and local campaigns whose success is reflected to a certain extent in the range of recycling rates throughout the UK.

'Wake up to Waste' is the central information campaign running in Northern Ireland aimed at increasing awareness of the issues and initiating a change in values and behaviour. The first phase of the multi-media campaign was timed to coincide with the consultation on the sub-regional waste management plans.

Using the slogan 'Your waste, your problem, your say', the public had the opportunity to comment on the proposals for waste management in their area, encouraging them to take ownership of the problem. The reaction to the awareness phase was very encouraging with almost 10% of all households in Northern Ireland responding.

The second 'Action' phase focused on reduce, re-use and recycle and the 'daily do-ables' simple steps everyone can take every day to reduce waste. The Wake up to Waste TV advert was the second most spontaneously recalled advert ahead of Guinness and Coca-Cola. 94% of respondents in a survey consider it important that households reduce, re-use and recycle their waste while contractors have reported increase of up to 100% in materials presented for recycling.

The third phase of the campaign is in development and will build on the success of Phases 1 and 2, to consolidate the message and help to deliver real and substantial changes in attitude and, more importantly, behaviour.

European and Government recycling targets represent a substantial challenge for local authorities and technological solutions, including the provision of facilities for collection, sorting and reprocessing are just the start. Success in meeting objectives set in national strategies and local plans requires a real change in attitudes to ensure waste materials are perceived as resources with real value. Effecting the substantial changes in attitude and behaviour needed to raise the levels of participation in recycling and minimisation programmes requires a long-term strategy combining education and information, in addition to exploring the possibility of both positive and negative financial incentives.

A REVIEW OF
SOLIDIFICATION/STABILISATION TECHNOLOGY
AND ITS INDUSTRIAL APPLICATION IN THE UK

P J Carey

L Barnard	**C D Hills**	**B Bone**
Arup	University of Greenwich	Environment Agency
	United Kingdom	

ABSTRACT. The Environment Agency guidance on stabilisation/solidification (s/s) technology has recently been published. This document has been prepared for use by consultants, contractors and regulators involved in the remediation of contaminated land and the treatment of waste. The primary role of s/s is to immobilise contaminants both chemically and physically, by the addition of a binder and thereby break the source-pathway-receptor linkage and reduce the risk associated with the contaminated material. In addition to immobilising the contaminants, the use of cementitious binders during the treatment, will also enhance the engineering properties of the material. S/s treatment also has significant potential to address the changing requirements for waste disposed of to landfill, following the introduction of the EU Landfill Directive. The development of best-practice guidance for s/s technology will assist in the development of an indigenous s/s industry for managing contaminated land and pre-treating hazardous wastes prior to landfill disposal.

Keywords: Solidification, Stabilisation, Heavy metals, Contaminated soil, Hazardous wastes, Best practice guidance.

Dr Paula Carey, Principal Lecturer, Centre for Contaminated Land Remediation, University of Greenwich. Research interests the engineering behaviour of geomaterials; the durability of engineering materials and the application of petrography to the study of the durability of materials.

Lindsay Barnard, Engineer, Arup. Seconded to CCLR University of Greenwich to assist with the preparation of the EA guidance and Industrial guidance documents.

Dr Colin Hills, Reader in cementitious systems, Centre for Contaminated Land Remediation, University of Greenwich. Research interests in s/s using cement based systems, reuse of waste materials in environmental engineering; applications of accelerated carbonation to waste management.

Brian Bone, Scientific officer at the National Contaminated Land and Groundwater Centre of the Environment Agency, responsible for the overall management of the production of the guidance for s/s.

INTRODUCTION

Solidification/stabilisation (s/s) systems have been used in the UK to treat wastes for at least 15 years, but in a much more limited way than in North America, Japan or some EU member states. In these countries, s/s technology is well established through 20-30 years of use to treat a variety of wastes and contaminated soil. The United States Environmental Protection Agency (USEPA) has designated s/s as the Best Demonstrated Available Technology (BDAT) for the treatment of waste contaminated with metals [1], but the technology has also been used to treat organics, radioactive metals and mixtures of metals and organics. (Table 1)

Table 1   Contaminant types treated by s/s: after USEPA, [2]

CONTAMINANT TYPE	NUMBER OF PROJECTS, %
Metals only	92 (56%)
Organics only	10 (6%)
Radioactive metals	3 (2%)
Radioactive metals and metals	4 (2%)
Metals and organics	50 (31%)
Radioactive metals and organics	1 (1%)
Non-metals only	2 (1%)
Organics and non-metals	1 (1%)

Stabilisation and solidification are separate, but complementary processes that can be used for both in-situ and ex-situ treatment of contaminated soils as well as the treatment of industrial waste streams, which reduce the availability of contaminants and/or change the physical state of contaminated material. Solidification and stabilisation are defined in CIRIA [3] as:

Stabilisation - involves adding chemicals to the contaminated material to produce more chemically stable constituents, for example by precipitating soluble metal ions out of solution. It may not result in an improvement in the physical characteristics of the material, e.g. it may remain a relatively mobile sludge, but the process will have reduced the toxicity or mobility of the hazardous constituents.

Solidification - involves adding reagents to the contaminated material to reduce its fluidity/friability and reduce access by external mobilising agents, such as wind or water, to contaminants contained in the solid product. It does not necessarily require a chemical reaction between contaminants and the solidification agent, although such reactions may take place depending on the nature of the reagent.

The effectiveness of s/s in practice will depend upon the following:

- Good characterisation of the material to be treated.
- Selection of the most appropriate binder formulation.
- Effective contact between the contaminants and treatment reagents.
- A high degree of chemical and physical consistency of the feedstock.
- The use of appropriate mixing equipment and good working practice.
- Control over external factors such as temperature, humidity, and the amount of mixing since these affect setting, strength development and durability of the product.
- The absence (or control) of substances that inhibit the s/s process and affect the product properties.

The reluctance of many clients to choose s/s as a remediation method is partially due to the lack of information on where and how s/s has been used in the past. In the UK there has been a relatively poor uptake of s/s technologies, or other novel remediation technologies, due to a number of factors that include:

- The relatively low cost and widespread disposal of contaminated soils to landfill;
- Inconsistent interpretation of the regulatory requirement for the redeposit of treated waste on-site following ex-situ remediation;
- Lack of authoritative technical guidance;
- Uncertainty over the medium and long-term stability of disposed waste forms; and
- Residual liability associated with contamination remaining on-site albeit in an immobilised state.

This problem is currently being addressed by a number of organisations, working to improve information dissemination with respect to s/s technology. A European project involving UK participants has involved the compilation of a database of published data on s/s [4]. STARNET, an EPSRC funded network to promote the development of co-ordinated research work on the implementation of UK s/s technologies has also been established [5,6].

Although contaminants are not removed or destroyed during s/s, the technology can have a number of advantages over alternative risk management options, including:

- Treatment can be completed in a short time period;
- Diverse and recalcitrant contaminants, such as heavy metals and dioxins can be treated;
- The process can be performed in-situ or ex-situ and occupy a relatively small footprint on-site; and
- The geotechnical properties of the soil may be significantly enhanced creating the potential for re-use as an engineering material.

The CASSST initiative was initiated by the British Cement Association (BCA) and University of Greenwich in 2000 (www.cassst.co.uk). It has since become a pan-industry project to produce guidance for UK contractors, consulting engineers, contaminated land stakeholders and regulators on the use of s/s, particularly in connection with urban regeneration.

The CASSST guidance promotes and encourages the use of s/s as an effective and sustainable remediation technology. The guidance provides advice on:

- Screening procedures to assess the viability of s/s;
- Design of optimum binder formulation;
- Selection of chemical and physical tests;
- Selection of plant and equipment; and
- Post closure monitoring.

The approach taken throughout the guidance is based on the development of a site specific conceptual model and the use of a site specific testing regime based on the risks posed for that site and is supported by a science review.

TREATMENT OF CONTAMINATED SOIL

Outside the UK, a large amount of work on the application of s/s to contaminated soil has been undertaken by the USEPA. Figure 1 illustrates the percentage of contaminated soil treated by various remediation techniques under the Superfund programme. It can be seen that s/s is one technology amongst a number of routinely used remediation techniques and it has been used for both inorganic and organic contamination [7]. Soil vapour extraction (SVE) has been selected for sites contaminated with VOCs, BTEX, and non-halogenated VOCs, bioremediation has been implemented for non-halogenated SVOCs, whilst PCBs and halogenated SVOCs have been treated by incineration.

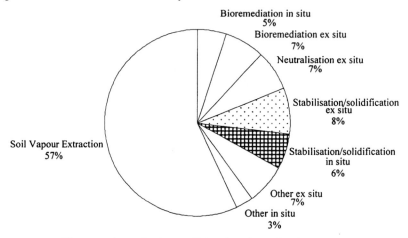

Figure 1   Superfund remedial actions by technology type [7]

The USEPA reported that at Superfund sites, s/s has been implemented (or is currently planned for use) in 183 instances, 137 of which were ex situ, with the remaining 46 in situ [7]. The binders used in Superfund projects included cement, phosphate, lime, pH, buffering agents, proprietary additives and other inorganic and organic components, including polymers, iron salts, silicates and clays (Table 2).

Table 2   Binders and reagents used for s/s projects [2,7]

BINDER OR REAGENT	NUMBER OF INSTANCES OF USE
Cement	50
Proprietary reagents	22
Other inorganic	20
Phosphate	14
pH controls*	12
Other inorganic	6
Fly ash	10
Lime	10
Sulphur	4
Asphalt	4

*pH buffering and adjustment agents, such as sodium hydroxide are sometimes added during stabilisation processes to decrease post-treatment contaminant leachability.

At many contaminated sites, the range of contaminants present may prove difficult to remediate. In some cases, the use of 'treatment trains' may be practicable, where two or more remediation processes are used in sequence to achieve the desired objectives.

WASTE TREATMENT PROCESSES

There are two routes for the treatment of waste streams by s/s:

- Treatment at the site of production as part of the waste-generating process, dealing with a specific waste stream, such as ash generated by incineration plants; or
- Treatment at a central processing plant, where a variety of wastes are treated, for example the former processing plant at Thurrock, Essex.

Although disposal is normally to landfill, exceptionally, the treated wastes may be suitable for use as general fill or road capping. In addition, s/s can be used for the treatment of contaminated dredgings where large volumes of material are treated and are generally re-used on or near the site, as fill.

The key issues associated with the implementation of s/s for waste treatment are:

- Landfill waste acceptance criteria and the implementation of the landfill directive;
- Can the s/s treated material be classified as a product for re-use or is it still designated a waste? and
- Suitability or waste for treatment by s/s.

There are a large number of s/s waste treatment processes available including at least seventy processes, which have been patented in the USA [8]. More than half of these processes involve combinations of cement, lime and PFA hydraulic binders. The remaining processes involve the use of organic compounds to form organic polymers or inorganic materials that rely on the formation of silicate polymers. The widespread use of organic processes for the s/s of hazardous wastes is hampered by their high cost, high-energy consumption or their specialised application [9]. Many of the s/s processes remain as 'blackbox' solutions to the treatment of wastes and the formulations are unknown, which has lead to a lack of confidence in the technology.

The six main proprietary systems that are best known and representative of the processes that use hydraulic binders are described in Table 3. Most s/s operators can refer to an extensive database of binder formulations previously used to treat characterised waste streams.

Lime and cement kiln dusts have also been used extensively at central waste management plants as absorbents, bulking agents or neutralizing agents for acidic wastes. Kiln dust and fly ash solidification techniques result in a stabilised product suitable for landfill. These systems are also very efficient for organic waste streams or for improving materials handling. The s/s treated products may take the form of:

- Granular materials which can be reused as fill or road sub-base or landfilled; and
- Monoliths which are subsequently landfilled.

Table 3   Typical cement based stabilisation processes
(modified from Clements and Griffiths, [8])

PROCESS	PRINCIPAL BINDER	OTHER ADDITIVES/PRE-TREATMENTS	APPLICATION
CHEMFIX	Sodium silicate	Calcium chloride Calcium sulphate Cement	America Canada Japan France UK (Mobile plant)
SOLIROC	Cement	Number of silicate waste materials Lime or alkali metal hydroxide	Belgium UK - Re-Chem Southampton France – Mitry Paris Norway Canada
PETRIFIX	Blast furnace slag	Alkaline activator	France UK - CHEMSAVE
IUCS system	PFA, coal ash and lime	cement	France US
STABLEX	Cement/fly ash	variety	Canada Japan UK – Birmingham, Thurrock
Geodur	Mixture of organic & non-organic chemicals	Variety of reducing agents or complexing agents as required	Switzerland UK (Mobile plant)

STABILISATION/SOLIDIFICATION OF INORGANIC CONTAMINANTS

Inorganic contaminants are commonly found associated with contaminated soil and hazardous wastes. Stabilisation and solidification processes can be used together to encapsulate the contaminants in the treated product and chemically immobilise them in the matrix of the treated waste form by adsorption, hydrolysis and/or precipitation reactions.

The effective use of s/s has been demonstrated with a range of inorganic contaminants, [9], including:

- Volatile metals
- Non-volatile metals
- Radioactive materials
- Asbestos
- Inorganic corrosives
- Inorganic cyanides

The speciation of heavy metals in the contaminated medium to be treated by s/s is important. Metals are immobilised through a number of mechanisms including- pH control, sorption, precipitation and by physical encapsulation in pores. The binder system used for s/s can be tailored for the contaminants to be treated. The redox environment of a binder-system can be altered by design to optimise s/s performance. Pre-treatment steps for difficult inorganic contaminants such as organo-metallic compounds and amphoteric metals are used to facilitate treatment by s/s. Interference effects from some inorganic or combinations of inorganic contaminants can affect the performance of s/s, but knowledge of contaminants to be treated and their likely interactions with available binders can remove any threat of interference on binder setting and hardening.

Chemical fixation is the most important factor in the successful long-term treatment of inorganic contaminants. [10] Natural analogues to cement hydrates, namely C-S-H, exist and are known to be stable over geological timescales. [11] This provides strong evidence that s/s systems will be persistent in the long term. Moreover, recent studies support the view that materials treated by s/s may provide an effective environment for the immobilisation of contaminants for thousands of years. [12]

STABILISATION/SOLIDIFICATION OF ORGANIC CONTAMINANTS

The s/s of organics presents more of a challenge than the s/s of metal-contaminated soils and wastes. Many authors have reported difficulty in the use of cement or lime alone in the treatment of organic-contaminated soils and wastes (e.g. [13]). This is because many organic compounds interfere with the hydration of cement/lime, resulting in retardation of set and a reduction in waste-form strength. Mixed wastes are typically presented for s/s and the direct consequences of interference from one contaminant or another are difficult to isolate. However, it should be noted that the successful treatment of difficult organic compounds in contaminated soil, has been carried out. For example, Bates *et al.* [14] describes the successful treatment of soil contaminated with PCP, Dioxin, Creosote NAPL and metals.

In general, the s/s of organic compounds may present greater challenges than the s/s of inorganic contaminants. The retarding effect of organics on the hydration of cement paste has been correlated with the number of hydroxyl, carboxylic, and carbonyl groups in the organic molecules, since the ability to form hydrogen bonds with cementitious constituents is thought to be a pre-requisite for interference in hydration mechanisms. As a rule of thumb, compounds that are volatile, water soluble, and have partition coefficients below 12 are likely to readily leach from cement or lime-based binder systems. The use of additives may improve the immobilisation of organic contaminants in s/s waste forms. Additives such as activated carbon, rice husk ash, shredded tyre particles and organoclays (sorbents) can increase the chemical containment of the contaminant. Additives such as silica fume and fly ash can improve the physical containment of organic compounds through the reduction of waste form porosity and permeability.

REGULATORY FRAMEWORK FOR THE TREATMENT OF WASTE AND CONTAMINATED SOILS

In the UK, the controls on the use of s/s depend on whether treatment is at a fixed plant which was in operation either prior to October 1999, when it would be licensed under the Waste Management Licensing Regulations 1994 (WMLR), or after October 1999, when it would be permitted under the Pollution Prevention and Control Regulations, 1999 (IPPC). For the treatment of contaminated soils on site using a mobile plant, a mobile plant license will regulate the treatment process but the redeposition of treated soil may be regulated under IPPC if the waste is not fully recovered or subject to an exemption under the WMLR.

A significant proportion of the UK industrial waste streams is currently co-disposed with bio-degradable waste without prior treatment. The Landfill Directive (Council Directive 1999/31/EEC [15]) requires member states to landfill only waste that has been subject to some form of treatment to reduce their hazard to human health or the environment. It will also prohibit the co-disposal of hazardous wastes with bio-degradable municipal wastes and require that such wastes be disposed of in a hazardous waste landfill or in a non-hazardous waste cell following treatment to satisfactory acceptance criteria.

The Urban Task Force report (1999) "Towards an Urban Renaissance", considers the current legislative position for remediation technologies such as s/s. It made a number of recommendations:

- Site owners should only have one set of standards to work to when resolving problems of site contamination
- Conflicts and inconsistencies between different environmental regulation systems covering land, water and waste should be resolved at the first opportunity
- The Environment Agency should provide a one-stop-shop service for regulatory and licensing requirements, moving quickly to a single remediation license to cover all the regulatory requirements for cleaning up a site.

DEFRA are currently reviewing the scope and detail for a single remediation license. The Single Remediation Permit working group has proposed that recovery and reuse of material on-site should be encouraged, that technical information about different remediation processes should be shared more widely and that codes of practice for different types of remediation activity should be developed.

Combined with the effects of increasing landfill taxes and a reduction in the available landfill space in the UK, disposal costs are likely to rise significantly. This will make process-based technologies more viable for on-site remediation (and re-use) and as a pre-treatment before landfilling. A considerable market is anticipated for the treatment of such waste streams by s/s technology.

CONCLUSIONS

The publication of industrial guidance document, together with a supporting science review, for the use of s/s through CASSST, has draw together the technical information about s/s processes for treating contaminated soils and wastes. The documents have also highlighted the opportunities to reuse s/s treated material on site or elsewhere as fill as well as reducing the hazardous nature of materials for which there is no alternative but landfill. The guidance, together with the development of a single remediation licence that should make reuse on site an option within the regulatory regime, should lead to the more wide spread industrial application of s/s technology.

REFERENCES

1. ALPERIN, E.S., LEAR, P.R. AND STINE, E.F., JR.. Stabilisation/Solidification of Wastes. BCA symposium University of Greenwich. 2000.

2. USEPA. Solidification/Stabilization Use at Superfund Sites, EPA-542-R-00-010, 2000.

3. CONSTRUCTION INDUSTRY RESEARCH AND INFORMATION ASSOCIATION (CIRIA). Remedial Treatment for Contaminated Land, Volume VII, *Ex Situ* Remedial Methods for Soils, Sludges and Sediments, CIRIA Special Publication 107. 1995.

4. STEGEMANN, J.A., BUTCHER, E.J, IRABIEN, A., JOHNSON, P., DE MIGUEL, R., OUKI, S.K., POLETTINI, A. AND SASSAROLI, G. Neural Network Analysis for Prediction of Interactions in Cement/Waste Systems (NNAPICS) Final Report 2001. (www.concrete.cv.is.ac.uk/iscowaa/nnapics/intro.html).

5. AL-TABBAA, A. AND PERERA, A.S.R. State of Practice Report UK Stabilisation/ Solidification Treatment and Remediation: Binders and Technologies – Basic principles. STARNET REPORT. 2002a

6. AL-TABBAA, A. AND PERERA, A.S.R. State of Practice Report UK Stabilisation/ Solidification Treatment and Remediation: Binders and Technologies – Applications. STARNET REPORT 2002b.

7. USEPA.. National Risk Management Research Laboratory. Treatabilty Database. http://www.epa.gov/tdnrmrl 2001.

8. CLEMENT, J.A. AND GRIFFITHS, C.M. Solidification Processes. In: Porteous, A. (Ed.) Waste Management Handbook. 1985. Butterworths, London.

9. USEPA. Stabilisation/Solidification of CERCLA and RCRA Wastes. Physical Tests, Chemical Testing Procedures. Technology Screening and Field Activities. EPA/625/6-89/022. 1989.

10. HILLS, C.D., JONES, H.M., BONE, B., CAREY, P.J., BARNARD, L.H., BOARDMAN, D.I., and MACLEOD, C.L. Validation of Remedial Performance of Stabilisation/Solidification. Science review accompanying EA guidance for stabilisation/solidification. 2003 (Contract Reference P5D (99) 02)

11. MCCONNELL, J.D.C. The Hydration of Larnite (B-Ca_2SiO_4) and Bredigite (a1-Ca_2SiO_4) and the Properties of the Resulting Gelatinous Plombierite. Mineral Magazine. 1955. 30, pp. 672-680

12. CATALAN, L.J.J., MERLIERE, E. and CHEZICK, C. Study of the Physical and Chemical Mechanisms Influencing the Long-term Environmental Stability of Natrojarosite Waste Treated by Stabilization/Solidification. Journal of Hazardous Materials 2002. B94, pp. 63-88.

13. USEPA. Solidification/Stabilisation of Organics and Inorganics. Engineering Bulletin. EPA/540/S-92/015, 1993.

14. BATES, E.R., SAHLE-DEMESSIE, E. AND GROSSE, D.W. Solidification /Stabilization for Remediation of Wood Preserving Sites: Treatment for Dioxins, PCP, Creosote and Metals. Portland Cement Association Publication RP371, Skokie, Il. 1999.

15. COUNCIL DIRECTIVE 1999/31/EC The Landfill of Waste. The Council of the European Union.

INDEX OF AUTHORS

SUBJECT INDEX

This index has been compiled from the keywords assigned to the papers, edited and extended as appropriate. The page references are to the first page of the relevant paper.

405

Printed in the United Kingdom
by Lightning Source UK Ltd.
126958UK00001B/101/A